Lecture Notes in Computer Science 4892

Commenced Publication in 1973
Founding and Former Series Editors:
Gerhard Goos, Juris Hartmanis, and Jan van Leeuwen

T0223303

Andrei Popescu-Belis   Steve Renals
Hervé Bourlard (Eds.)

Machine Learning for Multimodal Interaction

4th International Workshop, MLMI 2007
Brno, Czech Republic, June 28-30, 2007
Revised Selected Papers

 Springer

Volume Editors

Andrei Popescu-Belis
Hervé Bourlard
IDIAP Research Institute, Centre du Parc
Av. des Prés-Beudin 20, Case Postale 592, 1920 Martigny, Switzerland
E-mail: {andrei.popescu-belis, herve.bourlard}@idiap.ch

Steve Renals
University of Edinburgh, Centre for Speech Technology Research
2 Buccleuch Place, Edinburgh EH8 9LW, UK
E-mail: s.renals@ed.ac.uk

Library of Congress Control Number: 2008920951

CR Subject Classification (1998): H.5.2-3, H.5, I.2.6, I.2.10, I.2, D.2, K.4, I.4

LNCS Sublibrary: SL 3 – Information Systems and Application, incl. Internet/Web and HCI

ISSN   0302-9743
ISBN-10   3-540-78154-4 Springer Berlin Heidelberg New York
ISBN-13   978-3-540-78154-7 Springer Berlin Heidelberg New York

Springer is a part of Springer Science+Business Media

springer.com

© Springer-Verlag Berlin Heidelberg 2008
Printed in Germany

Typesetting: Camera-ready by author, data conversion by Scientific Publishing Services, Chennai, India
Printed on acid-free paper   SPIN: 12225712   06/3180   5 4 3 2 1 0

Preface

This book contains a selection of revised papers from the 4th Workshop on Machine Learning for Multimodal Interaction (MLMI 2007), which took place in Brno, Czech Republic, during June 28–30, 2007. As in the previous editions of the MLMI series, the 26 chapters of this book cover a large area of topics, from multimodal processing and human–computer interaction to video, audio, speech and language processing. The application of machine learning techniques to problems arising in these fields and the design and analysis of software supporting multimodal human–human and human–computer interaction are the two overarching themes of this post-workshop book.

The MLMI 2007 workshop featured 18 oral presentations—two invited talks, 14 regular talks and two special session talks—and 42 poster presentations. The participants were not only related to the sponsoring projects, AMI/AMIDA (http://www.amiproject.org) and IM2 (http://www.im2.ch), but also to other large research projects on multimodal processing and multimedia browsing, such as CALO and CHIL. Local universities were well represented, as well as other European, US and Japanese universities, research institutions and private companies, from a dozen countries overall.

The invited talks were given by Nick Campbell from the ATR Spoken Language Communication Research Labs in Kyoto, Japan—published as the opening chapter of this book—and by Václav Hlaváč from the Czech Technical University in Prague. The first day of the workshop included a special session with results from the second PASCAL Speech Separation Challenge (Part VII of the book), while the third day constituted the AMIDA Training Day. During the week following MLMI 2007, the Summer School of the European Masters in Speech and Language was also held at the Brno University of Technology. Oral presentations at the workshop were recorded using IDIAP's Presentation Acquisition System, thus making audio, video and slides publicly available at http://mmm.idiap.ch/talks.

The reviewing and revision process specific to the MLMI series ensured that high-quality chapters based on oral and poster presentations appear in this book. Prior to the workshop, full paper submissions were reviewed by at least three members of the Program Committee under the supervision of one of the five Area Chairs, some of the papers being accepted as oral presentations and some as posters. Submitted at a later deadline, additional abstracts for posters were reviewed by Area Chairs. After the workshop, all poster presenters were invited to (re)submit full papers, which underwent a second round of reviewing, while authors of oral presentations were asked to revise their full papers based on feedback received during the first round. Overall, about two thirds of all submitted full papers appear in this book, which is also about one third of all submitted papers and abstracts.

The editors would like to acknowledge the sponsorship of the AMI and AMIDA Integrated Projects, supported by the European Commission under the Information Society Technologies priority of the sixth Framework Programme, and of the IM2 National Center of Competence in Research, supported by the Swiss National Science Foundation. We also thank very warmly the members of the Program Committee, in particular the Area Chairs, as well as all those involved in the workshop organization, for making this fourth edition of MLMI a success, resulting in the present book.

To conclude, we would like to remind the reader that selected papers from previous editions of MLMI were published as Springer's LNCS 3361, 3869 and 4299, and that the fifth MLMI (http://www.mlmi.info) will take place in Utrecht, The Netherlands, on September 8–10, 2008.

October 2007   Andrei Popescu-Belis
                                                                Steve Renals
                                                              Hervé Bourlard

Organization

Organizing Committee

Hervé Bourlard	IDIAP Research Institute
Jan "Honza" Černocký	Brno University of Technology (Co-chair)
Andrei Popescu-Belis	IDIAP Research Institute (Program Chair)
Steve Renals	University of Edinburgh (Special Sessions)
Pavel Zemčík	Brno University of Technology (Co-chair)

Workshop Organization

Jonathan Kilgour	University of Edinburgh (Webmaster)
Sylva Otáhalová	Brno University of Technology (Secretary)
Jana Slámová	Brno University of Technology (Secretary)
Josef Žižka	Brno University of Technology (Webmaster)

Program Committee

Marc Al-Hames	Munich University of Technology
Jan Alexandersson	DFKI
Tilman Becker	DFKI
Samy Bengio	Google
Hervé Bourlard	IDIAP Research Institute
Nick Campbell	ATR Laboratories
Jean Carletta	University of Edinburgh
Jan "Honza" Černocký	Brno University of Technology
John Dines	IDIAP Research Institute
Joe Frankel	University of Edinburgh
Sadaoki Furui	Tokyo Institute of Technology
John Garofolo	NIST
Daniel Gatica-Perez	IDIAP Research Institute
Luc van Gool	ETHZ
Thomas Hain	University of Sheffield (Area Chair)
James Henderson	University of Edinburgh
Hynek Hermansky	IDIAP Research Institute
Václav Hlaváč	Czech Technical University Prague (Area Chair)
Alejandro Jaimes	IDIAP Research Institute
Samuel Kaski	Helsinki University of Technology
Simon King	University of Edinburgh
Denis Lalanne	University of Fribourg

Yang Liu	University of Texas at Dallas (Area Chair)
Stéphane Marchand-Maillet	University of Geneva
Jean-Claude Martin	LIMSI-CNRS
Helen Meng	Chinese University of Hong Kong
Nelson Morgan	ICSI
Petr Motlicek	IDIAP Research Institute
Luděk Müller	University of West Bohemia
Roderick Murray-Smith	University of Glasgow
Sharon Oviatt	Adapx and OGI/OHSU (Area Chair)
Andrei Popescu-Belis	University of Geneva (Program Chair)
Ganesh Ramaswamy	IBM T.J. Watson Research Center
Steve Renals	University of Edinburgh
Jan Šedivý	IBM Prague
Elizabeth Shriberg	SRI and ICSI
Rainer Stiefelhagen	University of Karlsruhe (Area Chair)
Jean-Philippe Thiran	EPFL
Pierre Wellner	Spiderphone
Dekai Wu	Hong Kong University of Science and Technology
Pavel Zemčík	Brno University of Technology

Sponsoring Programs, Projects and Institutions

Programs

- European Commission, through the Multimodal Interfaces objective of the Information Society Technologies (IST) priority of the sixth Framework Program.
- Swiss National Science Foundation, through the National Center of Competence in Research (NCCR) Program.

Projects

- AMI, Augmented Multiparty Interaction, and AMIDA, Augmented Multiparty Interaction with Distance Access, http://www.amiproject.org.
- IM2, Interactive Multimodal Information Management, http://www.im2.ch.

Institutions

- FIT BUT, Faculty of Information Technology, Brno University of Technology, http://www.fit.vutbr.cz.

Table of Contents

Invited Paper

Multimodal Processing

HCI, User Studies and Applications

Image and Video Processing

Discourse and Dialogue Processing

Speech and Audio Processing

PASCAL Speech Separation Challenge II

Robust Real Time Face Tracking for the Analysis of Human Behaviour

Damien Douxchamps[1] and Nick Campbell[2]

[1] Image Processing Laboratory,
Nara Institute for Science and Technology,
Nara 630-0192, Japan
[2] National Institute of Information and Communications Technology
& ATR Spoken Language Communication Research Labs
Keihanna Science City, Kyoto 619-0288, Japan
`nick@nict.go.jp, ddouxcha@is.naist.jp`

Abstract. We present a real-time system for face detection, tracking and characterisation from omni-directional video. Viola-Jones is used as a basis for face detection, then various filters are applied to eliminate false positives. Gaps between two detection of a face by the Viola-Jones algorithms are filled using a colour-based tracking. This system reliably detects more than 97% of the faces across several one-hour videos of unconstrained meetings, both indoor and outdoor, while keeping a very low false-positive rate ($<0.05\%$) and without changes in parameters. Diverse measurements such as head motion and body activity are extracted to provide input to further research on human behaviour and for tracking participant activites at round-table meetings and similar discourse environments.

1 Introduction

The analysis of the relation between human behaviour and speech has been the subject of numerous research in the past and has recently formed the core of integrated research on meetings activity. One particular case of interest is the analysis of discourse processes and human interactions in meetings because those are common, easy to setup and provide a relatively controlled environment while encouraging people to express themselves [1,2,3,4]. However, the various approaches used to track the people's behaviour in these circumstances often use intrusive equipment, like individual cameras and microphones. As intrusions into the discourse will inevitably change the behaviour of people, less invasive techniques are sought [5].

In this context, we have developed a real-time video system that relies on a single small omnidirectional camera to retrieve information about the attendants' motion and activity level. No specific lighting is required. Given the relatively low resolution of our video it is not possible to extract fine information such as eye gaze but we can still detect heads and calculate the person's motion and activity. In a later stage only briefly mentioned here, this data is then correlated with verbal and non-verbal speech to infer higher level information about behaviour of discourse participants.

A. Popescu-Belis, S. Renals, and H. Bourlard (Eds.): MLMI 2007, LNCS 4892, pp. 1–10, 2007.

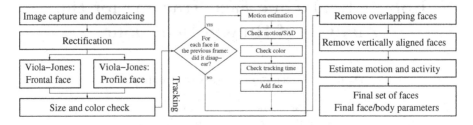

Fig. 1. Processing flow for each frame of the video

The detection and tracking of faces is a well covered subject in the literature. Among the different techniques available, the one proposed by Viola and Jones [6] [7] has the best results for low-resolution video and has been used in many situations. However, the number of studies reporting on their actual detection rate is surprisingly small, or they are limited to short video sequences. Examples include Fröba (90% detection rate, 0.5% false positive rate) [8], Kawato (89%, 1%) [9] and Castrillón-Santana [10]. In this paper we will show how to achieve a very high detection rate (>97%, <0.05%) in the case of unconstrained meetings lasting over an hour.

2   Video Processing

The processing techniques (Fig. 1) used in our system are standard and well documented, such as face detection and block matching (BMA). However, it is not trivial to build a real-time processing chain from these building blocks, especially when a high level of detection is to be achieved without any constraints given to the participants of the meeting.

Visual clues of the behaviour of discourse participants are extracted from the streaming video image by combining standard tools to form a more specialized video processing chain. Much of the processing is aimed towards a proper face detection since the face is a human feature that is relatively easy to detect and contains a lot of information concerning the behaviour of the person. Detecting hands is also an option but these are more difficult to track as their shape can vary greatly and they also move much faster. This in turn requires a higher video framerate, which weighs heavily on the processing speed. Our process for detecting and characterizing faces is as follows:

2.1   Video Capture, Demozaicing and Rectification

The video signal from a digital camera is decoded from a raw Bayer format to a full RGBI image. The demozaicing is performed using the 'Edge Sense II' algorithm presented in [12]. This algorithm provides good quality output while still being able to run at a reasonable speed. Other algorithms have been used [13] [14] but did not provide a significant advantage while being considerably slower. The circular, 360° image of Fig. 2 is then rectified with a linear subpixel

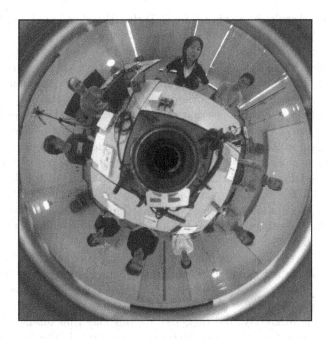

Fig. 2. Circular 360 degrees image captured by the camera (1040×1040)

Fig. 3. Rectified 360 degrees image (2048×260). Note the limited vertical resolution.

resampling before the face detection. To ensure a proper scaling of the faces, the horizontal size of the rectified image is set around $2\pi r$ where r is the average radial position of faces in the circular image. This rectification is necessary because the Viola-Jones face detection cannot detect faces in any orientation without a significant additional computational cost or a loss in accuracy. The resulting rectified image (Fig. 3) is now ready to be used for the face detection.

2.2 Face Detection

Face detection can be performed in a number of ways. The first technique that we tried was based on background subtraction and colour segmentation [15]. It has the advantage of not requiring an image rectification, but it failed to provide satisfactory results due to illumination changes and colour variability. A better approach is to use the Viola-Jones face detection [6] [7] which is based on pattern matching. One drawback of this approach is that the algorithm must be trained on a large number of images, but standard software packages such as OpenCV [11] provide example training data (in the form of Haar cascades) that we found to be very effective to detect the two patterns that we are most interested in: profile faces and frontal faces. In fact, using the Viola-Jones detection alone more than 50% of faces can be found during our round-table meetings.

To filter out the few non-faces that were detected we use two filters. The first one limits the size of the head within a reasonable range. The second one verifies that the face region contains a minimum of 25% of skin-coloured pixels. We found that the skin tone could be defined with sufficient accuracy in the RGB color space using the following criteria: $0.55 < R < 0.85$, $1.15 < R/G < 1.9$, $1.15 < R/B < 1.5$ and $0.2 < (R+G+B)/3 < 0.6$. We have successfully used the same criteria for both indoor and outdoor scenes. At last, the binary mask of skin-coloured pixels is eroded and dilated using mathematical morphology before counting the number of skin-coloured pixels.

After this filtering, overlapping face-regions can still exist but they are removed easily by verifying that their overlapping region is not greater than around 20%. If so, the smallest face is discarded. Note that removing all faces that have the slightest overlap is not appropriate because people may be approaching each other for talking discretely, and their face regions may thus intersect slightly.

The Viola-Jones face detection is strictly frame-based. The lack of time integration means that the detection is not guaranteed to be continuous. In fact it can oscillate even with small image variations: a face can be detected in one frame, disappear in the next frame and then reappear again. To avoid these instabilities, we introduce a method to track the faces and bridge the gaps between two successive detections.

2.3   Face Tracking

If a face region in one frame intersects with a face region in the next frame, they will be considered to be from the same physical face and tracking is not necessary. If no such face can be found in the next frame, we will attempt to bridge the Viola-Jones detection gap by looking for an instance of the older face in the newer frame, using a classic block-matching algorithm (BMA) based on the Sum of Absolute Differences (SAD). This matching can drift in time so it is necessary to limit it with some safeguards. The first one consists in limiting the time during which this gap-bridging will be performed. Given the very low false-positive rate of the face detection (see Section 3) we can allow a long maximal tracking time of 30 seconds. Secondly, we verify that the tracked face still contains a minimal amount of skin-coloured area, as we did after the Viola-Jones detection. Thirdly, the image difference between the old and tracked faces should be limited b y a threshold. Finally, the amount of face motion is also limited by the size of the search zone of the BMA.

At this point we have not yet included any situation-specific verifications that may help to filter out the last outlying faces. To remain as generic as possible

Fig. 4. Typical output from the program showing two 180 degrees sections on top of each other. Detected faces are shown with a white or black square.

we only include one: if two faces are overlapping vertically (i.e., if they belong to the same image column) then only the highest face is kept. This is a small restriction that remains valid for most meeting situations. A visual output of a final set of detected faces is shown in Fig. 4.

2.4   Motion and Activity Estimation

Once faces are properly detected a number of measurements are performed to identify their position, motion, and surface. The motion estimation cannot be performed on the positions of the detected faces because they are too unstable; parasitic motions of +/- 5 pixels are not uncommon with the Viola-Jones detection. The motion estimation is therefore performed using subpixel block matching (BMA) on the image content. Two measures of a person's activity are also computed as the mean SAD between the previous and the current image: one is computed on the face region, and the other on the body region, the latter being defined as the area below the face with a width three times that of the face.

The graphs in Fig. 5 show a small section of five minutes of a few head and body measures for the nine persons attending a meeting. These graphs show that the vertical and horizontal motion estimation of the face is able to resolve small details. For example persons mimicking a 'yes' or 'no' head movement are visible as small sinewave bursts in the vertical or horizontal head motion. Activity measures also correlate well with the global movements of a person. These measurements are now being correlated more systematically with manually labeled audiovisual data to provide clues about the link between physical activity and both verbal and nonverbal discourse events. This work, however, is beyond the scope of the present paper.

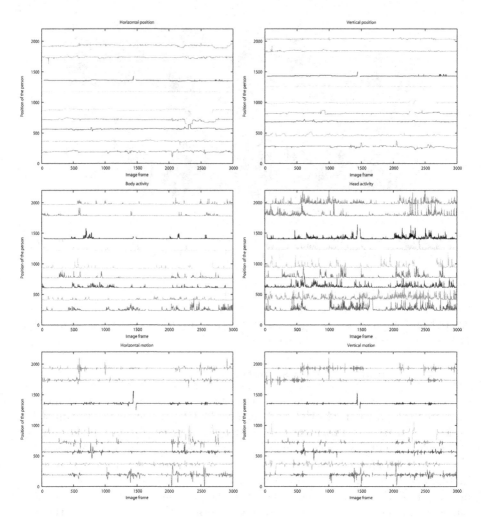

Fig. 5. The body and head activity (*top row*) and the head horizontal and vertical movement (*bottom row*) of the 9 participants found in the third sequence

3   Experiments

The proposed system has been tested under several conditions of lighting, image resolution and human activity. The minimum resolution for the circular image was found to be around 1000×1000; of the tests meeting this criteria five will be shown and discussed here. All tests were performed with the same hardware and processed with the same software and parameters. Typical output images are presented for each sequence in Fig. 6 together with a histogram of the frames of each sequence according to the number of faces detected in each of them. Ideally, each histogram should show a single bar for the bin corresponding to the number of people attending the meeting. Due to various errors, detection will sometimes fail and frames with fewer people detected than expected will exist. Similarly, false positives may lead to frames with a higher number of detected people than expected.

The first sequence is a one-hour meeting recorded at 10 frames per second. It has a raw detection rate of 95%. Some of the faces are not present at all during some periods of time, for instance when a person leaves her seat to write on a white-board. If we take these long events into account the detection rate climbs to 97%. The unrestricted movements of people also leads to other numerous small undetectability events that are more difficult to take into account, such as looking at the ceiling, looking back, face obscured by a sheet of paper, and so on .... The 97% figure may thus be an underestimate. At the same time, the amount of false positives is less than 2%.

The second sequence shows limits in our approach, with a poor detection rate of 57%. This is explained by three factors: 1) a high contrast video with strong shadows is not optimal for our Viola-Jones detection; 2) the rectangular table means that people far from the camera will appear too small, which is also difficult for the Viola-Jones algorithm to detect and 3) the sharpness of the sequence was poor, washing face features away. Consequently, further tests were performed with a lower contrast and a square table, the latter leading to more homogenous face sizes than the rectangular table used in this test.

The third sequence has a similar detection rate to the first one: 96%. This test suffers a high false positive rate of 2% due to a high colour noise and poor white balance, as lights were switched on and off during the meeting to allow a video projection to be displayed. Many of the non-detections during this sequence are due to people looking away from the camera at table centre towards the presentation screen instead. Their faces are then strongly tilted or hidden, presenting angles that the Viola-Jones algorithm was not trained to detect.

The fourth sequence was taken outdoors while using identical processing parameters. Surprisingly, the detectability is also good (93%) but could without doubt benefit from fine tuning of the parameters. However, the strong directionality of the sunlight results in a set of brighter faces (facing towards the sun) and darker faces (back to the sun) which cannot be simultaneously optimized. This difference in exposure poses problems both for the Viola-Jones and for the colour-tracking.

Finally, the fifth sequence has a poor detection rate of 33%. The conditions were that of the second test but the camera was in the way of the projector beam which strongly influenced the white balance, lowered contrast and added a significant amount of flare. The histogram shows two distributions which correspond to the projector being on or off. The large number of tilted heads also partially explain the lower detection rate.

Overall, the face detection appears to work very well even in very different situations, provided that the scene has a reasonable focus, white-balance, sharpness and dynamic range.

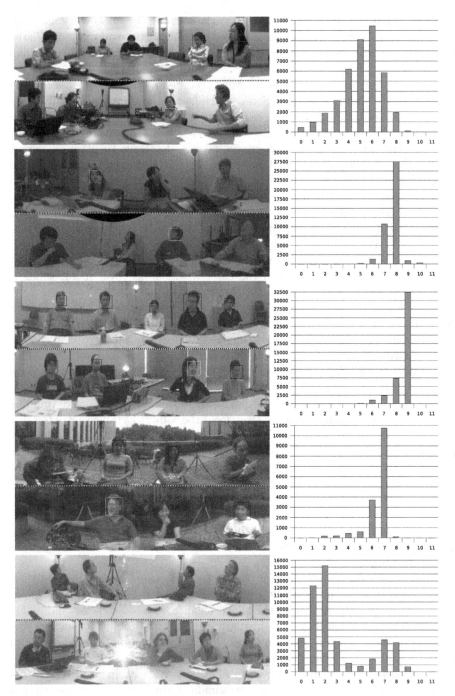

Fig. 6. Detection results for five test sequences. A representative image from the sequence is shown in the (*left column*). The (*right column*) contains the histogram of the number of images (*vertical axis*) per number of people detected (*horizontal axis*).

4   Conclusions

We have presented an image processing technique that is able to reliably extract faces from hour-long recordings of unconstrained meetings. Our technique is able to achieve a very good detection rate (>95%) while keeping the false positives to a negligible level (<1.5%). Conditions for which our approach has problems have been identified but can be easily avoided so that they do not limit its scope of application.

Acknowledgements

The second author is supported by NiCT, the National Institute for Communications and Information Technology. This work was partially funded under the SCOPE initiative. Both are under the Japanese Ministry of Internal Affairs and Communications,

References

1. Dielman, A., Renals, S.: Multistream Recognition of dilogue acts in meetings in Renals. In: Renals, S., Bengio, S., Fiscus, J.G. (eds.) MLMI 2006. LNCS, vol. 4299, pp. 178–189. Springer, Heidelberg (2006)
2. Burger, S., MacLaren, V., Yu, H.: Meeting corpus: The impact of meeting type on speech style. In: Proc. International Conference on Spoken Language Processing (ICSLP), Denver (September 2002)
3. Janin, A., Baron, D., Edwards, J., Ellis, D., Gelbart, D., Morgan, N., Peskin, B., Pfau, T., Shriberg, E., Stolcke, A., Wooters, C.: The ICSI meeting corpus. In: Proc. IEEE Int. Conf. on Acoustics, Speech and Signal Processing (ICASSP), Hong-Kong (April 2003)
4. McCowan, I., Gatica-Perez, D., Bengio, S., Lathoud, G., Barnard, M., Zhang, D.: Automatic analysis of multimodal group actions in meetings. IEEE Trans. on Pattern Analysis and Machine Intelligence 27(3), 305–317 (2005)
5. Campbell, W.N.: A multimedia database of meetings and informal interactions for tracking participant involvement and discourse flow. In: Proc. LREC 2006, Genoa, Italy (May 2006)
6. Viola, P., Jones, M.: Rapid Object Detection Using a Boosted Cascade of Simple Features. In: Proc. Intl. Conf. on Computer Vision and Pattern Recognition (CVPR), vol. 1, pp. 511–518 (2001)
7. Viola, P., Jones, M.: Robust Real-Time Face Detection. International Journal of Computer Vision 57(2), 137–154 (2004)
8. Froöba, B., Küblbeck, C.: Face Tracking by Means of Continuous Detection. In: Proc. of the IEEE 2004 Conf. on Computer Vision and Pat. Rec. Workshops (CVPRW 2004), 27, 02, 65–65 (June 2004)
9. Kawato, S., Tetsutani, N.: Scale Adaptive Face Detection and Tracking in Real Time with SSR filter and Support Vector Machine. IEICE - Transactions on Information and Systems, E88-D 12, 2857–2863 (2005)
10. Castrillón-Santana, M., Déniz-Suárez, O., Guerra-Artal, C., Isern-González, J.: Cue Combination for Robust Real-Time Multiple Face Detection at Different Resolutions. In: Moreno Díaz, R., Pichler, F., Quesada Arencibia, A. (eds.) EUROCAST 2005. LNCS, vol. 3643, pp. 398–403. Springer, Heidelberg (2005)

11. OpenCV, http://www.sourceforge.net/projects/opencvlibrary
12. Chen, T.: A Study of Spatial Color Interpolation Algorithms for Single-Detector Digital Cameras, http://www-ise.stanford.edu/tingchen/
13. Hirakawa, K., Parks, T.W.: Adaptive Homogeneity-Directed Demosaicing Algorithm. IEEE Trans. on Image Processing 14(3), 360–369 (2005)
14. Chang, E., Cheung, S., Pan, D.: Color filter array recovery using a threshold-based variable number of gradients. In: Proc. of the SPIE Conference, vol. 3650, pp. 36–43 (1999)
15. Hsu, R.L., Abdel-Mottaleb, M.: Face Detection in Color Images. IEEE Trans. on Pattern Analysis and Machine Intelligence 24(5), 696–706 (2002)

Conditional Sequence Model for Context-Based Recognition of Gaze Aversion

Louis-Philippe Morency and Trevor Darrell

MIT Computer Science and Artificial Intelligence Laboratory
Cambridge, MA 02139
{lmorency,trevor}@csail.mit.edu

Abstract. Eye gaze and gesture form key conversational grounding cues that are used extensively in face-to-face interaction among people. To accurately recognize visual feedback during interaction, people often use contextual knowledge from previous and current events to anticipate when feedback is most likely to occur. In this paper, we investigate how dialog context from an embodied conversational agent (ECA) can improve visual recognition of eye gestures. We propose a new framework for contextual recognition based on Latent-Dynamic Conditional Random Field (LDCRF) models to learn the sub-structure and external dynamics of contextual cues. Our experiments show that adding contextual information improves visual recognition of eye gestures and demonstrate that the LDCRF model for context-based recognition of gaze aversion gestures outperforms Support Vector Machines, Hidden Markov Models, and Conditional Random Fields.

Keywords: Contextual information, Conditional Random Fields, Eye gesture recognition, gaze aversion.

1 Introduction

In face to face interaction, eye gaze is known to be an important aspect of discourse and turn-taking. To create effective conversational human-computer interfaces, it is desirable to have computers which can sense a user's gaze and infer appropriate conversational cues. Embodied conversational agents, either in robotic form or implemented as virtual avatars, have the ability to demonstrate conversational gestures through eye gaze and body gesture, and should also be able to perceive similar displays as expressed by a human user.

Previous work has shown that human participants avert their gaze (i.e. perform "look-away" or "thinking" gestures) to hold the conversational floor even while answering relatively simple questions [1]. A gaze aversion gesture while a person is thinking may indicate that the person is not finished with their conversational turn. If the ECA senses the aversion gesture, it can correctly wait for mutual gaze to be re-established before taking its turn.

When recognizing visual feedback, people use more than their visual perception. Knowledge about the current topic and expectations from previous utterances help guide our visual perception in recognizing nonverbal cues. Context

A. Popescu-Belis, S. Renals, and H. Bourlard (Eds.): MLMI 2007, LNCS 4892, pp. 11–23, 2007.

information can be found from cues like the words and prosody/punctuation (e.g., word pair "do you" with question mark) of the current sentence but the real meaning and structure of these cues can sometimes be hidden (e.g., this is a yes/no question). The dynamic between these contextual cues (e.g., "do you" before the question mark) is also relevant information. An important challenge for context-based recognition is to learn these hidden sub-structures and external dynamics from the contextual cues.

In this paper, we present a framework for context-based recognition that uses Latent-Dynamic Conditional Random Field (LDCRF) models [2] to learn the hidden sub-structure and external dynamic of contextual information. The main two contributions of this paper are that we are the first to (1) show that dialog context can improve gaze aversion recognition and (2) demonstrate that LDCRF models are superior to other learning methods (i.e., SVM, CRF, and HMM) at learning relevant context and integrating it with visual observations for gaze aversion recognition. The power of LDCRFs comes from the fact that it learns the extrinsic dynamics by modeling a continuous stream of class labels, and learns internal sub-structure by utilizing intermediate hidden states.

The remainder of this paper is organized as follows. In Section 2 we review relevant related work, and in Section 3 we present our LDCRF context-based recognition framework. The details of our three set of experiments including information about the dataset, the compared models and the methodology are described in Section 4. We present and discuss the results of our experiments in Section 5. Finally, a summary and discussion of future work are provided in Section 6.

2   Related Work

Eye gaze plays an important role in face-to-face interactions. Kendon proposed that eye gaze in two-person conversation offers different functions: monitoring visual feedback, expressing emotion and information, regulating the flow of the conversation (turn-taking), and improving concentration by restricting visual input [3]. Many of these functions have been studied to create more realistic ECAs [4,5,6], but they have tended to explore only gaze directed towards individual conversational partners or objects.

A considerable body of work has been carried out regarding eye gaze and eye motion patterns for perceptive user interfaces. Velichkovsky suggested the use of eye motion to replace the mouse as a pointing device [7]. Qvarfordt and Zhai used eye-gaze patterns to sense the user interest with a map-based interactive system [8]. Li and Selker developed the InVision system which responded to a user's eye fixation patterns in a kitchen environment [9].

Context has been previously used in computer vision to disambiguate recognition of individual objects given the current overall scene category [10]. Fujie *et al.* also used HMMs to perform head nod recognition [11]. In their paper, they combined head gesture detection with prosodic low-level features computed

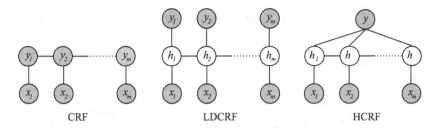

Fig. 1. Comparison of the LDCRF model [2] with two related models: CRF [14] and HCRF [16,17]. In these graphical models, x_j represents the jth observation (corresponding to the jth frame of the video sequence), h_j is a hidden state assigned to x_j , and y_j the class label of x_j (i.e. head-nod or other-gesture). Gray circles are observed variables. The LDCRF model combines the strengths of CRFs and HCRFs in that it captures both extrinsic dynamics and intrinsic structure and can be naturally applied to predict labels over unsegmented sequences.

from Japanese spoken utterances to determine strongly positive, weak positive and negative responses to yes/no type utterances.

The use of dialogue context for visual gesture recognition was first explored in [12]. In [12] they propose a late-fusion framework for incorporating dialog context in head gesture recognition. This framework was later extended to include context from conventional graphical user interfaces [13]. In both papers, the experiments were performed on head gesture recognition. This paper is the first to extend the idea of context-based recognition to recognize eye gesture. Also, the approach presented in [12,13] used multi-class SVMs to train the context-based recognizer. Unlike LDCRFs, SVMs do not model the external dynamics between classes and do not explicitly model hidden sub-structure.

LDCRF models offer several advantages over previous discriminative models (see Figure 1). In contrast to Conditional Random Fields (CRFs) [14], our method incorporates hidden state variables which model the sub-structure of gesture sequences. The CRF approach models the transitions between gestures, thus capturing extrinsic dynamics, but lacks the ability to learn the internal sub-structure. In contrast to Hidden-state Conditional Random Fields (HCRFs) [15], our method can learn the dynamics between gesture labels and can be directly applied to label unsegmented sequences. The results reported in [2] demonstrate that LDCRF outperforms models based on Support Vector Machines (SVMs), HMMs, CRFs and HCRFs on visual gesture recognition task. In this paper, we demonstrate that LDCRF models are superior to other learning methods at learning relevant context and integrating it with visual observations.

3   Context-Based Recognition Framework Using LDCRF

For reliable recognition of visual feedback during face-to-face conversational interactions, people use knowledge about the current dialogue to anticipate gestures from their interlocutors.

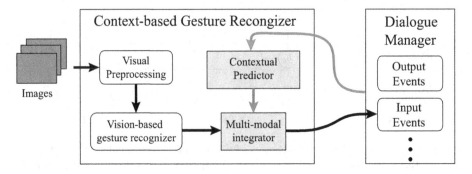

Fig. 2. Framework for context-based gesture recognition. The contextual predictor translates contextual features into a likelihood measure, similar to the visual recognizer output. The multi-modal integrator fuses these visual and contextual likelihood measures.

We can use a conversational agent's knowledge about the current dialogue to improve recognition of visual feedback (i.e., eye gestures). The dialogue manager merges information from the input devices with the history and the discourse model [18,19]. The dialogue manager contains two main sub-components, an agenda and a history: the agenda keeps a list of all the possible actions the agent and the user (i.e., human participant) can do next. This list is updated by the dialogue manager based on its discourse model (prior knowledge) and on the history. Dialogue managers generally exploit contextual information to produce output for the speech and gesture synthesizer, and we can use similar cues to predict when user visual feedback is most likely.

Following [12], we use three types of contextual features easily available from the dialogue manager: lexical features, prosody and punctuation, and timing. The contextual information is extracted from the dialogue manager rather than directly accessing internal ECA states. This strategy makes it possible to extract dialogue context without any knowledge of the internal representation and therefore makes it appliable to most ECA architectures. Figure 2 shows the general architecture of the context-based recognition framework.

In the following subsections we first give a formal description of the LD-CRF and then show how LDCRF is integrated in the context-based recognition framework.

3.1   LDCRF Model

As described in [2], the task of the LDCRF model is to learn a mapping between a sequence of observations $\mathbf{x} = \{x_1, x_2, ..., x_m\}$ and a sequence of labels $\mathbf{y} = \{y_1, y_2, ..., y_m\}$. Each y_j is a class label for the jth frame of a video sequence and is a member of a set $\mathcal{Y}$ of possible class labels, for example, $\mathcal{Y} = \{\texttt{gaze-aversion}, \texttt{other-gesture}\}$. Each frame observation x_j is represented by a feature vector $\phi(x_j) \in \mathbf{R}^d$, for example, the head velocities at each frame. For each sequence, we also assume a vector of "sub-structure" variables

$\mathbf{h} = \{h_1, h_2, ..., h_m\}$. These variables are not observed in the training examples and will therefore form a set of hidden variables in the model.

Given the above definitions, we define a latent conditional model:

$$P(\mathbf{y} \mid \mathbf{x}, \theta) = \sum_{\mathbf{h}} P(\mathbf{y} \mid \mathbf{h}, \mathbf{x}, \theta) P(\mathbf{h} \mid \mathbf{x}, \theta). \tag{1}$$

where θ are the parameters of the model.

To keep training and inference tractable, we restrict the LDCRF model to have disjoint sets of hidden states associated with each class label. Each h_j is a member of a set $\mathcal{H}_{y_j}$ of possible hidden states for the class label y_j. We define $\mathcal{H}$, the set of all possible hidden states, to be the union of all $\mathcal{H}_y$ sets. Since sequences which have any $h_j \notin \mathcal{H}_{y_j}$ will by definition have $P(\mathbf{y} \mid \mathbf{h}, \mathbf{x}, \theta) = 0$, we can express the LDCRF model as:

$$P(\mathbf{y} \mid \mathbf{x}, \theta) = \sum_{\mathbf{h}:\forall h_j \in \mathcal{H}_{y_j}} P(\mathbf{h} \mid \mathbf{x}, \theta). \tag{2}$$

Given a training set consisting of n labeled sequences $(\mathbf{x_i}, \mathbf{y_i})$ for $i = 1...n$, training is done following [20,14] using this objective function to learn the parameter θ^*:

$$L(\theta) = \sum_{i=1}^{n} \log P(\mathbf{y}_i \mid \mathbf{x}_i, \theta) - \frac{1}{2\sigma^2}||\theta||^2 \tag{3}$$

The first term in Eq. 3 is the conditional log-likelihood of the training data. The second term is the log of a Gaussian prior with variance σ^2, i.e., $P(\theta) \sim \exp\left(\frac{1}{2\sigma^2}||\theta||^2\right)$.

For testing, given a new test sequence $\mathbf{x}$, we want to estimate the most probable sequence labels $\mathbf{y}^*$ that maximizes our conditional model:

$$\mathbf{y}^* = \arg\max_{\mathbf{y}} \sum_{\mathbf{h}:\forall h_i \in \mathcal{H}_{y_i}} P(\mathbf{h} \mid \mathbf{x}, \theta^*) \tag{4}$$

For a more detailed discussion of LDCRF training and inference see [2].

3.2   LDCRF Context-Based Recognition

The contextual predictor outputs a likelihood measurement at the same frame rate as the vision-based recognizer so that the multi-modal integrator can merge both measurements. For this reason, feature vectors $\mathbf{x}_j$ are computed at every frame j (even though the contextual features do not directly depend on the input images).

For the LDCRF model, the likelihood measurement for a specific gesture g is equal to the marginal probability $P(y_j = g \mid \mathbf{x}, \theta^*)$. This probability is equal to the sum of the marginal probabilities for the hidden states part of the subset $\mathcal{H}_g$:

$$P(y_j = g \mid \mathbf{x}, \theta^*) = \sum_{\mathbf{h}:\forall h_j \in \mathcal{H}_g} P(\mathbf{h} \mid \mathbf{x}, \theta^*) \tag{5}$$

where $\mathbf{x}$ is the concatenation of all the feature vectors $\mathbf{x}_j$ for the entire sequence and θ^* are the model parameters learned during training. When testing offline, the marginal probabilities are computed using a forward-backward belief propagation algorithm. To estimate the marginal probabilities online, it is possible to define $\mathbf{x}$ as the concatenation of all feature vectors up to frame j and use the forward-only belief propagation algorithm.

Our integration component takes as input the likelihood measurement from the contextual predictor and the visual observations from the vision-based head gesture recognizer, and recognizes whether a head gesture has been expressed by the human participant. The output from the integrator is further sent to the dialogue manager or the window manager so it can be used to decide the next action of the ECA.

4   Experiments

We designed our experiments to demonstrate how contextual information can improve eye gesture recognition and to demonstrate the superior performance of LDCRF on context-based recognition compared to baseline methods. We performed three series of experiments:

Experiment 1. Where we compare the vision-only approach with the context-based recognition using LDCRF models. The goal of this experiment is to show that dialog context can improve eye gesture recognition

Experiment 2. Where we compare the LDCRF model to SVM, CRF and HMM models for context-based recognition of gaze aversion. In this set of experiments, the contextual predictor and the multimodal integrator are both trained using the same model (either LDRCF, SVM, CRF or HMM). The goal of this experiment is to show the superiority of LDCRF for context-based recognition.

Experiment 3. Where we first train the contextual predictor with the LDCRF model and then train the multimodal integrator with one of the four model. The goal of this experiment is to analyze the relative importance of LDCRF for contextual prediction and multimodal integration.

In the following subsections, we first describe our dataset used in our experiments, then present the models used to compare the performance of the LDCRF model, and finally describe our experimental methodology.

4.1   Eye Gesture Dataset

Our dataset came from a user study that shown that human participants naturally perform gaze aversion gestures when interacting with an avatar [1]. The goal of this dataset is to differentiate gaze aversion gestures from all other type of eye gestures (e.g., eye contact or deictic gestures). Our dataset consist of 6 human participants interacting with a virtual embodied agent. Each video sequence lasted approximately 10-12 minutes, and was recorded at 30 frames/sec, for a total of 105,743 frames. During these interactions, human participants would rotate their head up to +/-70 degrees around the Y axis and +/-20 degrees around the X axis, and would also occasionally move their head, mostly along the Z axis.

The dataset was labeled with the start and end points of each gaze aversion gestures. Each frame was labeled either as gaze-aversion or as other-gesture which included sections of video where people were looking at the avatar or performing deictic gestures. The contextual cues from the dialogue manager (spoken utterances with start time and duration) were recorded during each interaction and were later automatically processed to create the contextual features necessary for the contextual predictor. The previous section showed how the contextual features are automatically computed.

For each video sequence, the eye gaze was estimated using the view-based appearance model described in [1] and for each frame a 2-dimensional eye gaze estimate was obtained. The eye gaze estimates were logged online with the contextual cues. For this dataset, the vision-based recognizer is a LDCRF model trained and validated offline on the same training and validation sets used for the contextual predictor and the multi-modal integrator.

4.2 Models

In our experiments, the LDCRF model is compared with three models: Conditional Random Field (CRF), Hidden Markov Model (HMM) and Support Vector Machine (SVM).

Conditional Random Field. As a first baseline, we trained a single CRF chain model where every gesture class has a corresponding state label. During evaluation, marginal probabilities were computed for each state label and each frame of the sequence using belief propagation. The optimal label for a specific frame is typically selected as the label with the highest marginal probability. In our case, to be able to plot ROC curves of our results, the marginal probability of the primary label (i.e. `gaze-aversion`) was thresholded at each frame, and the frame was given a positive label if the marginal probability was larger than the threshold. The objective function of the CRF model contains a regularization term similar to the regularization shown in Equation 3 for the LDCRF model. During training and validation, this regularization term was validated with values $10^k, k = -3..3$.

Support Vector Machine. As a second baseline, a multi-class SVM was trained with one label per gesture using a Radial Basis Function (RBF) kernel. Since the SVM does not encode the dynamics between frames, the training set was decomposed into frame-based samples, where the input to the SVM is the head velocity or eye gaze at a specific frame. The output of the SVM is a margin for each class. This SVM margin measures how close a sample is to the SVM decision boundary [21]. The margin was used to plot the ROC curves. During training and validation, two parameters were validated: C, the penalty parameter of the error term in the SVM objective function, and γ, the RBF kernel parameter. Both parameters were validated with values $10^k, k = -3..3$.

Hidden Markov Model. As a third baseline, an HMM was trained for each gesture class. We trained each HMM with segmented subsequences where the frames of each subsequence all belong to the same gesture class. This training set contained the same number of frames as the one used for training the other models except frames were grouped into subsequences according to their label. The HMMs trained on subsequences are concatenated into a single HMM with the number of hidden states equal to the sum of hidden states from each individual HMM. For example, if the recognition problem has two labels and each individual HMM is trained using 3 hidden states, then the concatenated HMM will have 6 hidden states. To estimate the transition matrix of the concatenated HMM, we compute the Viterbi path of each training subsequence, concatenate the subsequences into their original order, and then count the number of transitions between hidden states. The resulting transition matrix is then normalized so that its rows sum to one. At testing, we apply the forward-backward algorithm on the new sequence, and then sum at each frame the hidden states associated with each class label. The resulting HMM can seen as a generative version of our LDCRF model. During training and validation, we varied the number of states from 1 to 6 and the number of Gaussians per mixture from 1 to 3.

Latent-Dynamic Conditional Random Field. Our LDCRF model was trained using the objective function described in [2]. During evaluation, we compute ROC curves using the maximal marginal probabilities of Equation 4. During training and validation, we varied the number of hidden states per label (from 2 to 6 states per label) and the regularization term (with values $10^k, k = -3..3$).

4.3   Methodology

In our experiments, the vision-based recognizer was trained and tested using LDCRF since this model gave the best performance for the visual recognition task (see [2] for details). The contextual predictors and multi-modal integrator (also referred as "Fusion" in the result section) were trained using one of the four models described in the previous subsection. The contextual features were computed from the dialog context of the avatar using the technique described in [12].

The experiments were performed using a leave-one-out testing approach. For validation, we did holdout cross-validation where a subject is randomly picked from the training set and kept for validation. The optimal validation parameters were picked based on the equal error rate for the validation set.

The dataset contained an unbalanced number of other-gesture frames. To have a balanced training set and reduce the training time, the training dataset was preprocessed to create a smaller training dataset containing an equal number of other-gesture and *transition* subsequences. Each *transition* subsequence includes frames from one complete gesture subsequence and frames before and after the gesture labeled as other-gesture. The size of the other-gesture gap before and after the gesture was randomly picked between 2 and 50 frames. The number of transition subsequences was equal to the number of ground truth gestures in the original training set. Other-gesture subsequences were randomly extracted from the original sequences with length varying between 30-60 frames.

5   Results and Discussion

For the ROC curves shown in this section, the true positive rate is computed by dividing the number of recognized frames by the total number of ground truth frames. Similarly, the false positive rate is computed by dividing the number of falsely recognized frames by the total number of other-gesture frames.

Figure 3 shows the results of Experiment 1 where we compare the LDCRF vision-only approach with the LDRCF context-based approach. We can see in this figure that context information does improve recognition of eye gesture. The ROC curve of LDCRF combining both vision and context is higher than that of LDCRF using only vision without context. Using t-test analysis, the difference between the two curves, calculated based on the equal error rates, is statistically significant (one-tail $p = 0.043$).

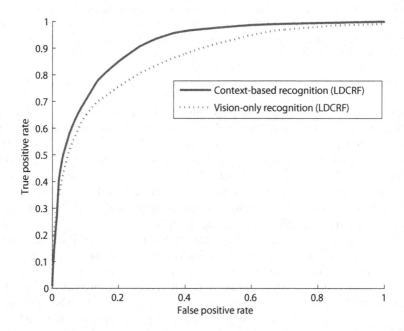

Fig. 3. Results from Experiment 1 comparing vision-only approach with context-based recognition using LDCRF models. We can see that dialog context significantly improves (p-value = 0.043) the gaze aversion recognition performance.

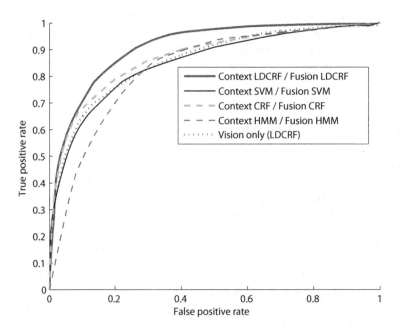

Fig. 4. Results from Experiment 2 comparing the LDCRF model to SVM, CRF and HMM models for context-based recognition of gaze aversion. Both the contextual predictor and the multimodal integrator were trained using the same model. The ROC curves show the performance of each trained multimodal integrator. LDCRF outperforms all three other models with statistically significant differences for SVM and HMM (p-values equal to 0.0329, and 0.0343 respectively).

Figure 4 shows the results from Experiment 2 where we compare the LDCRF model to SVM, CRF and HMM models for context-based recognition of gaze aversion. LDCRF outperforms all three other models (SVM, CRF and HMM) for context-based recognition. A paired t-test analysis over all tested subjects returns a one-tail p-value of 0.0329, 0.0717 and 0.0343 when comparing the equal error rate performance of LDCRF with SVM, CRF and HMM respectively. This analysis shows statistically significant improvement using the LDCRF model when compared to both SVM and HMM models.

Figure 5 shows the results of Experiment 3 where we analyze the relative importance of LDCRF for contextual prediction and multimodal integration by running a new set of experiments where only the multimodal integrator changes. The ROC curves in this figure show that LDCRF model outperforms all three other models. This demonstrates the superiority of LDCRF for the multimodal integration task. Also, by comparing Figures 4 and 5, we can see that both SVM and HMM curves improve, confirming the utility of LDCRF as a contextual predictor.

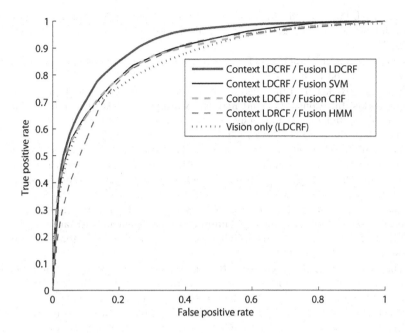

Fig. 5. Results from Experiment 3 analyzing the relative importance of LDCRF for contextual prediction and multimodal integration. Note that the contextual predictor is the same for all four cases and only the multimodal integrator changes in each case. This result demonstrates the superiority of LDCRF for the multimodal integration task and by comparing with Figures 4, we can see that both SVM and HMM curves improve, confirming the utility of LDCRF as a contextual predictor.

6   Conclusion

In this paper, we investigated how dialog context from an embodied conversational agent (ECA) can improve visual recognition of eye gestures. We proposed a new framework for contextual recognition based on Latent-Dynamic Conditional Random Field (LDCRF) models to learn the sub-structure and external dynamic of contextual cues. Our experiments showed that adding contextual information improves visual recognition of eye gestures and demonstrated that LDCRF models for context-based recognition outperform Support Vector Machines, Hidden Markov Models, and Conditional Random Fields for our visual feedback recognition tasks.

References

1. Morency, L.P., Christoudias, C.M., Darrell, T.: Recognizing gaze aversion gestures in embodied conversational discourse. In: Proceedings of the International Conference on Multi-modal Interfaces, Banff, Canada (2006)

2. Morency, L.P., Quattoni, A., Darrell, T.: Latent-dynamic discriminative models for continuous gesture recognition. Technical Report MIT-CSAIL-TR-2007-002, MIT CSAIL (2007)
3. Kendon, A.: Some functions of gaze direction in social interaction. Acta Psyghologica 26, 22–63 (1967)
4. Traum, D., Rickel, J.: Embodied agents for multi-party dialogue in immersive virtual worlds. In: Alonso, E., Kudenko, D., Kazakov, D. (eds.) Embodied agents for multi-party dialogue in immersive virtual worlds. LNCS (LNAI), vol. 2636, pp. 766–773. Springer, Heidelberg (2003)
5. Vertegaal, R., Slagter, R., van der Veer, G., Nijholt, A.: Eye gaze patterns in conversations: There is more to conversational agents than meets the eyes. In: CHI 2001. Proceedings of the SIGCHI conference on Human factors in computing systems, pp. 301–308 (2001)
6. Fukayama, A., Ohno, T., Mukawa, N., Sawaki, M., Hagita, N.: Messages embedded in gaze of interface agents — impression management with agent's gaze. In: CHI 2002. Proceedings of the SIGCHI conference on Human factors in computing systems, pp. 41–48 (2002)
7. Velichkovsky, B.M., Hansen, J.P.: New technological windows in mind: There is more in eyes and brains for human-computer interaction. In: CHI 1996. Proceedings of the SIGCHI conference on Human factors in computing systems (1996)
8. Qvarfordt, P., Zhai, S.: Conversing with the user based on eye-gaze patterns. In: CHI 2005. Proceedings of the SIGCHI conference on Human factors in computing systems, pp. 221–230 (2005)
9. Li, M., Selker, T.: Eye pattern analysis in intelligent virtual agents. In: (IVA 2002). Conference on Intelligent Virutal Agents, pp. 23–35 (2001)
10. Torralba, A., Murphy, K.P., Freeman, W.T., Rubin, M.A.: Context-based vision system for place and object recognition. In: IEEE Intl. Conference on Computer Vision (ICCV), Nice, France (2003)
11. Fujie, S., Ejiri, Y., Nakajima, K., Matsusaka, Y., Kobayashi, T.: A conversation robot using head gesture recognition as para-linguistic information. In: Proceedings of 13th IEEE International Workshop on Robot and Human Communication, RO-MAN 2004, pp. 159–164 (2004)
12. Morency, L.-P., Sidner, C., Lee, C., Darrell, T.: Contextual recognition of head gestures. In: Proceedings of the International Conference on Multi-modal Interfaces (2005)
13. Morency, L.-P., Darrell, T.: Head gesture recognition in intelligent interfaces: The role of context in improving recognition. In: Proceedings of Intelligent User Interfaces, Australia (2006)
14. Lafferty, J., McCallum, A., Pereira, F.: Conditional random fields: probabilistic models for segmenting and labelling sequence data. In: ICML (2001)
15. Quattoni, A., Collins, M., Darrell, T.: Conditional random fields for object recognition. In: NIPS (2004)
16. Gunawardana, A., Mahajan, M., Acero, A., Platt, J.C.: Hidden conditional random fields for phone classification. In: INTERSPEECH (2005)
17. Wang, S., Quattoni, A., Morency, L., Demirdjian, D., Darrell, T.: Hidden conditional random fields for gesture recognition. In: CVPR (2006)
18. Nakano, Reinstein, Stocky, Cassell, J.: Towards a model of face-to-face grounding. In: Proceedings of the Annual Meeting of the Association for Computational Linguistics, Sapporo, Japan (2003)

19. Rich, N., Sidner, Lesh: Collagen: Applying collaborative discourse theory to human–computer interaction. AI Magazine, Special Issue on Intelligent User Interfaces 22(4), 15–25 (2001)
20. Kumar, S., Herbert., M.: Discriminative random fields: A framework for contextual interaction in classification. In: ICCV (2003)
21. Vapnik, V.: The nature of statistical learning theory. Springer, Heidelberg (1995)

Meeting State Recognition from Visual and Aural Labels

Jan Curín, Pascal Fleury, Jan Kleindienst, and Robert Kessl

IBM, Prague, Czech Republic
{jan_curin,pascal.fleury,jankle,kesslr}@cz.ibm.com

Abstract. In this paper we present a meeting state recognizer based on a combination of multi-modal sensor data in a smart room. Our approach is based on the training of a statistical model to use semantical cues generated by perceptual components. These perceptual components generate these cues in processing the output of one or multiple sensors. The presented recognizer is designed to work with an arbitrary combination of multi-modal input sensors. We have defined a set of states representing both meeting and non-meeting situations, and a set of features we base our classification on. Thus, we can model situations like *presentation* or *break* which are important information for many applications. We have hand-annotated a set of meeting recordings to verify our statistical classification, as appropriate multi-modal corpora are currently very sparse. We have also used several statistical classification methods for the best classification, which we validated on the hand-annotated corpus of real meeting data.

1 Introduction

Today's applications and interaction systems tend to interact with users at inappropriate times as they have no way to determine when such an appropriate time for the interaction is. Sensor-based statistical models recognizing human activity in smart environments offer a potential solution to this problem. For example, a smart room equipped with audio and video sensors may automatically block phone calls for a person engaged in interruption-sensitive tasks such as presenting a lecture.

In this article, we present an approach to sensor-based statistical meeting state recognition in smart environments. Multiple aspects of this challenge have been tackled in past works, based on single modalities like speech features [1] or video features [2], simple on/off meeting states classification [3], or solving the aspect of data collection [4]. A recurring issue for such works is the limited sets of data available for training. Most of the works presented previously consider intra-state meetings, whereas our approach also classifies breaks and non-meeting times (extra-meeting data). We needed multi-modal data also presenting such extra-meeting material, presenting an additional limitation to the usability of existing corpora like AMI [5], M4 [6] or VACE [7]. We used a meeting corpus from the CLEAR evaluation [8], that contained all we need but needed annotation. This is presented in Section 4.1.

A. Popescu-Belis, S. Renals, and H. Bourlard (Eds.): MLMI 2007, LNCS 4892, pp. 24–35, 2007.

The next section introduces some of the terminology used in our meeting state detector. Section 3 presents the considered meeting states, the feature extraction and their selection for the meeting state detection. Section 4 describes the data we used, what annotations we had to create and how we used it for training and validation. Section 5 presents the meeting state detector's performance, and Section 6 concludes the paper's findings.

2   The Meeting Recognition Framework

We first introduce the architecture context, in which we carried the meeting recognition task. Let us describe the framework on a connector scenario proposed and exploited in [9]. The connector service is responsible for detecting acceptable interruptions (phone call, SMS, targeted audio . . .) of a particular person in the smart room. During the meeting, for example, a member of the audience might be interrupted by a message during the presentation, whereas the service blocks any calls for the meeting presenter.

We define the following abstraction levels: sensors, perceptual components, situation modeling, and services. We assume an information flow from sensors (e.g., cameras, microphones) to perceptual components (e.g., body trackers, speech recognition engines) and finally to situation model and services in a pipelined fashion. Figure 1 shows a schema of setup used in our experiments.

2.1   Perceptual Components

We suppose that the smart room is equipped with multiple cameras and microphones on the **sensor** level. The audio and video data are streamed into the following **perceptual components**:

- **Body Tracker** is a video-based component (theoretically it might be also audio or audio/video based tracking) for tracking 3D position of participants. It provides 3D coordinates of the head centroid for each person in the room.
- **Facial Features Tracker** is a video-based component providing information about face visibility for each participant. Multiple Facial Feature Trackers are running for each camera stream. We use the nose visibility feature in our setup.
- **Automatic Speech Recognition** is an audio-based component providing speech transcription for each participant in the room. In our scenario we expect the participant's label with speech transcriptions, through the use of wearable microphones or speaker identification for far-field microphones.

2.2   Situation Modeling

The **situation model** is the level transforming a set of facts about the environment into a set of situations [10]. As shown in Figure 1, the main information flows from the lower levels to the upper levels. Higher levels represent semantically higher abstractions.

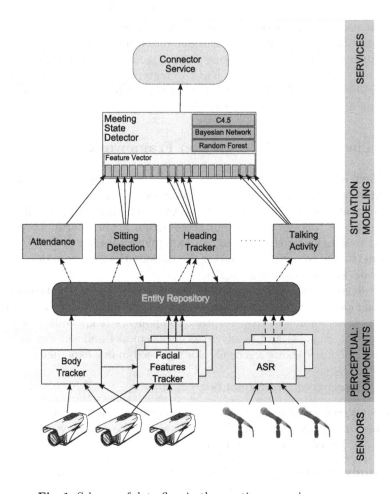

Fig. 1. Schema of data flow in the meeting scenario

The set of facts comes from the *perceptual components*, which collect and aggregate sensor information into *entities*. Entities—the base representation for the situation model—are stored in the **Entity Repository**. In the meeting scenario, the entities are objects from offices or meeting rooms (table, chair, whiteboard, PDA, etc.) and the people. Each entity may have a variable number of *attributes*, which describe the entity. Examples are PERSON ID and LOCATION for a person in the meeting room, or LOCATION and HEADING for a (movable) whiteboard. For each of these entities, the perceptual components send a sequence of attribute update *events*, which we call *streams*. A stream is relative to a single attribute of a single entity. The situation model considers these entities as having a one-to-one mapping to the real world in the modeled environment.

The situation model will use the attributes of the entities provided by the perceptual components, and infer the current state of a set of *situations*. Example of a situation is the *Meeting*, and its states may be PRESENTATION, DISCUSSION, or BREAK. The situation model will then generate an event, indicating that a particular situation has changed its state.

The goal of the situation model is to extract semantically higher information from these facts. This is achieved through *situation machines* (SM). A situation machine models a single situation which consists of a set of *states*, the state of the situation machine is detected by observing entities in the entity repository or current states of other situation machines.

In the connector scenario the following situation machines are used:

- **Attendance** tracks the number of participants (person entities) in the room.
- **Motion Detection** reports how many participants have moved over a certain distance threshold in a given time period.
- **Heading Tracker** infers the heading (head orientation) for each person from her position and face visibility for all cameras (provided by *Facial Features Tracker*).
- **Attention Direction** tracks the number of participants looking towards the same spot, using the heading information provided by the *Heading Tracker*.
- **Sitting Detection** infers whether a particular person is sitting or not from the Z (height) coordinate provided by the *Body Tracker*.
- **Crowd Detection** searches for crowds of people. A crowd is a group of at least 3 people where the distance between neighboring participants is less than a given threshold.
- **Talking Activity** tracks duration of speech activity and number of speaker changes in a given period.
- **Meeting State Detection** infers the current state of the meeting in the room. It uses information provided by other situation machines with the use of statistical methods for the meeting state detection described later this article. The information about the current meeting state and the current speaker is provided to the *Connector* service.

Certain situation machines can update attributes of entities in the entity repository. For example the *Heading Tracker SM* updates the HEADING attribute of person entities, whereas the *Sitting Detection SM* propagates the observed sitting or standing position of a participant back to the repository. These information may be used by other situation machines or can be useful for the visualization of the current scene.

3   Meeting State Recognition Task

The statistical-model presented in this article is trained to recognize meeting activity in smart rooms, i.e. a meeting environment equipped with sensors. We make no assumptions about the shape and dimensions of the meeting space, it could be a small meeting room or a large lecture hall. Similarly, we set no bounds on meeting attendance, it may have any number of participants.

The meeting recognizer works on features extracted from the stream of sensor data. We make the assumption that the smart room's video and audio sensors, through the perceptual components, provide at least one of the following streams:

1. The location of people in the room (2D or 3D coordinates);
2. The head orientation of each person in the room (angle);
3. The speech activity of a person or a group of persons (speech vs. silence, not transcription).

We designed the recognizer to work with any subset and/or combination of these sensor streams. This is based on the observed practical consideration that the above described outputs are generated by independent sensors that may or may not run at the same time. Thus we wanted our recognizer to work in the situation where only a body tracker is running in a smart room (case 1), as well as support the case when a head pose detector and a speech-activity detector are also available (case 1+2+3).

The meeting states which are of our interest are described in Table 1.

Table 1. Recognized Meeting States

State	Description
no meeting	empty room, people entering/leaving, drinking coffee
presentation	presentation of lecture or seminar is going on
discussion	discussion or around-the-table meeting

3.1 Feature Extraction

For the designing, debugging and running the situation modeling we have used SITCOM tool [11]. SITCOM (Situation Composer) is a simulator tool and runtime for the development of context-aware applications and services. SITCOM is able to simulate perceptual and other context acquisition components and allows the composition of situation models into hierarchies to provide event filtering, aggregation, and combination.

The SITCOM tool was also used for feature extraction; the data coming from various sensors (e.g. body tracker, speech-silence detector) where proceed by the set of situation machines producing the corresponding feature vectors, sampled at defined regular time intervals.

For statistical model training, tuning, and evaluation, we have used the WEKA tool [12], evaluating various statistical modules for classification and feature selection.

To compute the features, we wrote several *situation machines* (SM). We have used all 7 situation machines presented in Section 2.2. Then, the output of each such SM will be concatenated into a feature vector, that is provided to the meeting state classifier. Note that some of the features from the SMs are fed back into the entities they emanate from, as they are information that may be reused by other SMs. Depending on the availability of sensors and perceptual components, the feature vector may be incomplete, therefore the statistical classification used in the meeting classifier has to deal with such incomplete information.

SMs are also useful when missing information needs to be inferred. For example, the heading of the people was not explicitly included in the meeting recordings, so we used each camera's nose visibility information to compute an approximate heading.

The feature vectors are generated at regular sampling intervals. We have used one second interval as it is the frequency of 3D position and facial features annotations.

3.2 Feature Selection

Feature selection is an iterative process, as both the set of features and the domain of values have to be selected. We have experimented with continuous values such as the number of participants, or discrete values such as relative or proportional number of moving, talking, or sitting participants (*none, more than a third, more than half, all*). Parameterized features with boolean values, such as "*Is the number of moving participants higher than half the participants?*", typically generated better scores than features with proportional value such as "*What is the proportion of moving participants?*".

We also found useful to "discretize" some of the continuous values, for example *the time the current talker speaks* has one of the following values: *no speech, 10 seconds, 1 minute, 3 minutes, more than 5 minutes*.

Finally, trying several attribute selection methods available in WEKA, we have selected 8 of 22 features. Table 2 lists the selected features. The impact of feature selection is discussed in the next section.

Table 2. Feature Selection

ID	mod	feature description	values	selected
1	P	Number of participants	number	yes
2	P	Number of moving participants > 1	true/false	no
3	P	Proportion of moving participants $>^1/_3$	true/false	no
4	P	Proportion of moving participants $>^1/_2$	true/false	yes
5	P	Number of sitting participants > 1	true/false	no
6	P	Proportion of sitting participants $>^1/_3$	true/false	no
7	P	Proportion of sitting participants $>^1/_2$	true/false	yes
8	P	Number of people crowds	number	no
9	H	Number of attention directions	number	no
10	H	Proportion of people looking same direction $>^1/_3$	true/false	no
11	H	Proportion of people looking same direction $>^1/_2$	true/false	yes
12	H	Proportion of people looking same direction $> 90\%$	true/false	no
13	H	Proportion of people looking same direction $= 100\%$	true/false	no
14	H	Crowds Concentration	true/false	no
15	S	Number of talking participants	number	no
16	S	Number of speaker changes in last 2s	rel. num.	no
17	S	Number of speaker changes in last 5s	rel. num.	no
18	S	Number of speaker changes in last 20s	rel. num.	yes
19	S	Did the speaker changed in last 2s	true/false	yes
20	S	Did the speaker changed in last 5s	true/false	yes
21	S	Did the speaker changed in last 20s	true/false	no
22	S	Time the current talker speaks	time intervals	yes

The features and results are separated into three categories, based on the available perceptual components (or annotations in our case), i.e. position (**P**), heading (**H**), and speech activity (**S**).

4 Data Collection and Annotation

4.1 CLEAR Corpus

The corpus we use in our experiments was created for the purpose of the CLEAR evaluation task [8]. We found that this corpus is the only data set that provides multi-modal, multi-sensory recordings of realistic human behavior and interaction in the meeting scenarios. We have used 5 seminars prepared for the CLEAR 2007 evaluation campain and 1 seminar from CLEAR evaluation 2006. The total length of the recordings is 1 hour 51 minutes, and it includes rich manual annotations of both audio and visual modalities. In particular, it contains a detailed multi-channel transcription of the audio recordings that includes speaker identification, and acoustic condition information. Video labels provide multi-person head location in the 3D space, as well as information about the 2D face bounding box and facial feature locations visible in all camera views.

We have also checked other recently collected corpora, such as the AMI project, M4 corpus, and VACE project (reference in the Introduction Section), but these corpora are either limited to a single data collection site or contain scripted scenarios with static interaction among the meeting participants. They do not contain enough non-meeting material either, therefore, these data sets were not suitable for our task.

4.2 Meeting State Annotation

As the CLEAR recording does not include the meeting state annotation, we ask three human annotators to label the corpus by the appropriate meeting states. They were provided with sequences of camera pictures and schematic visualization figures of the observed attributes provided directly by the manual annotation or inferred by special situation machines (*Heading Tracker SM* and *Sitting Detection SM*), i.e.:

1. The location of people in the room (2D or 3D coordinates);
2. The head orientation of each person in the room (angle);
3. Information if the person is sitting or standing;
4. The speech activity of a person or a group of persons (speech vs. silence, transcription not used).

Their task was to identify one of the meeting state described in Table 1. Figure 2 shows the schematic visualization and corresponding pictures provided to the annotators (for presentation and break state). We can see the position and heading of each participant, visible legs for sitting persons, and highlighted bubbles for talking persons. Bubbles display persons IDs or names, if they are

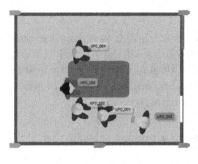 

PRESENTATION

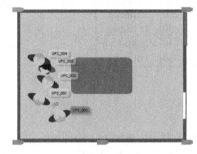 

NO MEETING / BREAK

Fig. 2. Screenshots of presentation and break in the UPC'06 recording

Table 3. Inter-anotator agreement (in %) for the first-round annotations

	A1 x A2	A3 x A2	A3 x A1
AIT'07	43.00	71.75	60.50
IBM'07	87.61	78.96	72.98
ITC'07	74.25	75.75	97.75
UKA'07	62.50	57.83	89.33
UPC'06	78.08	65.20	84.19
UPC'07	87.89	82.12	81.69
average	**72.22**	**71.60**	**81.07**

know to the system. For annotators convenience, objects like table, whiteboard, or door are also displayed in the schematic view, though they are not used for the experiments.

We found that the annotation of meeting states was not straightforward, as it can be seen on the inter-annotator agreement measure for individual recordings presented in the Table 3. Therefore we have decided to create a *gold annotation* by reviewing the disagreed sections. The *gold annotation* was then used in our experiments.

4.3   Data Split

We have divided the data set into two parts of similar size, one for training and other for evaluation. The training part consists of the *ITC'07, UKA'07 & UPC'07* seminars. The *IBM07, AIT07 & UPC06* seminars are in the evaluation part. Table 4 shows the length of the individual recordings, number of generated feature vectors and the data set they are belonging to.

Table 4. Data sizes of recordings used for training or evaluation

	length	feature vectors	data set
AIT'07	20 min	1200	evaluation
IBM'07	20 min	1200	evaluation
ITC'07	20 min	1200	training
UKA'07	20 min	1200	training
UPC'06	13 min	780	evaluation
UPC'07	23 min	1380	training

5   The Resulting Model

We have decided to present the results separately for several categories of "modality", assuming that all the perceptual components or sensors may not be available in a particular room configuration. The modalities are: position (**P**), heading (**H**), and speech activity (**S**).

5.1   Selecting the Model

In parallel with the feature extraction and selection process we have also been searching for the statistical model which provides the most relevant results for our classification task. Table 5 shows the percentage of correctly classified instances (10-fold cross-validation) using the selected features on training scenarios for the following classifiers: a *zero rule* classifier, a simple classifier that always predicts the most likely meeting state, as observed in training data, a *Bayesian network* with the structure trained from data, a decision-tree based classifier (*C4.5*), and a *random forest* classifier. For details about these classifiers and their parameters please refer to [12].

Among the tested methods we have identified the *random forest* classifier as the most reliable for our task. This was an expected result, as the *random forest*

Table 5. Comparison of Different Classification Methods

modality	P	PH	PS	PHS
Zero Rule	72.72	72.72	72.72	72.72
Bayesian Network	77.07	78.39	82.77	82.98
C4.5 Classifier	77.65	78.53	82.32	83.35
Random Forest	77.86	79.48	83.14	84.09

(forest of decision trees) can better adapt to sparse data than the single decision tree in a *C4.5* classifier or the fixed network structure in a *Bayesian network*.

5.2 Evaluation of Results

We evaluate the results on the evaluation part of CLEAR data (*IBM07, AIT07 & UPC06* seminars). In Table 6, we present the results for model (*random forest* classifier) build on training data and applied to the evaluation data with the selection of features (*A*) listed in Table 2. For the comparison, we also present results with the full set of features (*B*), figures for simple *zero rule* classifier (*C*) and for a 10-fold cross-validation of the evaluation data for selected features (*D*). Numbers are the percentage of correctly classified instances for different types of perceptual input.

Table 6. Results on evaluation scenarios

model/evaluation	P	PH	PS	PHS	2 states
trained & evaluated on selected features (*A*)	72.23	71.50	76.60	**76.96**	95.91
trained & evaluated on all features (*B*)	73.31	71.22	76.86	**75.68**	95.84
zero rule for evaluation data (*C*)	67.06	67.06	67.06	67.06	87.32
cross-validation on evaluation data (*D*)	78.06	78.76	81.96	82.94	98.92

The numbers for the zero rule classifier on evaluation data (*C*) and for the cross-validation on evaluation data (*D*) can be seen as lower and upper bounds for this task. The feature selection shows improvement for the system with all perceptual inputs available, i.e. modalities: position + heading + speech (**PHS**). Decrease of reliability for other combinations of modalities is due the fact that the selection of features was performed on the full set of modalities.

Adding the speech modality consistently improves the results on real data thus confirming our expectations. The resulting model yields a system performing significantly better than the simple classifier that always predicts the most likely value.

Both the classification method and the human annotator were uncertain in distinguishing between *presentation* and *discussion* states. The last column (labeled '2 states') of Table 6 shows figures for two states only (*meeting / no meeting*) with a full set of modalities (**PHS**). In this case, *presentation* and *discussion* states are labeled and evaluated as one state called *meeting*.

We believe that the performance is reasonable for the intended use.

6 Conclusions

A step in the direction of full-automatic meeting annotation has been presented in this document. We have developed a technique to use multiple modalities for the automatic segmentation of activities in a smart room into several meeting and non-meeting states.

Our approach differs from previous works in that it bases the state detection on the output of perceptual components, which do some processing and combination of sensor outputs. Such perceptual components produce facts about people's presence, location, pose, and voice activity in the room. These facts may themselves be the result of the analysis of multiple sensors. Then, we used a situation model to statistically detect the current activity in the room from the set of facts delivered by these perceptual components. Finally, from the room activities we compute the meeting state.

Annotated meetings providing both meeting and non-meeting material in multiple modalities are currently very sparse, so we had to annotate ourselves some data to verify our algorithms. This has been done on the CLEAR dataset, that is the only one we found that provides multimodal non-meeting material. This new corpus has been used to validate our results, and has shown to be able to classify the meeting status correctly in 77% of the cases, for a data set where the inter-annotator agreement is between 71% and 81% depending on the annotators.

Acknowledgements

We would like to acknowledge support of this work by the European Commission under IST FP6 integrated project CHIL (Computers in the Human Interaction Loop), contract number 506909.

References

1. Banerjee, S., Rudnicky, A.I.: Using simple speech-based features to detect the state of a meeting and the roles of the meeting participants. In: Proceedings of ICSLP 2004, Jeju Island, Korea (2004)
2. Hakeem, A., Shah, M.: Ontology and taxonomy collaborated framework for meeting classification. In: ICPR 2004. Proceedings of the 17th International Conference on Pattern Recognition (2004)
3. Wang, J., Chen, G., Kotz, D.: A meeting detector and its applications. Technical Report TR2004-486, Dept. of Computer Science, Dartmouth College Hanover, NH, Hanover, NH, USA (2004)
4. Campbell, N., Suzuki, N.: Working with very sparse data to detect speaker and listener participation in a meetings corpus. In: Proceedings of Multimodal Behaviour Theory to Usable Models, Genova, Italy (2006)
5. Carletta, J., Ashby, S., Bourban, S., Flynn, M., Guillemot, M., Hain, T., Kadlec, J., Kasairos, V., Kraaij, W., Kronenthal, M., Lathoud, G., Lincoln, M., Lisowska, A., McCowan, I., Post, W., Reidsma, D., Wellner, P.: The AMI meeting corpus: a pre–announcement. In: Renals, S., Bengio, S. (eds.) MLMI 2005. LNCS, vol. 3869, pp. 28–39. Springer, Heidelberg (2005)
6. McCowan, I., Gatica-Perez, D., Bengio, S., Lathoud, G., Barnard, M., Zhang, D.: Automatic analysis of multimodal group action in meetings. IEEE Transactions on Pattern Analysis and Machine Intelligence 27(3), 305–317 (2005)

7. Chen, L., Rose, R.T., Parrill, F., Han, X., Tu, J., Huang, Z., Harper, M., Quek, F., McNeill, D., Tuttle, R.: (VACE) multimodal meeting corpus. In: Renals, S., Bengio, S. (eds.) MLMI 2005. LNCS, vol. 3869, pp. 40–51. Springer, Heidelberg (2005)

8. Stiefelhagen, R., Bowers, R.: CLEAR (Classification of Events, Activities and Relationships) Evaluation Campaign and Workshop (May, Baltimore, MD, USA (2007), http://isl.ira.uka.de/clear07/

9. Danninger, M., Robles, E., Takayama, L., Wang, Q., Kluge, T., Nass, C., Stiefelhagen, R.: The connector service - predicting availability in mobile contexts. In: Renals, S., Bengio, S., Fiscus, J.G. (eds.) MLMI 2006. LNCS, vol. 4299, pp. 129–141. Springer, Heidelberg (2006)

10. Crowley, J.L., Coutaz, J., Rey, G., Reignier, P.: Perceptual components for context aware computing. In: Borriello, G., Holmquist, L.E. (eds.) UbiComp 2002. LNCS, vol. 2498, Springer, Heidelberg (2002)

11. Fleury, P., Cuřín, J., Kleindienst, J.: SitCom - development platform for multimodal perceptual services. In: Marik, V., Vyatkin, V., Colombo, A.W. (eds.) HoloMAS 2007. LNCS (LNAI), vol. 4659, pp. 104–113. Springer, Heidelberg (2007)

12. Witten, I.H., Frank, E.: Data mining. Practical machine learning tools and techniques, 2nd edn. Morgan Kaufmann, San Francisco (2005)

Object Category Recognition Using Probabilistic Fusion of Speech and Image Classifiers

Kate Saenko and Trevor Darrell

Computer Science and Artificial Intelligence Laboratory
Massachusetts Institute of Technology
32 Vassar Street, Cambridge, MA 02139, USA
{saenko,trevor}@csail.mit.edu

Abstract. Multimodal scene understanding is an integral part of human-robot interaction (HRI) in situated environments. Especially useful is category-level recognition, where the the system can recognize classes of objects of scenes rather than specific instances (e.g., any chair vs. this particular chair.) Humans use multiple modalities to understand which object category is being referred to, simultaneously interpreting gesture, speech and visual appearance, and using one modality to disambiguate the information contained in the others. In this paper, we address the problem of fusing visual and acoustic information to predict object categories, when an image of the object and speech input from the user is available to the HRI system. Using probabilistic decision fusion, we show improved classification rates on a dataset containing a wide variety of object categories, compared to using either modality alone.

Keywords: multimodal fusion, object recognition, human-computer interaction.

1 Introduction

Multimodal recognition of object categories in situated environments is useful for robotic systems and other applications. Information about object identity can be conveyed in both speech and image. For example, if the user takes a picture of a cylindrical object and says: "This is my pen," a machine should be able to recognize the object as belonging to the class "pen", and not "pan", even if the acoustic signal was too ambiguous to make that distinction. Conventional approaches to object recognition rely either on visual input or on speech input alone, and therefore can be brittle in noisy conditions. Humans use multiple modalities for robust scene understanding, and artificial systems should be able to do the same.

The conventional approach to *image*-based category recognition is to train a classifier for each category offline, using labeled images. Note that *category-level* recognition allows the system to recognize a class of objects, not just single instances. To date, automatic image-based category recognition performance has only reached a fraction of human capability, especially in terms of the variety of

A. Popescu-Belis, S. Renals, and H. Bourlard (Eds.): MLMI 2007, LNCS 4892, pp. 36–47, 2007.

Fig. 1. Examples of the most visually confusable categories in our dataset (see Section 4 for a description of the experiments). The image-based classifier most often misclassified the category on the left as the category on the right.

recognized categories, partly due to lack of labeled data. Accurate and efficient off-the-shelf recognizers are only available for a handful of objects, such as faces and cars. In an assistant robot scenario, the user would have to collect and manually annotate a database of sample images to enable a robot to accurately recognize the objects in the home.

A *speech*-only approach to multimodal object recognition relies on speech recognition results to interpret the categories being referred to by the user. This approach can be used, for example, to have the user "train" a robot by providing it with speech-labeled images of objects. Such a system is described in [9], where a user can point at objects and describe them using natural dialogue, enabling the system to automatically extract sample images of specific objects and to bind them to recognized words. However, this system uses speech-only object category recognition, i.e. it uses the output of a speech recognizer to determine object-referring words, and then maps them directly to object categories. It does not use any prior knowledge of object category appearance. Thus, if the spoken description is misrecognized, there is no way to recover, and an incorrect object label may be assigned to the input image (e.g., "pan" instead of "pen".) Also, the robot can only model object *instances* that the user has pointed out. This places a burden on the user to show the robot every possible object, since it cannot generalize to unseen objects of the same category.

We propose a new approach, which combines speech and visual object category recognition. Rather than rely completely on one modality, which can be error-prone, we propose to use both speech- and image-based classifiers to help determine the category of the object. The intuition behind this approach is that, when the categories are acoustically ambiguous due to noise, or highly confusable (e.g., "budda" and "gouda"), their visual characteristics may be distinct enough to allow an image-based classifier to correct the speech recognition errors. Even if the visual classifier is not accurate enough to choose the correct category from the set of all possible categories, it may be good enough to choose between a few *acoustically* similar categories. The same intuition applies in the other direction, with speech disambiguating confusable visual categories. For example, Figure 1 shows the categories that the visual classifier confused the most in our experiments.

There are many cases in the human-computer interaction literature where multimodal fusion helps recognition (e.g. [12], [10]). Although visual object *category* recognition is a well-studied problem, to the best of our knowledge, it has not been combined with speech-based category recognition. In the experimental section, we use real images, as well as speech waveforms from users describing objects depicted in those images, to see whether there is complementary information in the two channels. We propose a fusion algorithm based on probabilistic fusion of the speech and image classifier outputs. We show that it is feasible, using state-of-the-art recognition methods, to benefit from fusion on this task. The current implementation is limited to recognizing about one hundred objects, a limitation due to the number of categories in the labeled image database. In the future, we will explore extensions to allow arbitrary vocabularies and numbers of object categories.

2   Related Work

Multimodal interaction using speech and gesture dates back to Bolt's Put-That-There system [1]. Since that pioneering work, there have been a number of projects on virtual and augmented-reality interaction combining multiple modalities for reference resolution. For example, Kaiser, et. al. [10] use mutual disambiguation of gesture and speech modalities to interpret which object the user is referring to in an immersive virtual environment. Our proposed method is complimentary to these approaches, as it allows multimodal reference to objects in real environments, where, unlike in the virtual reality and game environments, the identity of surrounding objects is unknown and must be determined based on visual appearance.

Haasch, et. al. [9] describe a robotic home tour system called BIRON that can learn about simple objects by interacting with a human. The robot has many capabilities, including navigation, recognizing intent-to-speak, person tracking, automatic speech recognition, dialogue management, pointing gesture recognition, and simple object detection. Interactive object learning works as follows: the user points to an object and describes what it is (e.g., "this is my cup"). The system selects a region of the image based on the recognized pointing gesture and simple salient visual feature extraction, and binds that region to the object-referring word. Object detection is performed by matching previously learned object images to the new image using cross-correlation. The system does not use pre-existing visual models to determine the object category, but rather assumes that the dialogue component has provided it with the correct words. Note that the object recognition component is very simple, as this work focuses more on a human-robot interaction (HRI) model for object learning than on object recognition.

The idea of disambiguating which object the user is referring to using speech and image recognition is not new. In [13], the authors describe a visually-grounded spoken language understanding system, an embodied robot situated on top of a table with several solid-colored blocks placed in front of it on a green tablecloth. The robot learns by pointing to one of the blocks, prompting the user to provide a verbal description of the object, for example: "horizontal blue rectangle". The paired visual observations and transcribed words are used to learn concepts like the meaning of "blue", "above", "square". The key difference between this work and [13] is that we focus on a realistic object categorization task, and on disambiguating among many arbitrary categories using prior visual models.

There is a large body of work on object recognition in the computer vision literature, a comprehensive review of which is beyond the scope of this paper. Here we only review several recent publications that focus on object presence detection, where, given an image, the task is to determine if a particular object category is present, and on object classification, where the task is to determine which one out of C categories is present in the image.

Murphy, et. al. present a context-sensitive object presence detection method [11]. The overall image context gives the probability of the object being present in the image, which is used to correct the probability of detection based on the local image features. The authors show that the combination of experts based on local and global image features performs better than either expert alone. Our proposed disambiguation method is somewhat similar to this, except that, in our case, the two experts operate on speech and global image features.

The current best-performing object classification methods on *Caltech 101* [3], the image database we use in our experiments, are based on discriminative multi-class classifiers. In [5], a nearest-neighbor classifier is used in combination with a perceptual distance function. This distance function is learned for each individual training image as a combination of distances between various visual features. The authors of [15] use a multi-class support vector machine (SVM) classifier with local interest point descriptors as visual features. We use the method of [6], which is also based on a multi-class SVM, but in combination with a kernel that computes distances between pyramids of visual feature histograms.

There has been some recent interest in using weakly supervised cross-modal learning for object recognition. For example, Fergus et al. [4] learn object categories from images obtained using a Google search for keywords describing the categories of interest. While, in this paper, we describe a supervised approach, we are also interested in exploring the idea of learning visual category classifiers in an unsupervised fashion, perhaps using web-based image search for keywords corresponding to the top speech hypotheses. This would allow an arbitrary vocabulary of object-referring words to be used, without requiring that a labeled image database exists for each word in the vocabulary.

3   Speech and Image-Based Category Recognition

In this section, we describe an algorithm for speech and image-based recognition of object categories. We assume a fixed set of C categories, and a set W of nouns (or compound nouns), where W_k corresponds to the name of the kth object category, where $k = 1, ..., C$.

The inputs to the algorithm consist of a visual observation x_1, derived from the image containing the object of category k, and the acoustic observation x_2, derived from the speech waveform corresponding to W_k. In this paper, we assume that the user always uses the same name for an object category (e.g., "car" and not "automobile".) Future work will address an extension to multiple object names. A simple extension would involve mapping each category to a list of synonyms using a dictionary or an ontology such as WordNet.

The disambiguation algorithm consists of decision-level fusion of the outputs of the visual and speech category classifiers. In this work, the speech classifier is a general-purpose recognizer, but its vocabulary is limited to the set of phrases defined by W. Decision-level fusion means that, rather than fusing information at the observation level and training a new classifier on the fused features $x = x_1, x_2$, the observations are kept separate and the decision of the visual-only classifier, $f_1(x_1)$, is fused with the decision of the speech-only classifier, $f_2(x_2)$. In general, decisions can be in the form of the class label k, posterior probabilities $p(c = k | x_i)$, or a ranked list of the top N hypotheses.

There are several methods for fusing multiple classifiers at the decision level, such as letting the classifiers vote on the best class. We propose to use the probabilistic method of combining the posterior class probabilities output by each classifier. We investigate two combination rules. The first one, the weighted mean rule, is specified as:

$$p(c | x_1, ..., x_m) = \sum_{i=1}^{m} p(c | x_i) \lambda_i, \tag{1}$$

where m is the number of modalities, and the weights λ_i sum to 1 and indicate the "reliability" of each modality. This rule can be thought of as a mixture of experts. The second rule is the weighted version of the product rule,

$$p(c | x_1, ..., x_m) = \prod_{i=1}^{m} p(c | x_i)^{\lambda_i} \tag{2}$$

which assumes that the observations are independent given the class, which is a valid assumption in our case. The weights are estimated experimentally by enumerating a range of values and choosing the one that gives the best performance. Using one of the above combination rules, we compute new probabilities for all categories, and pick the one with the maximum score as the final category output by the classifier.

Note that our visual classifier is a multi-class SVM, which returns margin scores rather than probabilities. To obtain posterior probabilities $p(c = k|x_2)$ from decision values, a logistic function is trained using cross-validation on the training set. Further details can be found in [2].

4 Experiments

If there is complementary information in the visual and spoken modalities, then using both for recognition should achieve better accuracy than using either one in isolation. The goal of the following experiments is to use real images, as well as recordings of users describing the objects depicted in those images, to see if such complementarity exists. Since we are not aware of any publicly available databases that contain paired images and spoken descriptions, we augmented a subset of an image-only database with speech by asking subjects to view each image and to speak the name of the object category it belongs to. Using this data, we evaluate our probabilistic fusion model. We investigate whether weighting the modalities is advantageous, and compare the mean and product combination rules.

Image Dataset. Most publicly available image databases suitable for category-level recognition contain very few object categories. The exceptions include the *PASCAL, LabelMe, Caltech101, ESP* and *Peekaboom* databases, which are described in [14]. We chose to use the *Caltech101* database, because it contains a large variety of categories, and because it is a standard benchmark in the object recognition field. The database has a total of 101 categories, with about 50 images per category. Although the categories are challenging for current object recognition methods, the task is made somewhat easier by the fact that most images have little or no background clutter, and the objects tend to be centered in the image and tend to be in a stereotypical pose. Sample images from each of the 101 categories are shown in Figure 4.

Speech Collection. We augmented a subset of the images with spoken utterances recorded in our lab, to produce a test set of image-utterance pairs on which to evaluate the fusion method. We chose the set of names W based on the names provided with the image database, changing a few of the names to more common words. For example, instead of "gerenuk", we used the word "gazelle", and so on. The exact set of names W is shown in Figure 4. A total of 6 subjects participated in the data collection, 4 male and 2 female, all native speakers of American English. Each subject was presented with 2 images from each category in the image test set, and asked to say the exact object name for each image, resulting in 12 utterances for each category, for a total of 1212 image-utterance pairs. The reason that the images were shown, as opposed to just prompting the subject with the category name, is that some names are homonyms (e.g., here "bass" refers to the fish, not the musical instrument), and also to make the experience more natural. The speech data collection took place in a quiet office, on a laptop computer, using its built-in microphone.

The nature of the category names in the *Caltech101* database, the controlled environment, and the small vocabulary makes this an easy speech recognition task. The speech recognizer, although it was trained on an unrelated phone-quality audio corpus, achieved a word error rate (WER) of around 10% when tested on the collected category utterances. In realistic human-computer interaction scenarios, the environment can be noisy, interfering with speech recognition. Also, the category names of everyday objects are shorter, more common words (e.g. "pen" or "pan", rather than "trilobite" or "mandolin"), and the their vocabulary is much larger, resulting in a lot more acoustic confusion. Our preliminary experiments with large-vocabulary recognition of everyday object names, using a 25K-phrase vocabulary, produced WERs closer to 50%. Thus, to simulate a more realistic speech task, we added "cocktail party" noise to the original wave-forms, using increasingly lower signal-to-noise ratios (SNRs): 10db, 4db, 0db, and -4db. For the last two SNRs, the audio-only WERs are in a more realistic range of around 30-60%.

Training of Classifiers. We trained the image-based classifier on a standard *Caltech101* training set, consisting of the first 15 images from each category, which are different from the test images mentioned above. The classification method is described in detail in [6], here we only give a brief overview. First, a set of feature vectors is extracted from the image at each point on a regular 8-by-8 grid. A gradient direction histogram is computed around each grid point, resulting in a 128-dimensional SIFT descriptor. The size of the descriptor is reduced to 10 dimensions using principal component analysis, and the x,y position of the point is also added, resulting in a 12-dimensional vector. Vector quantization is then performed on the feature space [7], and each feature vector (block) of the image is assigned to a visual "word". Each image is represented in terms of a bag (histogram) of words. Two images can then be matched using a special kernel (the pyramid match kernel) over the space of histograms of visual words. Classification is performed with a multi-class support vector machine (SVM) using the pyramid match kernel. Our implementation uses a one-vs-rest multi-class SVM formulation, with a total of C binary SVMs, each of which outputs the visual posterior probabilities $p(c = k|x_1)$ of the class given the test image.

The speech classifier is based on the Nuance speech recognizer, a commercial, state-of-the-art, large-vocabulary speech recognizer. The recognizer has pre-trained acoustic models, and is compiled using a grammar, which we set to be the set of object names W, thus creating an isolated phrase recognizer with a vocabulary of 101 phrases. This recognizer then acts as the speech-based classifier in our framework. The recognizer returns an N-best list, i.e. a list of N most likely phrase hypotheses $k = k_1, ..., k_N$, sorted by their confidence score. We use normalized confidence scores as an estimate of the posterior probability $p(c = k|x_2)$ in Equations 1, 2. For values of k not in the N-best list, the probability was set to 0. The size of the N-best was set to 101, however, due to pruning,

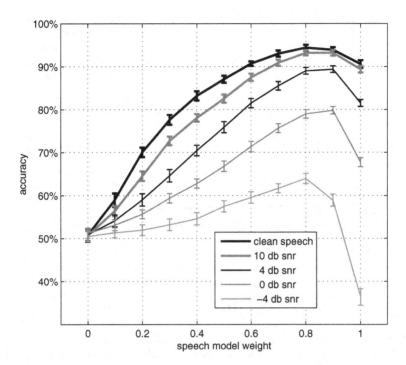

Fig. 2. Object classification using the mean rule, on the development set. Each line represents the performance on a different level of acoustic noise. The y-axis shows the percent of the samples classified correctly, the x-axis plots the speech weight used for the combined classifier.

most lists were much shorter. The accuracy is measured as the percentage of utterances assigned the correct category label.

Development and Test Sets. The test set of image-utterance pairs was further split randomly into a development set and test set. The development set was used to optimize the speech weight. All experiments were done by averaging the performance over 20 trials, each of which consisted of randomly choosing half of the data as the development set, optimizing the weight on it, and then computing the performance with that weight on the rest of the data.

Results. First, we report the single-modality results. The average accuracy obtained by the image-based classifier, measured as the percentage of correctly labeled images, was 50.7%. Chance performance on this task is around 1%. Note that it is possible to achieve better performance (58%) by using 30 training images per category [8], however, that would not leave enough test images for some of the categories. The average 1-best accuracy obtained by the speech classifier in the clean audio condition was 91.5%. The oracle N-best accuracy, i.e. the accuracy that would be obtained if we could choose the best hypothesis by hand from the N-best list, was 99.2%.

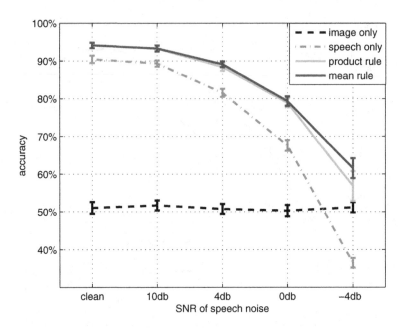

Fig. 3. Absolute improvement across noise conditions on the test set. The Y-axis shows the percent of the test samples classified correctly, the X-axis shows the SNR of the noise condition. Chance performance is around 1%.

Next, we see how the fused model performs on different noise levels. Figure 2 shows the results of the fusion algorithm on the development set, using the mean combination rule. The plot for the product rule, not shown here, is similar. Each line represents a different level of acoustic noise, with the top line being clean speech, and the bottom line being the noisiest speech with -4db SNR. The x-axis plots the speech model weight λ_2 in increments of 0.1, where $\lambda_1 + \lambda_2 = 1$. Thus, the leftmost point of each line is the average image-only accuracy, and the rightmost point is the speech-only accuracy. As expected, speech-only accuracy degrades with increasing noise. We can see that the fusion algorithm is able to do better than either single-modality classifier for some setting of the weights. The product combination rule gives similar performance to the mean rule. We also see that the weighted combination rule is better than not having weights (i.e. setting each weight to 0.5). The average accuracy on the test set, using the weight chosen on the development set for each noise condition, is plotted in Figure 3. The plot shows the gains that each combination rule achieved over the single modality classifiers. The mean rule (red line) does slightly better than the product rule (green line) on a number of noise conditions, and significantly better than the either speech or vision alone on all conditions.

Fig. 4. Sample images from the *Caltech101* database. The category name used in our experiments is shown at the top of each image

5   Conclusion and Future Work

We presented a multimodal object category classifier that combines image-only and speech-only hypotheses in a probabilistic way. The recognizer uses both the name of the object and its appearance to disambiguate what object category the user is referring to. We evaluated our algorithm on a standard image database of 101 object categories, augmented with recorded speech data of subjects saying the name of the objects in the images. We have simulated increasingly difficult speech recognition tasks by adding different levels of noise to the original speech data. Our results show that combining the modalities improves recognition across all noise levels, indicating that there is complementary information provided by the two classifiers. To avoid catastrophic fusion, we have proposed to use the weighted version of the mean rule to combine the posterior probabilities, and showed experimentally that there exists a single weight that works for a variety of audio noise conditions. We have thus shown that it may be advantageous for HRI systems to use both channels to recognize object references, as opposed to the conventional approach of relying only on speech or only on image recognition, when both are available.

We regard this work as a proof of concept for a larger system, the first step towards multimodal object category recognition in HRI systems. We plan to continue this line of research, extending the model to handle multiple words per category, and, eventually, to extract possible object-referring words from natural dialogue. A simple extension to handle multiple object names is to map each category to a list of synonyms using a dictionary or an ontology such as WordNet. We are also interested in enabling the use of arbitrary vocabularies by learning visual category classifiers in an unsupervised fashion, using methods similar to [4]. With this approach, web-based image search would be conducted for keywords corresponding to words in the N-best list output by the speech recognizer. The returned images could then be used to build visual models for disambiguation of arbitrary objects.

References

1. Bolt, R.: "Put-that-there": Voice and gesture at the graphics interface. In: SIG-GRAPH 1980. Proceedings of the 7th Annual Conference on Computer Graphics and interactive Techniques, pp. 262–270. ACM Press, New York (1980)
2. Chang, C., Lin, C.: LIBSVM: A library for support vector machines, Software (2001), http://www.csie.ntu.edu.tw/~cjlin/libsvm
3. Fei-Fei, L., Fergus, R., Perona, P.: Learning generative visual models from few training examples: In: an incremental Bayesian approach tested on 101 object categories. IEEE. CVPR, Workshop on Generative-Model Based Vision (2004)
4. Fergus, R., Fei-Fei, L., Perona, P., Zisserman, A.: Learning Object Categories from Google's Image Search. In: Proc. of the 10th Inter. Conf. on Computer Vision, ICCV (2005)
5. Frome, A., Singer, Y., Malik, J.: Image Retrieval and Recognition Using Local Distance Functions. In: Proceedings of Neural Information Processing Systems (NIPS) (2006)
6. Grauman, K., Darrell, T.: The Pyramid Match Kernel: Discriminative Classification with Sets of Image Features. In: Proceedings of the IEEE International Conference on Computer Vision (ICCV), Beijing, China (October 2005), http://people.csail.mit.edu/jjl/libpmk/
7. Grauman, K., Darrell, T.: Approximate Correspondences in High Dimensions. In: Proceedings of Advances in Neural Information Processing Systems (NIPS) (2006)

8. Grauman, K., Darrell, T.: Pyramid Match Kernels: Discriminative Classification with Sets of Image Features. MIT Technical Report MIT-CSAIL-TR-2006-020, 2006. The Journal of Machine Learning (2006) (to appear)
9. Haasch, A., Hofemann, N., Fritsch, J., Sagerer, G.: A multi-modal object attention system for a mobile robot, Intelligent Robots and Systems (2005)
10. Kaiser, E., Olwal, A., McGee, D., Benko, H., Corradini, A., Li, X., Cohen, P., Feiner, S.: Mutual disambiguation of 3D multimodal interaction in augmented and virtual reality. In: Proceedings of the 5th International Conference on Multimodal Interfaces (ICMI) (2003)
11. Murphy, K., Torralba, A., Eaton, D., Freeman, W.T.: Object detection and localization using local and global features. In: Sicily workshop on object recognition, LNCS(unrefered) (2005)
12. Potamianos, G., Neti, C., Gravier, G., Garg, A., Senior, A.: Recent Advances in the Automatic Recognition of Audio-Visual Speech. In: Proc. IEEE (2003)
13. Roy, D., Gorniak, P., Mukherjee, N., Juster, J.: A Trainable Spoken Language Understanding System for Visual Object Selection. In: Proceedings of the International Conference of Spoken Language Processing (2002)
14. Russell, B., Torralba, A., Murphy, K., Freeman, W.T.: LabelMe: a database and web-based tool for image annotation. MIT AI LAB MEMO AIM-2005-025 (2005)
15. Zhang, H., Berg, A., Maire, M., Malik, J.: SVM-KNN: Discriminative Nearest Neighbor Classification for Visual Category Recognition. In: proceedings of CVPR (2006)

Automatic Annotation of Dialogue Structure from Simple User Interaction

Matthew Purver, John Niekrasz, and Patrick Ehlen

CSLI, Stanford University, Stanford CA 94305, USA
{mpurver,niekrasz,ehlen}@stanford.edu
http://godel.stanford.edu/

Abstract. In [1,2], we presented a method for automatic detection of *action items* from natural conversation. This method relies on supervised classification techniques that are trained on data annotated according to a hierarchical notion of dialogue structure; data which are expensive and time-consuming to produce. In [3], we presented a meeting browser which allows users to view a set of automatically-produced action item summaries and give feedback on their accuracy. In this paper, we investigate methods of using this kind of feedback as implicit supervision, in order to bypass the costly annotation process and enable machine learning through use. We investigate, through the transformation of human annotations into hypothetical idealized user interactions, the relative utility of various modes of user interaction and techniques for their interpretation. We show that performance improvements are possible, even with interfaces that demand very little of their users' attention.

1 Introduction

Few communicative events in a working day are more important than group decisions committing to future action. These events mark concrete progress toward shared goals, and are the bread and butter of face-to-face meetings. However, information produced in conversation is frequently forgotten or mis-remembered due to the limited means of memory, attention, and supporting technologies. Organizations are unable to review their own internal decisions, and individuals forget their own commitments. This information loss seriously impacts productivity and causes enormous financial hardship to many organizations [4].

The primary objective of our research is to assist meeting participants by automatically identifying *action items* in meetings: public commitments to perform some concrete future action. Some related work has sought to classify individual utterances or sentences as action-item-related [5,6,7,8,9]. These approaches have limited success when applied to multi-party conversational speech: individual utterances often do not contain sufficient information, as commitment arises not through utterances in isolation but through the dialogue interaction as a whole.

In contrast, our approach [1,2] employs shallow local dialogue structure, identifying short subdialogues and classifying the utterances within them as to their role in defining the action item. This approach improves accuracy and allows

A. Popescu-Belis, S. Renals, and H. Bourlard (Eds.): MLMI 2007, LNCS 4892, pp. 48–59, 2007.

extraction of specific information about individual semantic properties of the action item (such as what is to be done and who has agreed to take responsibility for it). However, one negative consequence of this approach is that data needs to be annotated with this structure – a complex and costly process.

In this paper we investigate the use of user interaction (in combination with classifier outputs) to automatically annotate previously unseen data, thus producing new training data and enabling a system to learn through use alone.

1.1 Action Item Detection

Our methods for annotating action items and automatically detecting them rely on a two-level notion of dialogue structure. First, short sequences of utterances are identified as *action item discussions*, in which an action item is discussed and committed to. Second, the utterances within these subdialogues are identified as belonging to zero or more of a set of four specific dialogue act types that can be thought of as *properties* of the action item discussion: *task description* (proposal or discussion of the task to be performed), *timeframe* (proposal or discussion of when the task should be performed), *ownership* (assignment or acceptance of responsibility by one or more people), and *agreement* (commitment to the action item as a whole or to one of its properties). A description of the annotation schema and classification method may be found in [1].

1.2 Data Needs and Implicit User Supervision

Because dialogue annotation is resource-intensive, we are interested in methods for producing annotated data with minimal human effort. Additionally, the characteristics of action item discussion vary substantially across users and meeting types. We therefore would like the system to learn "in the wild" and adapt to new observations without the need for *any* human annotation.

Rather, we would prefer to use *implicit user supervision* by harnessing subtle user interactions with the system as feedback that helps to improve performance over time. Implicit supervision of this kind has proved effective for topic segmentation and identification in meetings [10].

Figure 1 shows a broad view of our architecture for using implicit supervision. First, a set of utterance classifiers detects the action-item-related dialogue acts in a meeting and tags them. Then a subdialogue classifier identifies patterns of these tagged utterances to hypothesize action items. The relevant utterances are then fed into a summarization algorithm suitable for presentation in a user interface. From the interface, a user's interactions with the summarized action items can be interpreted, providing feedback to a feedback interpreter that updates the hypothesized action items and utterance tags, which are ultimately treated as annotations that provide new training data for the classifiers.

1.3 User Interfaces for Meeting Assistance

The following question now arises: What kinds of interfaces could harness user feedback most effectively?

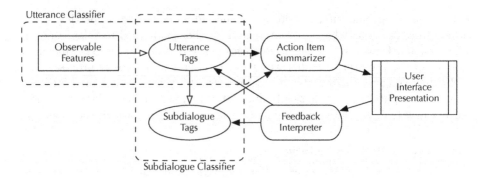

Fig. 1. An outline of the action item detection and feedback system

In [3], we described a meeting browser that allows participants to view a set of automatically hypothesized action item summaries after a meeting. Participants can confirm a hypothesized action item by adding it to a to-do list, or reject it by deleting it. They can also change the hypothesized properties (the who, what, and when for the action item) by selecting from alternate hypotheses or changing text directly. While a number of meeting browser tools allow users to inspect different facets of information that might be gleaned from a meeting (see [11] for an overview), our browser tool was specifically designed to harvest these ordinary user interactions for implicit user feedback.

Specifically, our browser is designed to verify two types of information: the *time* when an action item discussion occurred, and its description in *text*—gleaned from the language used in conversation—that contributes to our understanding of the action item's properties. Different feedback actions may give different information about each of these types. For example, overall confirmation tells us that the extracted text descriptions were adequate, and therefore implicitly confirms that the time period used to extract that text must also have been correct. Similarly, editing just one property of a hypothesized action item might tell us that the overall time period was correct, but that one aspect of the extracted text could have been better.

The distinction between temporal and textual information is important, since it is not clear which of these provides the most benefit for creating new accurate training data. And interfaces that emphasize one or the other may vary in their usefulness and cognitive demands. Certain common types of "meeting interfaces" (like handwritten notes or to-do lists) may provide information about the text of an action item, but not about the time it was discussed. Other in-meeting interfaces could provide temporal information instead of or in addition to text, such as flags or notes made during the meeting using a PDA, a digital pen, or electronic note-taking software like SmartNotes [12].

The distinction between user- and system-initiative is also important. A system-initiative approach might be to identify action items to the user while the meeting is ongoing, perhaps by popping up a button on a PDA which the user can confirm or reject. However, a user-initiative approach – allowing the user to

either take notes or flag parts of the meeting where action item discussions occur
– might provide more independent information, though perhaps at the cost of
higher cognitive demands on the user. So this third characteristic of *initiative*,
in addition to the temporal and textual characteristics, could have a profound
effect on the quality of supervision a user can provide as it relates to producing
valuable annotations.

Thus, we wish to address two basic questions in this paper. First, to what
extent do the two informational types of *time* and *text* contribute to producing
valuable data for implicit supervision of action item detection? Second, does
user- or computer- *initiative* produce more valuable information?

2   Method

2.1   Outline

Our goal is to investigate the efficacy of various methods of learning from implicit human supervision by comparing various possible user interfaces that offer
users different capabilities for providing feedback. But rather than implement
all of these possible interfaces to record real feedback data, we compare the best
possible results that each notional interface could achieve, given an "ideal" user.
We simulate this ideal user by using our existing gold-standard annotations and
positing them as user feedback. That is, we use varying amounts of the information provided by our gold-standard annotations in an attempt to reproduce
those annotations in their entirety.

The procedure involves the following steps. First, a baseline classifier is trained
on an *initial* set of human annotations of action items. That classifier then produces action item hypotheses for a second set of meetings, the *learning* set. Next,
for each notional user interface, we translate our gold-standard human annotations of the learning set into idealized user feedback interactions – rejecting,
adding, or otherwise editing the action items – while constraining the types of
feedback information (time, text, and initiative) to comply with the constraints
afforded by each interface.

These idealized interactions are then used to modify ("correct") our
automatically-generated action item hypotheses from the learning set, producing
a new updated set of hypotheses that are used as new training data. The accuracy of this updated set can be evaluated by comparing its corrected hypotheses
to the existing human annotations. We can also evaluate the effect of our feedback on the classifiers by retraining them on the union of the initial dataset and
the updated dataset, and then testing then again on a third, held-out *test* set.

2.2   Datasets

For our experiments we use 18 meetings selected from the ICSI Meeting Corpus [13]. The ICSI corpus contains unscripted natural meetings, most of which
are recordings of regularly occurring project group meetings. We have annotated a series of 12 meetings held by the Berkeley "Even Deeper Understanding"

project group (code Bed) and an additional 6 meetings selected randomly from the rest of the corpus. There are a total of 177 action items in the 18 meetings, with 944 total utterances tagged. This makes an average of 9.8 action items per meeting and 5.3 utterances per action item.

We distribute the meetings randomly into the three sets described above: an *initial* set of meetings for training the baseline classifier, a *learning* set of meetings which the system will encounter "in the wild" and learn from through user feedback, and a *test* set used for an overall performance evaluation. Each of these sets has an equal proportion of meetings from the Bed series. These sets and their human annotations can be summarized as:

– An *initial* set of meetings I, and the set of *human* annotations D_{IH}
– A *learning* set of meetings L, and the set of *human* annotations D_{LH}
– A *test* set of meetings T, and the set of *human* annotations D_{TH}

2.3   Baseline Experiments with No User Input

Using the test set for evaluation, we compare the performance of the re-trained classifiers to three baseline measures, in which the user is not a factor. The first baseline, *Initial-only*, assumes no means of obtaining user input on the L meetings at all, and thus provides an expected lower bound to the performance: we simply ignore L and train classifiers only on I. The second, *Optimal*, is a ceiling which provides an upper bound by assuming perfect annotation of the L meetings: we train the classifiers on human annotations of both I and L. Finally, the third baseline, *Naively-retrained*, examines the effect of retraining the classifier on its own output for L (the automatically-produced hypotheses D_{LA}) without modification by user feedback. The baseline experiments are therefore:

– *Initial-only*: Train on D_{IH}; test against D_{TH}
– *Optimal*: Train on $D_{IH} \cup D_{LH}$; test against D_{TH}
– *Naively-retrained*: Train on $D_{IH} \cup D_{LA}$; test against D_{TH}

2.4   Experiments Involving User Input

For each type of user interface, we perform the following experimental steps:

– *Step 1*: Train on D_{IH}
– *Step 2*: Produce automatic hypotheses D_{LA}
– *Step 3*: Update D_{LA} based on user feedback, producing a dataset D_{LU}
– *Step 4*: Test the updated D_{LU} against the gold-standard D_{LH}
– *Step 5*: Retrain on $D_{IH} \cup D_{LU}$
– *Step 6*: Test against D_{TH} and the baselines

We can test the effectiveness of user feedback in two ways. Results can be presented in terms of the accuracy of the updated hypotheses for L (i.e. directly measuring agreement between D_{LH} and D_{LU}). Results can also be presented in terms of the overall effect on classifier performance as tested on T (i.e. measuring agreement between D_{TH} and D_{TA}, and comparing with the above baselines).

3   User Interfaces and Artificial Feedback

We characterize potential interfaces along three dimensions: *temporal* (whether an interface provides information about *when* action items were discussed), *textual* (whether it provides information about the *properties* of the action items), and *initiative* (whether interaction is initiated by computer, user, or both).

Accordingly, we can define a set of hypothetical user interfaces, summarized in Table 1. The *Proactive Button* provides only user-initiated temporal information, simply allowing the user to signal that an action item has just occurred at any time during the meeting – this could be realized as a virtual button on a tablet PC or PDA, or perhaps digital paper. The *Reactive Button* provides computer-initiated temporal information: it tells the user during a meeting when an action item is detected, requiring the user to confirm or reject (or ignore) the hypothesis. Again, this could be realized on a PC, PDA, or phone. Our third interface, *Post-meeting Notes*, assumes only textual information supplied by the user after the meeting is finished, either via note-taking software or by scanning hand-written notes. Our *In-meeting Notes* interface provides both textual and temporal information, assuming that the user is willing to take descriptive notes when action items are discussed (perhaps via collaborative note-taking software).

Table 1. The set of user interfaces investigated

Interface	Temporal Information	Textual Information	Initiative
Proactive Button	Yes	No	User
Reactive Button	Yes	No	Computer
Post-Meeting Notes	No	Yes	User
In-Meeting Notes	Yes	Yes	User

3.1   Simulating Feedback from "Idealized" Users

To simulate feedback as it would be produced by an "ideal" (perfectly informative and correct) user, we use information from our existing human annotations.

To simulate **user-initiated** feedback, we need to provide the textual descriptions and/or discussion times of each action item. Times are taken as the end time of the final utterance in an annotated action item subdialogue. Text information is taken from the properties annotated for each individual action item (the task description, the timeframe description, and the identity of the responsible party). Note that annotators were allowed to specify these properties as free text, paraphrasing or rewording the actual discussion as needed: there was no requirement to copy the words or phrases actually used in the transcripts. While they showed a tendency to re-use important words from the utterances themselves, this seems entirely natural and we expect that users would behave similarly. Importantly, annotated information about which utterances belong to a subdialogue, or which utterances play which dialogue act roles, is not used.

To simulate **computer-initiated** feedback, we must compare the automatic hypotheses with the gold-standard annotations, and provide negative or positive feedback accordingly. This requires a criterion for acceptability, which must vary depending on the interface's information content. Where temporal information only is concerned, we class a hypothesis as correct if its corresponding subdialogue period overlaps by more than 50% with that of a gold-standard subdialogue. Where textual information is concerned, we compare each property description using a string similarity metric, and class a hypothesis as correct if the similarity is above a given threshold (see below for more details).

3.2   Interpreting Feedback as Annotation

Given these varying degrees of feedback information, our task is now to infer a complete set of structured action item annotations (both the action item subdialogue periods, and the individual utterances which perform the various dialogue acts within those periods). The inference method depends, of course, on the amount and type of information provided – a summary is shown in Table 2.

The simplest case is that of overall **confirmation or rejection** of a computer-generated hypothesis (as provided by computer-initiated feedback, whether determined on a temporal or textual basis). In this case, we already have a record of which utterances were involved in generating the hypothesis, and their hypothesized dialogue act types; if confirmed, we can use these types directly as (positive) annotation labels; if rejected, we can mark all these utterances as negative instances for all dialogue act types.

The most complex case is that of independent **creation** of a new action item (as provided by user-initiated feedback). In this case, we have no information as to which utterances are involved, but must find the most likely candidates for each relevant dialogue act type; of course, we may have an indicative time, or textual descriptions, or both, to help constrain this process. Given only the time, we must use what prior information we have about the characteristics of the various dialogue act types: given our approach, this means using the existing subclassifiers – using each one to assign a confidence to each utterance within a realistic time window, and labeling those above a given confidence threshold. However, textual information (if available) can provide more independent evidence, at least for semantically informative dialogue act types. For the purely textual *task description* and *timeframe* acts, we can use an independent similarity measure to score each utterance, and label accordingly;[1] for *owner* acts, where the owner's identity usually arises through discourse reference (e.g. personal pronouns) rather than lexical information, we can filter potential utterances based on their referential compatibility. However, as the *agreement* act type does not correspond to any textual information that a user might realistically provide, it must always be assigned using the existing subclassifier.

---

[1] Our current similarity measure uses a heuristic combination of lexical overlap and semantic distance (using WordNet [14]); we plan to investigate alternatives in future.

Table 2. Inference procedures for each feedback type. Here, "likely" refers to the use of subclassifier confidences, "relevant" to the use of similarity measures.

Feedback	Text Info.	Time Info.	Procedure
confirm	(Given)	(Given)	Label all utterances as hypothesized.
reject	(Given)	(Given)	Label all utterances as negative.
edit	Yes	(Given)	Label most relevant utterance(s) in time window.
create	No	Yes	Label most likely utterances in time window.
create	Yes	Yes	Label most relevant utterances in time window.

Other cases such as **editing** of an individual property of an action item fall in between these two cases: we use a relevant similarity measure (if text is available) or the relevant subclassifier (if not) to label the most likely corresponding utterance(s) – but in these cases we have more constraining information (knowledge of the part of the subdialogue/hypothesis *not* being edited).

Note that an alternative to purely computer- or user-initiated feedback exists: we can attempt to interpret user-initiated feedback as *implicitly* giving feedback on the computer's hypotheses. This way, a user specification of an action item could be interpreted as an implicit confirmation of a suitable overlapping hypothesis (and a rejection of an unsuitable or non-overlapping one) if one exists, and only as an independent creation otherwise. We investigate both approaches.

3.3   Research Questions

By analyzing these re-interpreted annotations as idealized feedback coming from different types of interfaces, we hope to answer a few questions.

First, in comparing the two types of *time* data and *text* data, will either of these types prove to be more informative than the other? Will either type prove itself not valuable at all? A "yes" to either of these questions could save us time that we might otherwise spend testing interfaces with no inherent promise.

Similarly, since relying on user initiative can burden the user in a way that a system-initiative interface doesn't (by requiring the user to keep another task "in mind" while doing other things), is there any value to a *user-initiative* system over and above a *system-initiative* one? And is there a notable benefit to combining both initiatives, by treating user-initiated actions as implicit confirmation for system-initiated ones?

Our final question is how the overall performances compare to the baseline cases. How close is the overall performance enabled by feedback to the ideal performance achieved when large amounts of gold-standard annotated data are available (our *Optimal* ceiling)? Does the use of feedback really provide better performance than naively using the classifier to re-annotate (the *Naively-retrained* baseline)? If not, we must consider the possibility that such semi-supervised feedback-based training offers little benefit over a totally unsupervised approach, and save users (and ourselves) some wasted effort.

4　Results

We evaluate the experimental results in two separate ways. The first evaluation directly evaluates the quality of the new training data inferred from feedback on the L dataset. Tables 3 & 4 report the accuracy of the updated annotations D_{LU} compared with the gold-standard human annotations D_{LH}, both in terms of the kappa metric (as widely used to assess inter-annotator agreement [15]) and as F-scores for the task of retrieving the utterances which should be annotated. Kappa figures are given for each of the four utterance classes, showing the accuracy in identifying whether utterances belong to the four separate classes of *description*, *timeframe*, *owner*, and *agreement*, together with the figure for all four classes taken together (i.e. agreement on whether an utterance should be annotated as action-item-related or not). Table 3 shows these results calculated over individual utterances; Table 4 shows the same, but calculated over 30-second intervals – note that training data which assigns, say, the agreement class to an incorrect utterance, but one which is within a correct subdialogue, may still be useful for training the overall classifier.

The second evaluation shows the effect on overall performance of re-training on this new inferred data. Table 5 reports the classifier performance on the test set T, i.e. the accuracy of the hypothesized D_{TA} compared with the gold-standard D_{TH}. We show results as F-scores for two retrieval tasks: firstly, identifying the individual component utterances of action items; and secondly, identifying the presence of action items within 30-second intervals (an approximation of the task of identifying action item subdialogues).

The training data quality results (Tables 3 & 4) suggest several possible conclusions. Firstly, we see that using either temporal or text information alone allows improvement over raw classifier accuracy, suggesting that either can be useful in training. Secondly, combining both types of information (in-meeting notes) does best. Thirdly, textual information seems to be more useful than temporal (post-meeting notes do better than either button). This is useful to

Table 3. Utterance-level accuracy for the training annotations inferred from user feedback (D_{LU}) in comparison to the gold-standard human annotations (D_{LH})

Interface (& implicit hyp use)	Utterance Classes				Average	
	agreement	description	timeframe	owner	Kappa	F_1
(Raw hypotheses)	0.06	0.13	0.03	0.12	0.13	0.15
Proactive Button	0.22	0.30	0.17	0.37	0.35	0.36
–"– (implicit)	0.21	0.35	0.16	0.35	0.39	0.41
Reactive Button	0.15	0.31	0.09	0.28	0.32	0.33
Post-meeting Notes	0.26	0.65	0.75	0.29	0.56	0.56
–"– (implicit)	0.10	0.32	0.13	0.15	0.25	0.27
In-meeting Notes	0.26	0.72	0.75	0.40	0.61	0.62
–"– (implicit)	0.21	0.61	0.32	0.26	0.52	0.53

Table 4. 30-second interval accuracy for the training annotations inferred from user feedback (D_{LU}) in comparison to the gold-standard human annotations (D_{LH})

Interface (& implicit hyp use)	Utterance Classes				Average	
	agreement	description	timeframe	owner	Kappa	F_1
(Raw hypotheses)	0.13	0.23	0.11	0.22	0.19	0.27
Proactive Button	0.46	0.71	0.30	0.67	0.65	0.68
–"– (implicit)	0.41	0.75	0.44	0.63	0.64	0.67
Reactive Button	0.34	0.51	0.35	0.46	0.42	0.45
Post-meeting Notes	0.39	0.80	0.89	0.43	0.75	0.77
–"– (implicit)	0.19	0.41	0.24	0.29	0.36	0.43
In-meeting Notes	0.43	0.91	0.89	0.53	0.82	0.84
–"– (implicit)	0.43	0.89	0.68	0.62	0.79	0.81

Table 5. The classification accuracy of the retrained classifiers' hypotheses on the test set (D_{TU}), and those of the baseline classifiers

Interface	Utterances			30-sec Windows		
	Prec.	Recall	F_1	Prec.	Recall	F_1
Initial-only	0.08	0.13	0.09	0.21	0.27	0.22
Naively-retrained	0.09	0.20	0.12	0.23	0.45	0.30
Optimal	0.19	0.26	0.22	0.47	0.44	0.45
Proactive Button	0.18	0.25	0.21	0.47	0.48	0.46
Reactive Button	0.20	0.26	0.22	0.49	0.55	0.50
Post-meeting Notes	0.14	0.20	0.17	0.41	0.41	0.40
In-meeting Notes	0.16	0.27	0.20	0.42	0.52	0.46

know for interface design; it seems likely that temporal synchrony of interface actions and actual discussion of action item may be less exact with real users, so a design which does not have to rely on this synchrony may be advantageous (see below for a discussion of future experiments to investigate this.)

We also note that user initiative does seem to provide extra information above that provided by a purely system-initiative approach (proactive beats reactive). Of course, this may be influenced by the current low overall performance of the classifiers themselves, and may change as performance improves; however, it does suggest that a new technology such as action item detection, with inherently high error rates, is best applied without system initiative until accuracy improves. Similarly, implicit use of computer hypotheses harms the performance of the text-based notes; although it seems to marginally help the proactive button.

We note that the absolute level of agreement for utterance-level annotations is poor, and well below that which might be expected from human annotators. However, some utterance classes do well in some cases, with task description and timeframe utterances giving good agreement with the notes-based interfaces (unsurprisingly, these are the utterance classes which convey most of the information which a text note might contain). And all interfaces achieve respectable levels when considered over 30-second intervals, allowing us to expect that these inferred data could to some extent replace purely human annotation.

The overall performance results (Table 5) show that all kinds of feedback improve the system. All interfaces outperformed the *Naively-retrained* and *Initial-only* baselines; in fact, at the 30-second interval level, all perform approximately as well as our *Optimal* ceiling. However, we hesitate to draw strong conclusions from this, as the test set is currently small – we also note that differences in training data accuracy do not seem to translate directly into the performance differences one might expect, and this may also be due to small test set size.

5   Conclusions and Further Work

These results now lead us directly to the important problem of balancing cognitive load with usefulness of feedback. As intuition predicts, interfaces which supply more information perform better, with synchronous note-taking the most useful. But there is a cognitive price to pay with such interfaces. For now, our results suggest that interfaces like the proactive or reactive buttons, which likely demand less of users' attention, can still significantly improve results, as can post-meeting note-taking, which avoids distractions during the meeting.

Of course, these idealized interactions only give us a Platonic glimpse of what can be expected from the behavior of actual people working in the shadows of the real world. So our next step is to run an experiment with human subjects using these types of interfaces during actual meetings, and compare actual use data with the results herein in order to understand how well our observations of these simulated interfaces will extend to actual interfaces, and to ascertain the level of cognitive effort each interface demands. Then we can hope to determine an optimal balance between the demands of a "meeting assistant" system and the demands of the meeting participants who use it.

Future research will also involve efforts to improve overall classifier performance, which is currently low. Although action item detection is a genuinely hard problem, we believe that improvement can be gained both for utterance and subdialogue classifiers by investigating new classification techniques and feature sets (for example, the current classifiers use only basic lexical features).

Our final set of future plans involve exploring new and better ways for interpreting user feedback, updating automatically-produced hypotheses, and making decisions about retraining. Individual meetings display a high degree of variability, and we believe that feedback on certain types of meetings (e.g. planning meetings) will benefit the system greatly, while other types (e.g. presentations) may not. We will therefore investigate using global qualities of the updated hypotheses to determine whether or not to retrain on certain meetings at all.

References

1. Purver, M., Ehlen, P., Niekrasz, J.: Detecting action items in multi-party meetings: Annotation and initial experiments. In: Renals, S., Bengio, S., Fiscus, J.G. (eds.) MLMI 2006. LNCS, vol. 4299, pp. 200–211. Springer, Heidelberg (2006)
2. Purver, M., Dowding, J., Niekrasz, J., Ehlen, P., Noorbaloochi, S., Peters, S.: Detecting and summarizing action items in multi-party dialogue. In: Proceedings of the 8th SIGdial Workshop on Discourse and Dialogue, Antwerp, Belgium (2007)
3. Ehlen, P., Purver, M., Niekrasz, J.: A meeting browser that learns. In: Proceedings of the AAAI Spring Symposium on Interaction Challenges for Intelligent Assistants (2007)
4. Romano, Jr. N.C., Nunamaker, Jr. J.F.: Meeting analysis: Findings from research and practice. In: Proceedings of the 34th Hawaii International Conference on System Sciences (2001)
5. Cohen, W., Carvalho, V., Mitchell, T.: Learning to classify email into "speech acts". In: Proceedings of Empirical Methods in Natural Language Processing, pp. 309–316 (2004)
6. Corston-Oliver, S., Ringger, E., Gamon, M., Campbell, R.: Task-focused summarization of email. In: Proceedings of the 2004 ACL Workshop Text Summarization Branches Out (2004)
7. Bennett, P.N., Carbonell, J.: Detecting action-items in e-mail. In: Proceedings of the 28th Annual International ACM SIGIR Conference on Research and Development in Information Retrieval, Salvador, Brazil, ACM Press, New York (2005)
8. Gruenstein, A., Niekrasz, J., Purver, M.: Meeting structure annotation: data and tools. In: Proceedings of the 6th SIGdial Workshop on Discourse and Dialogue, Lisbon, Portugal (2005)
9. Morgan, W., Chang, P.C., Gupta, S., Brenier, J.M.: Automatically detecting action items in audio meeting recordings. In: Proceedings of the 7th SIGdial Workshop on Discourse and Dialogue, Sydney, Australia. Association for Computational Linguistics, pp. 96–103 (2006)
10. Banerjee, S., Rudnicky, A.: Segmenting meetings into agenda items by extracting implicit supervision from human note-taking. In: IUI 2007. Prooceedings of the International Conference on Intelligent User Interfaces, Honolulu, Hawaii, ACM Press, New York (2007)
11. Tucker, S., Whittaker, S.: Accessing multimodal meeting data: Systems, problems and possibilities. In: Bengio, S., Bourlard, H. (eds.) MLMI 2004. LNCS, vol. 3361, pp. 1–11. Springer, Heidelberg (2005)
12. Banerjee, S., Rudnicky, A.: Smartnotes: Implicit labeling of meeting data through user note-taking and browsing. In: Proceedings of the Human Language Techonolgy Conference of the NAACL (2006) (companion volume)
13. Janin, A., Baron, D., Edwards, J., Ellis, D., Gelbart, D., Morgan, N., Peskin, B., Pfau, T., Shriberg, E., Stolcke, A., Wooters, C.: The ICSI meeting corpus. In: ICASSP. Proceedings of the 2003 International Conference on Acoustics, Speech, and Signal Processing (2003)
14. Miller, G.A.: WordNet: A lexical database for English. Communications of the ACM 38(11), 39–41 (1995)
15. Carletta, J.: Assessing agreement on classification tasks: The kappa statistic. Computational Linguistics 22(2), 249–255 (1996)

Interactive Pattern Recognition*

Enrique Vidal[2], Luis Rodríguez[1], Francisco Casacuberta[2],
and Ismael García-Varea[1]

[1] Departamento de Sistemas Informáticos, Universidad de Castilla La Mancha, Spain
[2] Departamento de Sistemas Informáticosy Computación,
Universidad Politécnica de Valencia, Spain

Abstract. Pattern Recognition systems are not error-free. Human intervention is typically needed to verify and/or correct the result of such systems. To formalize this fact, a new framework, which integrates the human activity into the recognition process taking advantage of the user's feedback, is described. Several applications, involving Interactive Speech Transcription and Multimodal Interactive Machine Translation, have recently been considered under this framework. These applications are reviewed in this paper, and some experiments, showing that the proposed framework can save significant amounts of human effort, are also presented.

1 Introduction

The idea of interaction between humans and machines is by no means new. In fact, historically, machines have mostly been developed with the aim of *assisting* human beings in their work. Since the introduction of computer machinery, however, the idea of fully automatic devices that would completely substitute the humans in certain types of tasks, has been gaining increasing popularity.

This is particularly the case in areas such as Pattern Recognition (PR), where only a very small fraction of the huge potential of the interactive framework has been exploited so far. Scientific and technical research in this area has followed the "full automation" paradigm traditionally, even though, in practice, full automation often proves elusive or unnatural and application developments typically end up in "semiautomatic systems" or systems for "computer assisted" operation. By neglecting the regular need for human feedback in the initial problem formulation, the resulting systems generally fail to take full advantage of the opportunities underlying the interactive framework [11,6].

The aim of the present article is to introduce a formal PR framework that explicitly includes the human activity in the recognition process. This framework entails several interesting features. First, it shows how using the *human feedback* information obtained in each interaction step directly allows to improve subsequent system performance. Second, it clearly suggests that human feedback implicitly entails some form of *multimodal interaction*, thereby promoting

---

* This work has been partially supported by the Spanish project iDoc TIN2006-15694-C02-01. Reviewer's comments have significantly improved the original manuscript.

A. Popescu-Belis, S. Renals, and H. Bourlard (Eds.): MLMI 2007, LNCS 4892, pp. 60–71, 2007.
© Springer-Verlag Berlin Heidelberg 2007

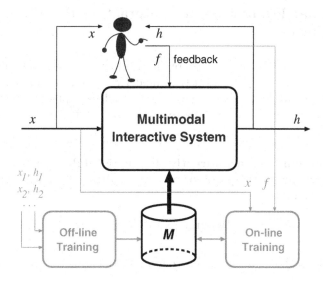

Fig. 1. Diagram of multimodal interaction in IPR

improved system ergonomics. And last but not least, since the successive results produced by the system at each interaction step are fully human-validated,
these results can be used as new and completely reliable training material for
adaptative learning of the models involved.

Figure 1 shows a schematic view of these ideas. Here, $x \in \mathcal{X}$ is an input stimulus, observation or signal and $h \in \mathcal{H}$ is a hypothesis or output, which the system
derives from x. By observing x and h the user provides some (perhaps null) feedback signal, $f \in \mathcal{F}$, which may iteratively help the system to refine or to improve
its hypothesis until it is finally accepted. Note that, typically, $\mathcal{F} \neq \mathcal{X}$; hence the
inherent multi-modality of the interactive framework. M is a *model* used by
the system to derive its hypotheses. It was initially obtained through classical
"batch" or "off-line" training from some initial *training sequence* (x_i, h_i). However, using the input data x and feedback signals f produced in each interaction
step, the model can also be *adapted* to the specific task and/or to the user's
interaction style. Note that (on-line, adaptive) learning issues are not developed
in this article, leaving them for consideration in future studies.

In the following sections we present a general formalization of these ideas
for *Interactive Pattern Recognition* (IPR). It is followed by a review of several
applications, involving interactive *Computer Assisted Speech Transcription* and
Multimodal Interactive Machine Translation, which we have recently considered
and which specifically fit under this framework. Results achieved in these applications are also summarized, empirically showing that significant amounts of
human effort can be saved in all the cases.

2 A Formal Framework for Interactive Pattern Recognition

Since adaptive learning is not considered in this paper, system operation is supposed to be driven by a *fixed* statistical model M. Under this assumption, we examine how human feedback can be directly used to improve the system performance and discuss the multimodal issues entailed by the resulting interactive framework.

Using the human feedback directly. In traditional PR [5] a best hypothesis is one which maximizes the posterior probability $\Pr(h \mid x)$. Using a fixed model M this is approximated as:

$$\hat{h} = \underset{h \in \mathcal{H}}{\operatorname{argmax}} \, P_M(h \mid x) \tag{1}$$

Now, interaction allows adding more *conditions*, that is:

$$\hat{h} = \underset{h \in \mathcal{H}}{\operatorname{argmax}} \, P_M(h \mid x, f) \tag{2}$$

where $f \equiv f_1 f_2 \ldots$ stands for the feedback, interaction-derived informations; e.g., in the form of *partial hypotheses* or *constraints* on $\mathcal{H}$. The new system hypothesis, $\hat{h}$, may prompt the user to provide further feedback informations, thereby starting a new interaction step. The process continues this way until the system output is acceptable by the user[1].

Clearly, the more f_i terms can be added in (2) the greater the opportunity to obtain better $\hat{h}$. But solving the maximization (2) may be more difficult than in the case of our familiar $P_M(h \mid x)$. Adequate solutions are discussed in the following sections for some specific applications.

Multimodality. As previously mentioned, in general, the interaction feedback informations f do not naturally belong to the original domain from which the main data, x, come from. This entails some sort of *multimodality*, apart from the possible multimodal nature of the input signals.

For the sake of simplicity, we assume that neither x nor f are multimodal themselves. Therefore, Eq.(2) corresponds to a fairly conventional *modality fusion* problem which can be straight-forwardly re-written as:

$$\hat{h} = \underset{h \in \mathcal{H}}{\operatorname{argmax}} \, P_M(x, f \mid h) \cdot P(h) \tag{3}$$

where $P(h)$ is the prior probability of the hypothesis. In many applications it is natural and/or convenient to assume conditional independence of x and f given

---

[1] This is a first-order approach, where $\hat{h}$ is derived using only the feedback obtained in the previous iteration step. More complex, higher-order models are not considered in this paper.

h. Consider for instance that x is an image and f the acoustic signal of a speech command. In this case, a *naïve* Bayes decomposition leads to:

$$\hat{h} = \underset{h \in \mathcal{H}}{\operatorname{argmax}} P_{M_X}(x \mid h) \cdot P_{M_F}(f \mid h) \cdot P(h) \qquad (4)$$

which allows for a separate estimation of independent models, M_X and M_F for the image and speech components respectively [2]. As will be discussed in Section 5, the maximization in Eq. (4) can often be approached by means of adequate extensions of available techniques to solve the corresponding search problems in the traditional non interactive/multimodal framework.

3   Computer Assisted Speech Transcription

In this section we discuss the application of the IPR framework to the transcription of spoken documents. This application is called Computer Assisted Speech Transcription (CAST) [10].

In conventional Automatic Speech Recognition (ASR) the input pattern v is a *speech utterance* and system hypotheses are *sequences of transcribed words, t, from a certain language*. This way, Eq. (1) is rewritten as:

$$\hat{t} = \underset{t}{\operatorname{argmax}} P(t \mid v) = \underset{t}{\operatorname{argmax}} P(v \mid t) \cdot P(t) \qquad (5)$$

where $P(v \mid t)$ is given by acoustic models (usually hidden Markov models [7]) and $P(t)$ by a target language model (usually a n-gram [7]).

In the CAST framework, the user is directly involved in the transcription process since he/she is responsible for validating and/or correcting the ASR output during the process. According to the IPR paradigm, the system should take into account the current state to improve the following predictions. The process starts when the system makes an initial prediction consisting in a whole transcription of (some adequate segment of) the input signal. Then, the user reads this prediction until a transcription error is found. At this point, the user corrects this error generating a new, extended prefix (the previous validated prefix plus the amendments introduced by the user). This new prefix is used by the ASR system to attempt a new prediction thereby starting a new cycle that is repeated until a final and successful transcription is reached. An example of this process is shown in Fig. 2.

Formally, the CAST framework can be seen as an instantiation of the problem formulated in Eq. (2) where, in addition to the given utterance v (x in Eq.(2)), a *prefix* t_p of the transcription is available. This prefix, which corresponds to the feedback f in Eq. (2), contains information from the last system's prediction plus user's actions, in the form of amendment keystrokes. The ASR should try to complete this prefix by searching for the most likely *suffix* $\hat{t}_s$ ($\hat{h}$ in Eq.(2)), according to:

$$\hat{t}_s = \underset{t_s}{\operatorname{argmax}} P(t_s \mid v, t_p) = \underset{t_s}{\operatorname{argmax}} P(v \mid t_p, t_s) \cdot P(t_s \mid t_p) \qquad (6)$$

---

[2] To simplify notation, $P_{M_Z}(z \mid \ldots)$ will be denoted as $P(z \mid \ldots)$ from now on.

	(v)	
ITER-0	(t_p)	( )
ITER-1	$(\hat{t}_s)$	(*Nine extra soul are planned half beam discovered these years*)
	(a)	(**Nine**)
	(k)	(extrasolar)
	(t_p)	(Nine extrasolar)
ITER-2	$(\hat{t}_s)$	(*planets have been discovered these years*)
	(a)	(**planets have been discovered**)
	(k)	(this)
	(t_p)	(Nine extrasolar planets have been discovered this)
FINAL	$(\hat{t}_s)$	(*year*)
	(k)	(#)
	$(t_p \equiv t)$	(**Nine** extrasolar **planets have been discovered** this **year**)

Fig. 2. Example of CAST interaction. Initially the prefix t_p is empty, and the system proposes a complete transcription (t) of the audio input (v). In each iteration the user reads this transcription, validating a prefix (a) of it. Then, he or she corrects some words (k) of the transcription provided by the system, generating so a new prefix t_p (the previous one plus the words added by the user). At this point, the system will suggest a suitable continuation to this prefix (t_p) and this process will be repeated until a complete and correct transcription (t) of the input signal is reached.

Eq. (6) is very similar to Eq. (5), t being the concatenation of t_p and t_s. The main difference is that t_p is given. Therefore, the search must be performed over all possible suffixes t_s of t_p and the language model probability $P(t_s \mid t_p)$ must account for the words that can be uttered *after the prefix* t_p.

In order to solve Eq. (6), the signal v can be considered split into two fragments, v_1^b and v_{b+1}^m, where m is the length of v. By further considering the boundary point b as a hidden variable, we can write:

$$\hat{t}_s = \operatorname*{argmax}_{t_s} \sum_{0 \leq b \leq m} P(v, b \mid t_s, t_p) \cdot P(t_s \mid t_p) \tag{7}$$

We can now make the *naïve* (but realistic) assumption that the probability of v_1^b given t_p does not depend on the suffix and the probability of v_{b+1}^m given t_s does not depend on the prefix and, approximating the sum over all the possible segmentation by the dominating term, Eq.(7) can be rewritten as:

$$\hat{t}_s \approx \operatorname*{argmax}_{t_s} \max_{0 \leq b \leq m} P(v_1^b \mid t_p) \cdot P(v_{b+1}^m \mid t_s) \cdot P(t_s \mid t_p) \tag{8}$$

This optimization problem entails finding an optimal boundary point, $\hat{b}$, associated with the optimal suffix decoding, $\hat{t}_s$. That is, the signal v is actually split into two segments, $v_p = v_1^{\hat{b}}$ and $v_s = v_{\hat{b}+1}^m$, the first one corresponding to the *prefix* and the second to the *suffix*. Therefore, the search can be performed just over segments of the signal corresponding to the possible suffixes and, on the other hand, we can take advantage of the information coming from the prefix to tune the language model constraints modelled by $P(t_s \mid t_p)$. Details of this language model adaptation using n-grams can be seen in [10].

3.1 Experimental Results

To assess the CAST approximation discussed above, different experiments were carried out. These experiments were performed on two different corpora, the EUTRANS tourist corpus and the ALBAYZIN geographic corpus. Description of both corpora can be found in [10].

Two kinds of assessment metrics were adopted: the *Word Error Rate* (WER) and the *Word Stroke Ratio* (WSR). WER is a well known measure which estimates the number of *off-line* word corrections needed to achieve a perfect transcription, divided by the total number of words in the correct transcription. WSR, on the other hand, estimates the number of *on-line* user interactions (in terms of whole typed words) which are necessary to reach the same perfect transcription, also relative to the total number of words in the correct transcription.

The comparison between WER and WSR estimates the amount of effort required by a CAST user with respect to the effort needed by using a classical speech recognition system followed by manual off-line post-editing.

Table 1 summarizes the results, empirically showing a clear effectiveness of the CAST approach in both corpora.

Table 1. CAST results (in %) obtained with two corpora

	EUTRANS	ALBAYZIN
WER	11.4	11.6
WSR	9.3	10.1
Relative Improvement	≈ 19	≈ 14

4 Computer Assisted Translation (CAT)

The statistical (pattern recognition) framework for Machine Translation (MT) can also be stated as a particular case of Eq. (1). Given a *text* sentence x from a source language, search for a sentence $\hat{t}$ ($\hat{h}$ in Eq. 1) from a target language for which the posterior probability is maximum, that is:

$$\hat{t} = \operatorname*{argmax}_{t} P(t \mid x) = \operatorname*{argmax}_{t} P(t) \cdot P(x \mid t) . \tag{9}$$

The models adopted for each factor of Eq. (9) play an important role. On the one hand, $P(t)$ is modeled by a *target language model* which gives high probability to well formed target sentences. On the other hand, models for $P(x \mid t)$ should give high probability for those sentences from the source language which are good translations for a given target sentence. These models generally consist of *stochastic dictionaries*, along with adequate models to account for *word alignments* [1,9].

An alternative is to transform Eq. (9) as:

$$\hat{t} = \underset{t}{\operatorname{argmax}} P(t, x) \ . \tag{10}$$

In this case, the joint probability distribution can be adequately modeled by means of *stochastic finite-state transducers (SFST)* [3], among other possible models.

Following the IPR paradigm, user intervention can be included into this framework. The process is quite similar to what was described for CAST in Section 3. Here, the difference is that the input is a source text rather than an audio signal.

Given a source text x and a user validated *prefix* t_p of the target sentence (the feedback f in Eq. (2)), search for a *suffix* t_s of the target sentence that maximizes the posterior probability over all possible suffixes:

$$\hat{t}_s = \underset{t_s}{\operatorname{argmax}} P(t_s \mid x, t_p) \ . \tag{11}$$

Taking into account that $P(t_p \mid x)$ does not depend on t_s, we can write:

$$\hat{t}_s = \underset{t_s}{\operatorname{argmax}} P(t_p t_s \mid x) \ , \tag{12}$$

where $t_p t_s$ is the concatenation of the given prefix t_p and a suffix t_s. Eq. (12) is similar to Eq. (9), but here the maximization is carried out over a (possibly infinite) set of suffixes, rather than full sentences as in Eq. (9).

The system proposed in [12] addresses the search problem stated in Eq. (12) by means of statistical phrase-based models. Alternatively, Eq. (12) can be rewritten as:

$$\hat{t}_s = \underset{t_s}{\operatorname{argmax}} P(t_p t_s, x) \ . \tag{13}$$

In this case, as in Eq. (10), this joint distribution can be adequately modeled by means of SFSTs and, the search problem is addressed by adequate extensions of the Viterbi algorithm. This approach is followed in the CAT systems presented in [4].

4.1   Experiments

Different experiments have been carried out to assess the proposed CAT approach. Apart from other assessment metrics commonly used in machine translation [2], an important performance criterion in this case is the amount of effort that the human translator would need, with respect to the effort needed to type the whole translation without the CAT system. This is estimated by the so called *Key Stroke Ratio* (KSR) which is related to WSR described in section 3 and measures the number of user interactions in terms of *keystrokes* needed to achieve a perfect translation of the source sentence. KSR can be automatically computed by using reference translations of the test source sentences [4].

The experiments were carried out on two corpora: the *Xerox printer manuals* and the *Bulletin of the European Union*. In both corpora, three language pairs were considered in both translation directions: English/Spanish, English/French and English/German [4].

According to the results reported in [3,4,12] the KSR ranged from 22% to 47%, depending on the language pair and corpus considered. In all the cases, the estimated saved effort was significant. In addition to these experiments, a CAT prototype was implemented and formal on-line tests with real (professional) users were carried out. Results reported in [8] confirmed that our CAT system can actually save significant amounts of human translator effort when high-quality translations are needed.

5   Multimodality in Computer Assisted Translation Speech-Enabled CAT

In CAT applications the user is repeatedly interacting with the system. Hence, the quality and ergonomics of the interaction process is crucial for the success of the system. Peripherals like keyboard and mouse can be used to validate and correct the successive predictions of the CAT system. Nevertheless, providing the system with a multimodal interface should result in an easier and more comfortable human-machine interaction. Different ways of communication can be explored; gaze and gesture tracking, touchscreens, etc.

In this section we focus on a *speech* interface for CAT. Speech has been selected for several reasons. On the one hand, it is a very natural and seamless way to convey information for humans. On the other, CAT is about natural language processing, where the use of natural spoken language seems to be most appropriate [11]. More specifically, the user has to read the system's predictions to search for possible mistakes and correct them if necessary. Allowing the user to read aloud and letting the application capture this information and use it as feedback to perform a new CAT interaction constitutes, a priori, a good premise to implement a CAT multimodal system in this way. Additionally, high speech recognition accuracy can be achieved by adopting a IPR approach to implement this multimodal interface. These ideas, which were first introduced in [13], are reviewed below.

Let x be the source text and t_p a validated prefix of the target sentence. The user is then allowed to utter some words v, generally related to the suffix suggested by the system in the previous iteration. This utterance is aimed at accepting or correcting parts of this suffix. It can be used to add more text as well. Moreover, the user may enter some keystrokes k in order to correct (other) parts of this suffix and/or to add more text. Using this information, the system has to suggest a new suffix t_s as a continuation of the previous prefix t_p, the voice v and the typed text k. That is, the problem is to find t_s given x and a feedback information composed of t_p, v and k, considering all possible *decodings* of v (i.e., letting the decoding of v be a hidden variable).

According to this very general discussion, it might be assumed that the user can type with independence of the result of the speech decoding process. However, it can be argued that this generality is not realistically useful in practical situations. Alternatively, it is much more natural that the user waits for a system outcome ($\hat{d}$) from the spoken utterance v, prior to start typing amendments (k) to the (remaining part of the previous) system hypothesis. Furthermore, this allows the user to fix possible speech recognition errors in $\hat{d}$.

In this more pragmatic and simpler scenario, the interaction process can be formulated in two steps. The first step is to rely on the source text x and the previous target prefix t_p, in order to search for a target suffix $\hat{t}_s$:

$$\hat{t}_s = \underset{t_s}{\operatorname{argmax}} \, Pr(t_s \mid x, t_p) \tag{14}$$

Once $\hat{t}_s$ is available, the user can produce some speech, v, and the system has to decode v into a target sequence of words, $\hat{d}$:

$$\hat{d} = \underset{d}{\operatorname{argmax}} \, P(d \mid x, t_p, \hat{t}_s, v) \tag{15}$$

Finally, the user can enter adequate amendment keystrokes k, if necessary, and produce a new consolidated prefix, t_p, based on the previous t_p, $\hat{d}$, k and parts of $\hat{t}_s$. The process will continue in this way until t_s is completely accepted by the user as a full target text which is an appropriate translation of x. An example of this combined speech and text interaction with a CAT system is shown in Fig. 3.

Since we have already dealt with Eq. (14) in section 4 (Eq. 11-13), we focus now on Eq. (15). As compared with Eq. (2), here the triplet $(x, t_p, \hat{t}_s)$ and v would correspond to two modalities: x (text) and f (voice). Therefore, assumptions and developments similar to those of Eq. (3-4) lead to:

$$\hat{d} = \underset{d}{\operatorname{argmax}} \, P(x, t_p, \hat{t}_s \mid d) \cdot P(v \mid d) \cdot P(d) \tag{16}$$

which can be strightforwardly re-written as:

$$\hat{d} = \underset{d}{\operatorname{argmax}} \, P(d \mid x, t_p, \hat{t}_s) \cdot P(v \mid d) \tag{17}$$

As in section 3, $P(v \mid d)$ is modelled by the acoustic models of the words in d. Here, $P(d \mid x, t_p, \hat{t}_s)$ can be provided by a target language model constrained by the source sentence x, by the previous prefix t_p and by the suffix $\hat{t}_s$ produced at the beginning of the current iteration.

Eq. (17) leads to different scenarios depending on the assumptions and constraints adopted for $P(d \mid x, t_p, \hat{t}_s)$. Three of them are discussed hereafter.

The first one consists in pure speech recognition of sentence-fragments. This scenario, called DEC, is considered here just as a *baseline*, where all the available conditions are ignored; i.e., $P(d \mid x, t_p, \hat{t}_s) \equiv P(d)$.

A second possibility would be to use only the given prefix; that is, $P(d \mid x, t_p, \hat{t}_s) \equiv P(d \mid t_p)$. This is similar to the CAST framework discussed in section 3 and is not considered here.

ITER-0	(t_p)	( )		
	$(\hat{t}_s)$	(*Haga clic para cerrar el diálogo de impresión*)		
	(v)	~~~~~~~~~~~~~~		
ITER-1	$(\hat{d})$	**(Haga clic a)**		
	(k)	**(en** ACEPTAR**)**		
	(t_p)	**(Haga clic en** ACEPTAR**)**		
	$(\hat{t}_s)$	(*para cerrar el diálogo de impresión*)		
	(v)	~~~~~~~~~~~~~~		
ITER-2	$(\hat{d})$	**(cerrar el cuadro)**		
	(k)	( )		
	(t_p)	**(Haga clic en** ACEPTAR **para cerrar el** cuadro)		
	$(\hat{t}_s)$	(*de diálogo de impresión*)		
FINAL	(k)	**(#)**		
	$(t_p \equiv t)$	**(Haga clic** en ACEPTAR **para cerrar el cuadro** de diálogo de impresión)		

Fig. 3. Example of keyboard and speech interaction with a CAT system, to translate the English sentence *"Click OK to close the print dialog"*. Each iteration starts with a target language prefix t_p that has been fixed in the previous iteration. First, the system suggests a suffix $\hat{t}_s$ and then, the user speaks (v) and/or types some keystrokes (k), possibly aimed to amend $\hat{t}_s$ (and maybe $\hat{d}$). A new prefix, t_p, is built from the previous prefix, along with (parts of) the system suggestion, $\hat{t}_s$, the decoded speech, $\hat{d}$, and the typed text in k. The process ends when the user types the special character "#". System suggestions are printed in cursive, text decoded from user speech in boldface and typed text in boldface typewriter font. In the final translation, t, text obtained from speech decoding is marked in boldface, while typed text is underlined.

If the information of the source sentence is additionally used, a new scenario, CAT-PREF, arises. In this case the speech recognition is constrained to find suitable continuations that are also partial translations for the source sentence; that is, $P(d \mid x, t_p, \hat{t}_s) \equiv P(d \mid x, t_p)$.

Finally, a most constrained scenario is CAT-SEL, where the human translator is only expected to utter exact prefixes of the suggestion made by the CAT system in the previous iteration; i.e. $P(d \mid x, t_p, \hat{t}_s) \equiv P(d \mid \hat{t}_s)$. This scenario aims at enabling highly accurate speech recognition as a way to validate correct prefixes of the current predictions of the translation engine.

Note that the different constraints underlying DEC, CAT-PREF and CAT-SEL, only affect the *language model* component of the speech decoding problem described by Eq. (17). Correspondingly, techniques similar to those used in CAST (section 3) also apply in this case. Details of these techniques can be found in [13].

5.1 Results

In the previous sections, experiments aimed at estimating the ability of the IPR applications to help the user achieve perfect results with the minimum effort.

In contrast, here our aim is to estimate the effectiveness of the multimodal (speech) feedback mechanism. To this end we are interested in measuring the degree of accuracy achieved when decoding the feedback signal with the help of other interaction-related informations which are assumed available in the different scenarios considered.

The test data for these experiments is composed of utterances of fragments of target-language (Spanish) sentences extracted from the test part of the XEROX corpus mentioned in Section 4. See [13] for details.

Speech decoding accuracy was assessed in terms of two well known measures, *Word Error Rate* (WER) and *Sentence Error Rate*. SER is the number of "sentences" (actually *sentence fragments*) correctly recognized divided by the overall number of sentences. Particularly in CAT-SEL, SER estimates how often the user will have to resort to non-speech (keyboard or mouse) interaction to amend speech-derived positioning errors.

As expected, the results of Table 2 show that increasing speech recognition performance is achieved as the language models become more constrained. By adding constraints derived from the source text and the target sentence prefix to DEC, a significant improvement is achieved in CAT-PREF: 8.0 points of WER and 19.8 points of SER. It is worth noting that the results that can be achieved in this scenario heavily depend on the quality of the underlying translation models employed and also on the difficulty of the translation task. Finally, the restrictions added in the most constrained scenario, CAT-SEL, are *directly* derived from the suggestions of the CAT system. In this case, the improvement with respect to the previous scenario (CAT-PREF) is even more important: 9.0 points of WER and 26.4 points of SER. The decoding computational demands of the different systems are also worth mentioning. A substantial reduction of memory and computing time is observed from DEC to CAT-PREF. Finally CAT-SEL only requires very light computing, which allows implementing this kind of speech-enabled CAT systems on low-end desktop or laptop computers.

Table 2. Speech decoding results (in %) for different scenarios

	WER	SER
DEC (baseline)	18.6	50.2
CAT-PREF	10.6	30.0
CAT-SEL	1.6	3.6

6 Conclusions

A new general framework to integrate human activity into pattern recognition systems, IPR, has been presented. In addition, two applications, CAST and CAT, have been reviewed and the inclusion of a multimodal interface in CAT has been discussed. This formulation and examples clearly suggest that IPR can be advantageously extended to new application fields. Furthermore, new ways to take advantage of the informations derived from the user interactions could be explored in the future.

References

1. Brown, P.F., Pietra, S.A.D., Pietra, V.J.D., Mercer, R.L.: The mathematics of statistical machine translation: Parameter estimation. Computational Linguistics 19(2), 263–311 (1993)
2. Casacuberta, F., Ney, H., Och, F.J., Vidal, E., Vilar, J.M., Barrachina, S., García-Varea, I., Llorens, D., Martínez, C., Molau, S., Nevado, F., Pastor, M., Picó, D., Sanchis, A., Tillmann, C.: Some approaches to statistical and finite-state speech-to-speech translation. Computer Speech and Language 18, 25–47 (2004)
3. Casacuberta, F., Vidal, E.: Learning finite-state models for machine translation. Machine Learning 66(1), 69–91 (2007)
4. Civera, J., Lagarda, A.L., Cubel, E., Casacuberta, F., Vidal, E., Vilar, J.M., Barrachina, S.: Computer-Assisted Translation Tool based on Finite-State Technology. In: Proc. of EAMT 2006, pp. 33–40 (2006)
5. Duda, R.O., Hart, P.E., Stork, D.G.: Pattern Classification, 2nd edn. John Wiley and Sons, New York, NY (2000)
6. Frederking, R., Rudnicky, A.I., Hogan, C.: Interactive speech translation in the DIPLOMAT project. In: Procs. of the ACL-97 Spoken Language Translation Workshop, pp. 61–66, Madrid, ACL (1997)
7. Jelinek, F.: Statistical Methods for Speech Recognition. The MIT Press, Cambridge, Massachusetts, USA (1998)
8. Macklovitch, E.: The contribution of end-users to the transtype2 project (TT2). In: Frederking, R.E., Taylor, K.B. (eds.) AMTA 2004. LNCS (LNAI), vol. 3265, pp. 197–207. Springer, Heidelberg (2004)
9. Ney, H., Niessen, S., Och, F.J., Sawaf, H., Tillmann, C., Vogel, S.: Algorithms for statistical translation of spoken language. IEEE Transactions on Speech and Audio Processing 8(1), 24–36 (2000)
10. Rodríguez, L., Casacuberta, F., Vidal, E.: Computer assisted transcription of speech. In: 3rd Iberian Conference on Pattern Recognition and Image Analysis, Girona (Spain) (June 2007)
11. Suhm, B., Myers, B., Waibel, A.: Multimodal error correction for speech user interfaces. ACM Trans. Comput.-Hum. Interact. 8(1), 60–98 (2001)
12. Tomás, J., Casacuberta, F.: Statistical phrase-based models for interactive computer-assisted translation. In: Proceedings of the Coling/ACL, pp. 835–841, Sydney, Australia (17th-21th July 2006),
http://acl.ldc.upenn.edu/P/P06/P06-2107.pdf
13. Vidal, E., Casacuberta, F., Rodríguez, L., Civera, J., Martínez, C.: Computer-assisted translation using speech recognition. IEEE Transaction on Audio, Speech and Language Processing 14(3), 941–951 (2006)

User Specific Training of a Music Search Engine

David Little, David Raffensperger, and Bryan Pardo

EECS Department
Northwestern University
Evanston, IL 60208
{d-little,d-raffensperger,pardo}@northwestern.edu

Abstract. Query-by-Humming (QBH) systems transcribe a sung or hummed query and search for related musical themes in a database, returning the most similar themes as a play list. A major obstacle to effective QBH is variation between user queries and the melodic targets used as database search keys. Since it is not possible to predict all individual singer profiles before system deployment, a robust QBH system should be able to adapt to different singers after deployment. Currently deployed systems do not have this capability. We describe a new QBH system that learns from user provided feedback on the search results, letting the system improve while deployed, after only a few queries. This is made possible by a trainable note segmentation system, an easily parameterized singer error model and a straight-forward genetic algorithm. Results show significant improvement in performance given only ten example queries from a particular user.

1 Introduction

Most currently deployed music search engines, such as Amazon.com and local libraries, make use of metadata about the song title and performer name in their indexing mechanism. Often, a person is able to sing a portion of the piece, but cannot specify the title, composer or performer. Query by humming (QBH) systems [1,2,3,4,5] solve this mismatch between database keys and user knowledge. This is done by transcribing a sung query and searching for related musical themes in a database, returning the most similar themes as a play list.

One of the main difficulties in building an effective QBH system is dealing with the variation between user queries and the melodic targets used as database search keys. Singers are error prone: they may go out of tune, sing at a different tempo than expected, or in a different key, and notes may be removed or added [1,6]. Further, singers differ in their error profiles. One singer may have poor pitch, while another may have poor rhythm. Similarly, different environments may introduce variation through different levels of background noise or availability of different microphones.

Since it is not possible to predict all individual singer profiles or use cases before deployment of a system, a robust QBH system should be able to adapt to different singers and circumstances after deployment. Currently deployed QBH systems do not have this capability.

A. Popescu-Belis, S. Renals, and H. Bourlard (Eds.): MLMI 2007, LNCS 4892, pp. 72–83, 2007.

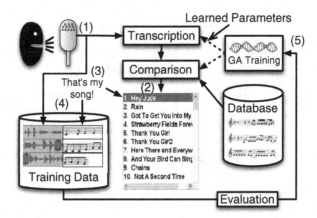

Fig. 1. System overview

While there has been significant prior work that addresses (or is applicable to) singer error modelling [6,7,8,5] for QBH, researchers have not focused on fully automated, ongoing QBH optimization after deployment. Thus, these approaches are unsuited for this task, requiring either hundreds of example queries [6,7], or training examples where the internal structure of each query is aligned to the structure of the correct target [8].

We are developing a QBH system (shown in Figure 1) that personalizes a singer model based on user feedback, learning the model on-line, after deployment without intervention from the system developers and after only a few example queries. The user sings a query (step 1 in the figure). The system returns a list of songs from the database, ranked by similarity (step 2). The user listens to the songs returned and selects the one he or she meant to sing (step 3). The more a person uses and corrects the system, the better the system performs. This is done by building a database of a user's paired queries and correct targets (step 4). These pairings are used to train optimal note segmentation and note interval similarity parameters for each specific user (step 5).

In this paper, we focus on how we automatically optimize backend QBH system performance, given a small set of example queries. We refer the reader to [9] for a description of the user interface and user interaction. Section 2 describes our query representation. Section 3 describes our optimizable note segmentation algorithm. Section 4 describes our melodic comparison algorithm. Section 5 describes our optimizable note interval similarity function. Section 6 describes our genetic algorithm learning approach. Section 7 outlines an empirical study of our system. Section 8 contains conclusions and future directions.

2   Query Representation

In a typical QBH system, a query is first transcribed into a time-frequency representation where the fundamental frequency and amplitude of the audio is

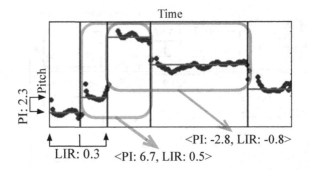

Fig. 2. Several unquantized pitch intervals built from a melodic contour

estimated at very short fixed intervals (on the order of 10 milliseconds). We call this sequence of fixed-frame estimates of fundamental frequency a melodic contour representation. Figure 2 shows the melodic contour of a sung query as a dotted line.

Finding queries represented by melodic contour imposes computational overhead on the search engine [10]. The melodic contour representation uses absolute frequency and tempo, making it sensitive to queries that vary in tempo and key from the target. To overcome these problems one can transform the contour through a number of keys and tempi [1] but this incurs significant overhead.

To improve system speed, we encode a query as a series of note intervals, created as follows:

1. Divide the melodic contour into segments corresponding to notes.
2. Find the median pitch and the length in frames of each note segment.
3. Encode segments as a sequence of note intervals.

Each note interval is represented by a pair of values: the pitch interval (PI) between adjacent note segments (encoded as un-quantized musical half-steps) and the log of the ratio between the length of a note segment and the length of the following segment (LIR) [10]. Since each LIR is a ratio, values do not have units.

This representation is both transposition and tempo invariant. It is also compact, only encoding salient points of change (note transitions), rather than every 10 millisecond frame. Figure 2 shows several note intervals.

We note that the use of unquantized PI and LIR values makes the representation insensitive to issues caused by a singer inadvertently singing in an unexpected tuning (A4 ≠ 440), or slowly changing tuning and tempo over the course of a query. This lets us avoid having to model these common singer errors.

The note-interval representation speeds search by a factor of roughly 100, when compared with a melodic contour representation on real-world queries [1]. This speedup comes at the price of introducing potential errors when dividing the query into note segments. We address note segmentation in Section 3.

3   Note Segmentation

In the initial transcription step, our system estimates the pitch, root-mean-squared amplitude and harmonicity (the relative strength of harmonic components compared to non-harmonic components) of the audio. This is done every 10 milliseconds, resulting in a sequence of fixed-length frames, each of which is a vector of these three features.

We assume significant changes in these three features occur at note boundaries. Thus, we wish to find a good way to determine what constitutes significant change. Our initial approach was to use a Euclidean distance metric on the sequence of frames. In this approach, a new note was assumed to begin whenever the distance between adjacent frames exceeds a fixed threshold.

Unfortunately, this introduces error when a single fixed threshold is not appropriate. For example, a singer may use vibrato at times and not at other times. Thus, the local pitch variation that might constitute a meaningful note boundary in one query may be insufficient to qualify as a note boundary in another query by the same singer. We wish to take local variance into account when determining whether or not a note boundary has occurred.

Note segmentation is related to the problem of visual edge detection [11,12]. Accounting for local variation has been successfully used in image processing to perform edge detection in cases where portions of the image may be blurry and other portions are sharp [13,11]. The Mahalanobis distance [14] differs from the Euclidean distance in that it normalizes distances over a covariance matrix M. Using the Mahalanobis lets one measure distance between frames relative to local variation. In a region of large variance, a sudden change will mean less than in a relatively stable region.

We find the distance between adjacent frames in the sequence using a Mahalanobis distance measure, shown in Equation 1. Recall that each frame is a three element vector containing values for pitch (p), amplitude (a) and harmonicity (h). Given a frame f_i, $f_i = \langle p_i, a_i, h_i \rangle$, we assume a new note has begun wherever the distance between two adjacent frames f_i and f_{i+1}, exceeds a threshold, T. This is shown in Equation 1.

$$\sqrt{\left(\mathbf{f}_i - \mathbf{f}_{i+1}\right) M^{-1} \left(\mathbf{f}_i - \mathbf{f}_{i+1}\right)'} \tag{1}$$

Our covariance matrix M depends on local variation in the three features of a frame. The matrix has three rows and three columns, with each row, ρ, corresponding to one feature and each column, η, corresponding to one feature (so $\rho, \eta \in \{p_i, a_i, h_i\}$ for f_i). We calculate each element of M for a frame f_i using Equation 2:

$$M_{\rho\eta} = \frac{1}{2\tau} \sum_{k=i-\tau}^{i+\tau} \left(\frac{\rho_k - \bar{\rho}}{w_\rho}\right) \left(\frac{\eta_k - \bar{\eta}}{w_\eta}\right) \tag{2}$$

In Equation 2, the w terms are weighting parameters that adjust the importance of the features of the frame (pitch, harmonicity and amplitude). The parameter τ is the size of a window (measured in frames) surrounding the current frame. The window size determines the number of frames considered in finding local variation. The averages for ρ and η ($\bar{\rho}$ and $\bar{\eta}$) are calculated over this window.

Thus, Equation 2 finds the covariance between a pair of features: for example with $\rho = a$, and $\eta = h$, Equation 2 would find the covariance between the amplitude and harmonicity for frames $i - \tau$ to $i + \tau$.

Our note segmenter has four tuneable parameters: the segmentation threshold (T), and the weights (w) for each of the three features (pitch, harmonicity and amplitude). We address tuning of these four parameters in Section 6. We leave the tuning of parameter τ for future work, setting it to a value of 25 frames (250 milliseconds) to either side of the current frame i.

Once we have estimated note segment boundaries, we build note intervals from these note segments.

4   Measuring Melodic Similarity

Once a query is encoded as a sequence of note intervals, we compare it to the melodies in our database. Each database melody is scored for similarity to the query using a classic dynamic-programming approach to performing string alignment [7]. The note interval matcher computes the similarity $Q(A, B)$ between two melodic sequences $A = a_1, a_2, \cdots a_m$ and $B = b_1, b_2, \cdots b_n$. by filling the matrix $Q = (q_{1\cdots m,1\cdots n})$. Each entry q_{ij} denotes the maximum melodic similarity between the two prefixes $a_1 \cdots ai$ and $b_h \cdots b_j$ where $1 \le h \le j$.

We use a standard calculation method for the algorithm, shown in Equation 3

$$q_{i,j} = max \begin{cases} 0 \\ q_{i-1,j-1} + s(a_i, b_i) \\ q_{i-1,j} - c_a \\ q_{i,j-1} - c_b \end{cases} \tag{3}$$

Here, $s(a_i, b_j)$ is the similarity reward for aligning note interval a_i to note interval b_j. We define $s(a_i, b_j)$ in Section 5. The costs for skipping a note interval from melody A or from melody B are given by c_a and c_b, respectively. Equation 3 is a local alignment method, so any portion of the query is allowed to match any portion of the target. The overall similarity is taken to be the maximum value in the matrix Q.

5   Modeling Singer Error

We now define the similarity function s for note intervals. Ideally we would like interval a_i to be similar to interval b_j if a_i likely to be sung when a singer intended to sing b_j. That is, likely errors should be considered similar to the correct interval, and unlikely errors should be less similar. Such a function lets a string-alignment algorithm correctly match error-prone singing examples to the correct targets, as long as the singer is relatively consistent with the kinds of errors produced.

In our previous work [7], the similarity function was represented with a table of 625 (25 by 25) values for 25 possible pitch intervals (from -12 to 12 half steps). The similarity between a perfect fifth (7 half steps) and a tritone (6 half steps) was determined by calculating the statistical likelihood of a person singing a perfect fifth when instructed to sing a tritone. The more likely this was, the higher the similarity. This required hours of directed training to learn all pairings in our table. This is something users outside the laboratory setting are unwilling to do.

For this work, we have developed a similarity function that captures singer variation by tuning only a few parameters, so that suitable values can be learned quickly. The normal function, $N(a, \mu, \sigma)$ returns the value for a given by a Gaussian function, centered on μ, with a standard deviation σ. Equation 4 shows a simple note-interval similarity function, based on the normal function.

$$s(x, y) = w_p N(y_p, x_p, \sigma_p) + w_r N(y_r, x_r, \sigma_r) \tag{4}$$

Let x and y be two note intervals. Here, x_p and y_p are the pitch intervals of x and y respectively, and x_r and y_r are the LIRs of x and y. The values w_p and w_r are the weights of pitch and rhythm. The sum of w_p and w_r is 1. Equation 4 is maximal when x and y are identical. As the difference between x and y grows, Equation 4 returns a value approaching 0.

Equation 4 assumes increasing distance between pitch intervals is equivalent to decreasing similarity between intervals. There is at least one way in which this is untrue for pitch: octaves. Shepard [15] proposes a pitch similarity measure that accounts for two dimensions: pitch chroma and pitch height. In this model, octaves are fairly close to each other. Criani [16] proposes a psychological model of pitch similarity based on the Pearson correlations of temporal codes which represent pitched stimuli. These correlations show strong peaks not only near like pitches, but also near pitches an octave apart. In previous work [10] we also found high likelihood of octave substitutions in sung queries. This suggests that, at a minimum, we should account for octave similarities in our measure of pitch similarity. We thus modify Equation 4 to give Equation 5.

$$s(x, y) = w_r N(y_r, x_r, \sigma_r) + w_p \sum_{i=-n}^{n} \lambda^{|i|} N(y_p, x_p + 12i, \sigma_p) \tag{5}$$

Here, the pitch similarity is modeled using $2n+1$ Gaussians, each one centered at one or more octaves above or below the pitch of x. The height of each Gaussian is determined by an octave decay parameter λ, in the range from than 1 to 0. This reward function provides us with five parameters to tune: the pitch and rhythm weight (w_p and w_r), the sensitivity to distances for pitch and rhythm (σ_p and σ_r), and the octave decay (λ). Figure 3 shows the positive portion of the pitch dimension of this function, given two example parameter settings, with two octaves shown.

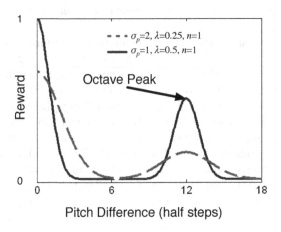

Fig. 3. The pitch dimension of the similarity function in Equation 5

6   System Training

We train the system by tuning the parameters of our note segmenter (Equations 1 and 2) and note similarity reward function (Equation 5). We measure improvement using the mean reciprocal rank (MRR) of a set of n queries. This is shown in Equation 6. Here, we define the rank of the ith query, r_i as the rank returned by the search engine for the correct song in the database.

$$MRR = \frac{\sum_{i=1}^{n} \frac{1}{r_i}}{n} \tag{6}$$

MRR emphasizes the importance of placing correct target songs near the top of the list while still rewarding improved rankings lower down on the returned list of songs [1]. Values for MRR range from 1 to 0, with higher numbers indicating better performance. Thus, MRR = 0.25 roughly corresponds to the correct answer being in the top four songs returned by the search engine, MRR = 0.05 indicates the right answer is in the top twenty, and so on.

We use a simple genetic algorithm [17,18] to tune system parameters. Each individual in the population is one set of parameter values for Equations 1, 2 and 5. The fitness function is the MRR of the parameter settings over a set of queries.

The genetic algorithm represents each parameter as a binary fraction of 7 bits, scaled to a range of 0 to 1. We allow crossover to occur between (not within) parameters.

During each generation, the fitness of an individual is found based on the Mean Reciprocal Rank (MRR) of the correct targets for a set of queries. Parameter settings (individuals) with high MRR values are given higher probability of reproduction (fitness proportional reproduction). We speed training by finding MRR over random subset of the database.

Given ten queries from each singer labeled with the correct target name and a set of 250 target melodies from our database, the system personalizes two singers per hour on our current hardware. These results are obtained with an iMac 2.16 GHz Intel Core Duo 2, with 2 GB of RAM. At this speed, our system can update models overnight for up to 16 singers who have provided new queries during the previous day.

7   Empirical Evaluation

Our empirical evaluation sought to compare user-specific training to training on a general set of queries. We also sought to verify the utility of two new design choices: the use of a Mahalanobis distance to account for local variation during note segmentation, and the use of unquantized, as opposed to quantized note-intervals.

Our query set was drawn from the QBSH corpus [19] used during the 2006 MIREX comparison of query-by-humming systems [20]. We used 10 singers, each singing the same 15 songs from this dataset. Our target database was composed of the 15 targets corresponding to these queries plus 1471 distracter melodies drawn from a selection of Beatles songs, folk songs and classical music, resulting in a database of 1486 melodies. Chance performance, on a database of this size would result in an $MRR \approx 0.005$, given a uniform distribution.

For the genetic algorithm, we chose a population size of 60. Initial tests showed learning on this task typically ceases by the 30th generation, thus results shown here report values from training runs of 40 generations. We used a mutation probability of 0.02 per parameter.

We ran two experiments: the first examines the overall performance of user specific training and the improvements the new system features introduced. The second compares our system with using user-specific to a version without user-specific training.

7.1   Experiment 1

For Experiment 1 we considered three different conditions. In the first condition we accounted for local variation during note segmentation using the Mahalanobis distance and used an unquantized note interval representation (Local). In the second condition, we used Euclidean distance for note segmentation and an unquantized note interval representation (Unquantized). In the third condition (Quantized) we used the Euclidean distance for note segmentation and a quantized note interval representation. The quantized condition is equivalent to a previous system we developed [7].

A single trial consists of selecting a condition (Local, Unquantized or Quantized) and performing a training run for one singer, selecting ten of the singers fifteen queries to train on and testing on the remaining five. To speed learning, training was done using a random sample of 250 target songs from the database. The same sample of 250 songs was used for training in each of the

Table 1. Mean (Standard Deviation) MRR over 30 trials in Experiment 1

Local	Unquantized	Quantized	Chance
0.228(0.14)	0.176(0.14)	0.089(0.09)	0.005

three conditions, so that results could be compared fairly. For each trial, the set of parameters with the best training performance was evaluated by finding the MRR of the five testing queries, searching over all 1486 melodies in the database.

We performed three-fold cross validation. Thus, there were three trials per singer for a total of 30 trials per condition. The mean MRR for each of the three conditions is shown in Table 1.

A paired-sample t-test was performed between each pair of conditions and all difference in means were found to be statistically significant ($p < 0.031$ in all cases). A t-test showed all three conditions also performed significantly better than chance ($p < 0.001$ in all cases). This indicates user-specific training does have a positive effect on search results. Further, the use of Mahalanobis distance for note segmentation and unquantized note interval representation significantly improves performance.

One cause for concern is the variance in the results. In a few trials the learned performance was below chance (< 0.005). In these cases it would make sense to back off to a more generalized set of parameters, learned from a larger population of singers. We explore this idea in Experiment 2.

7.2   Experiment 2

In practice, we would like to utilize user-specific training only when it improves performance relative to an un-personalized system. One simple option is to only use user-specific parameters if the user-specific performance (MRR_u) is superior to the performance using parameters learned on a general set of queries by multiple users (MRR_g).

To test this idea, we first trained the system on all queries from nine of the ten singers used in Experiment 1. For these trials we used the Mahalanobis distance for note segmentation and unquantized note intervals. We then tested on all the queries from the missing singer. Cross validation across singers was performed, thus the experiment was repeated ten times, testing with the queries from a different singer each time. This gave us parameters for each singer that were learned on the queries by the other nine singers. These are the General parameter settings for a singer. The mean MRR testing performance of the General parameters was 0.235 (Std. Dev. = 0.063).

We then repeated the user-specific training in Experiment 1 with the Local condition. For each trial we determined whether the singer-specific parameters found in a trial had an MRR on the training set (ten songs by one singer) that exceeded by the MRR that would result from using general parameters learned from the other nine singers. If, on the training set, $MRR_u > MRR_g + \epsilon$ we used the user-specific parameters. Else, we used the general parameters. For this experiment, ϵ is an error margin set to 0.04.

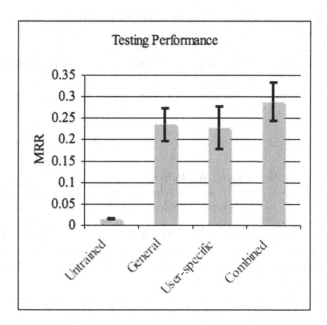

Fig. 4. Testing Performance on Experiment 2

Once the parameters (general or user-specific) were selected, we tested them on the testing set for that trial (the remaining five songs by that singer). We called this a combined trial. The combined trials had an average MRR of 0.289 (Std. Dev. = 0.086). A t-test indicated the improvement of the combined results over the general parameter settings is statistically significant ($p = 0.024$).

On 50% of the combined trials the user specific parameters were used and improved performance compared to general training. On 13% of the trials, user-specific parameters were selected, but made performance worse compared to general training. On the remaining 36% of trials, the general data set parameters were used. Figure 4 comaprers the results for conditions.

8   Conclusions

We have described a QBH system that can learn both general and singer-specific error models and note segmentation parameters from labeled user queries. This system can automatically customize parameters to individual users after deployment. This is done by taking pairings of queries to the correct targets (provided by the user) and using these queries' MRR as a fitness function to a genetic algorithm that optimizes the note segmentation and note similarity parameters of our system for that specific user. Our results show that combining user-specific training with more general training significantly improves mean search performance, resulting in a mean MRR of 0.289 on a database of 1486 melodies. This roughly corresponds to consistently placing the correct target in the top four results. Our results also show both unquantized note intervals and modeling local singer variation for note segmentation significantly improve performance.

In future work we plan to explore how performance varies with respect to the number of training examples provided and improve the information that can be used for training while maintaining user-specificity. We will also explore more sophisticated criteria to determine when user-specific training should be used.

References

1. Dannenberg, R., Birmingham, W., Pardo, B., Hu, N., Meek, C., Tzanetakis, G.: A comparative evaluation of search techniques for query-by-humming using the mustart testbed. Jounral of the American Society for Information Sicence and Technology 58(3) (2007)
2. Kosugi, N., Sakurai, Y., Morimoto, M.: Soundcompass: A practical query-by-humming system; normalization of scalable and shiftable time-series data and effective subsequence generation. In: International Conference on Management of Data, Paris, France, pp. 881–886 (2004)
3. Pauws, S.: Cubyhum: A fully operation query by humming system. In: ISMIR (2002)
4. Shifrin, J., Pardo, B., Meek, C., Birmingham, W.: Hmm-based musical query retrieval. In: Joint Conference on Digital Libraries, Portland, Oregon, USA (2002)
5. Unal, E., Narayanan, S., Chew, E.: A statistical approach to retrieval under user-dependent uncertainty in query-by-humming systems. In: Multimedia Information Retrieval Conference, New York, NY (2004)
6. Meek, C., Birmingham, W.: A comprehensive trainable error model for sung music queries. Journal of Artificial Intelligence Research 22, 57–91 (2004)
7. Pardo, B., Birmingham, W.P., Shifrin, J.: Name that tune: A pilot study in finding a melody from a sung query. Journal of the American Society for Information Science and Technology 55(4), 283–300 (2004)
8. Parker, C., Fern, A., Tadepalli, P.: Gradient boosting for sequence alignment. In: The Twenty-First National Confeerence on Artificial Intelligence (2006)
9. Pardo, B., Shamma, D.: Teaching a music search engine through play. In: CHI 2007. Computer/Human Interaction (2007)
10. Pardo, B., Birmingham, W.: Encoding timing information for music query matching. In: International Conference on Music Information Retrieval (2002)
11. Elder, J.H., Zucker, S.W.: Local scale control for edge detection and blur estimation. Pattern Analysis and Machine Intelligence, IEEE Transactions on 20(7), 699–716 (1998)
12. Meer, P., Georgescu, B.: Edge detection with embedded confidence. Pattern Analysis and Machine Intelligence, IEEE Transactions on 23(12), 1351–1365 (2001)
13. Ahmad, M.B., Tae-Sun, C.: Local threshold and boolean function based edge detection. Consumer Electronics, IEEE Transactions on 45(3), 674–679 (1999)
14. Tzanetakis, G., Cook, F.: A framework for audio analysis based on classification and temporal segmentation. In: EUROMICRO Confrence, Milan, vol. 2, pp. 61–67 (1999)
15. Shepard, R.: Geometrical approximations to the structure of musical pitch. Psychological Review 89(4), 305–309 (1982)

16. Criani, P.: Temporal codes, timing nets, and music perception. Journal of New Music Research 30(2), 107–135 (2001), Good/useful representation of pitch similarity here, also some interesting ideas on how to use temporal codes throughout computation.
17. Parker, J.: Genetic algorithms for continuous problems. In: 15th Conference of the Canadian Society for Computational Studies of Intelligence on Advances in Artificial Intelligence (2002)
18. Wright, A.: Genetic algorithms for real parameter optimization. In: The First workshop on the Foundations of Genetic Algorithms and Classier Systems (1990)
19. Jyh-Shing, J.R.: Qbsh: A corups for designing qbsh (query by singing/humming) systems, Available at the QBSH corpus for query by singing/humming link at the organizer's homepage (2006)
20. Downie, J.S., West, K., Ehmann, A., Vincent, E.: The 2005 music information retrieval evaluation exchange (mirex 2005): Preliminary overview. In: 6th International Conference on Music Information Retrieval (2005)

An Ego-Centric and Tangible Approach to Meeting Indexing and Browsing

Denis Lalanne, Florian Evequoz, Maurizio Rigamonti, Bruno Dumas, and Rolf Ingold

DIVA Group, Department of Informatics, University of Fribourg, Bd. de Pérolles 90
CH-1700 Fribourg, Switzerland
firstname.lastname@unifr.ch
http://diuf.unifr.ch/diva

Abstract. This article presents an ego-centric approach for indexing and browsing meetings. The method considers two concepts: meetings' data alignment with personal information to enable ego-centric browsing and live intentional annotation of meetings through tangible actions to enable ego-centric indexing. The article first motivates and introduces these concepts and further presents brief states-of-the-art of the domain of tangible user interaction, of document-centric multimedia browsing, a traditional tangible object to transport information, and of personal information management. The article then presents our approach in the context of meeting and details our methods to bridge the gap between meeting data and personal information. Finally the article reports the progress of the integration of this approach within Fribourg's meeting room.

1 Introduction

With the constant growth of information a person owns and handles, it is particularly important to find ways to support information organization as well as personalized access to information. Information is dematerializing in our daily and professional life and thus, people are often experiencing the "lost-in-infospace" effect, i.e. overloaded with information and tasks to do. Our documents are multiplying in very large file hierarchies, meetings attendance is increasing, emails are no longer organized due to lack of time, our pictures are no longer stored in photo-albums, our CDs are taking the form of mp3 files, etc. What we often miss in our daily-life and professional life are personal access to information, either tangible or digital, like used to be books in our shelves or the librarian who knew our interests.

Google and Microsoft recently tried to solve the "lost-in-infospace" issue by providing, respectively, a desktop search engine and a powerful email search engine, in order to minimize the effort made by people to organize their documents and access them later by browsing. However, in order to find a file, one has to remember a set of keywords or at least remember its "virtual" existence. If one does not remember to have a certain document, browsing could still be helpful. Browsing can reveal related keywords and documents that help you remember, since the process of browsing works by association, like our human memory does [15][20][28]. For this reason, information is generally easier to retrieve, and to "remember" if it is associated to **personal information**, either in a **digital** or **physical** form.

A. Popescu-Belis, S. Renals, and H. Bourlard (Eds.): MLMI 2007, LNCS 4892, pp. 84–95, 2007.

Meetings are central in our professional lives, not only formal meetings but also meetings at the coffee machines. Numerous recent works have tried to support the meeting process with recorders, analyzers and browsers, however most of those projects try to automate and generalize the whole process for every user, with the central long term goal to automate meeting minutes authoring [5, 6, 9, 16, 22, 26]. Moreover, all these projects try to hide technological support during the meeting, instead of exploring new ways to improve human-human communication with technology or new human-machine interaction to augment meeting rooms or enable intentional live annotations. Our claim is that replaying and browsing a meeting is personal in that it depends of each individual person's interests and that it is hard to find an agreement on what is interesting for everybody during a meeting. This claim is sustained by a recent survey we performed on 118 users [3]. This survey clearly shows that people have different needs from the meeting recordings and often personal needs, for instance what are the tasks they need to do or what do they need to prepare for the following meeting. Another interesting finding is that people use often their personal emails in order to access meeting information (reminders, outcomes, retrieve files).

For the above reasons, we propose in this article an ego-centric approach, complementary to other projects, which consider (a) meeting participants as potential live annotators and (b) persons who consult meetings afterwards as individuals, and thus with their own interests, derived from their personal information. The proof of concept of this tangible and personal access to information is currently assessed through two major applications:

- The control of our daily life information through the design, implementation and evaluation of tangible interfaces, called MeModules [1], allowing access, control, personalization and materialization of personal multimedia in-formation;
- The control of our professional information, meetings, emails and documents. This application, in the context of the NCCR on Interactive Multimodal Information Management (IM2), aims at recording meeting with audio/video devices, and further at analyzing the captured data in order to create indexes to retrieve interesting parts of meeting recordings.

Although these applications share common type of multimedia information, either personal or professional, the way information is accessed can not be done in a unique manner because users do not share the same roles and context of interaction in the two applications. Our assumption is that an organization of meetings centered on users could bridge the gap between these two applications and thus re-enforce user experience in meeting browsing. For this reason, this article discusses mainly the second application.

A brief state-of-the-art of tangible user interfaces is depicted in the next section. Further, the section presents a state-of-the-art on past and current document-centric browsers. The third section presents our ego-centric approach for indexing and browsing meetings, with tangible live annotations first, and secondly through the alignment with personal information structure. Finally, the last section presents our meeting room and the advancement of the integration of this ego-centric vision. The conclusion finally wrap-up the article and presents the future works.

2 State of the Art

We believe that tangible interfaces can be useful to establish links between our memory and information. They roost abstract information in the real world through tangible reminders, i.e. tiny tagged physical objects containing a link towards information sources that can be accessed by several devices. In the context of meetings, physical documents are well suited to play the role of such tangible hypermarks, that could provide entry points in the digital meeting minutes, or enable live personal annotation of meeting parts.

Brief states-of-the-art on tangible user interactions and document-centric tangible interfaces are presented in this section. Moreover, as our goal is to bridge the gap between the personal and professional information spaces by a unified ego-centric approach of both personal and professional fields, the domain of personal information management is introduced at the end of the section.

2.1 Tangible User Interfaces

Over the last couple of years, it has been demonstrated that Graspable, or Tangible User Interfaces (TUIs), a genre of human-computer interaction that uses physical objects as representations and controls for digital information [24], make up a promising alternative for the 'traditional' omnipresent graphical user interface (GUI). TUIs have also shown a high potential for supporting social interaction and collaboration. The majority of existing systems have targeted learning or playful learning, office environments, or collaborative planning tasks [25]. While quite a few systems have demonstrated the technical feasibility of associating digital media with tangible objects [12,24], these have often remained stand-alone proof of concept prototypes or applications of limited functionality.

Two existing TUIs are particularly close to our approach: MediaBlocks [23] and Phenom [12]. MediaBlocks consist of generic symbolic tangibles which act as containers for information. The system enables people to use tagged physical blocks to "copy and paste" (or move) digital media between specialized devices and computers and to physically compose and edit this content. The Phenom souvenirs are personal iconic tangibles embedded with digital ID tags to online content (an URL).

2.2 Document-Centric Tangible Interfaces

Several works explored tangible interface using printed or printable documents. The DigitalDesk [27] is centered on the co-existence of physical papers with digital information. It is built around an ordinary physical desk. It identifies and tracks documents on the table, and identifies user's pointing. Thanks to a computer-driven electronic projector above the desk, the system augments the real world with electronic objects onto the surface and onto real paper documents.

In fact, documents have properties that justify their users' acceptance. These properties are presented in this subsection, along with projects taking benefit of them. Unlike other augmented media, printed and written documents preserve their information. For instance, the RASA [17] project uses augmented maps and Post-it™ notes for military operations. In normal conditions, the system support users with speech and gesture. In case of blackout, militaries can continue their works in the classical way. People are able to associates pages topology and indexes to document message. Books with Voice [14] allows to hear oral histories thanks to books tagged with barcodes. This project was founded on observation that historians disposing of both printed document and audiotapes preferred to retrieve information in the transcript, because it is rich in visual information such as page numbers, indexes, etc. Moreover, printed documents are not intrusive. Palette [18] for example allows managing slideshows presentation thanks to printed index cards, identified by RFID chips, and its evaluation showed that users concentrated exclusively on the communication task. In addition, paper is a portable interface. Take for instance the interactive paper maps for tourists that have been studied in [19] and which allow to get multi-media information by pointing on interesting areas. Finally, printed document are not exclusively associated to working or unpleasant tasks. Thus, Listen Reader [2] allows consulting a book, while the atmosphere is adapted with music in respect to the text focused by readers.

2.3 Personal Information Management

Personal information management (PIM) supports the daily life activities people perform in order to manage information of all kind. Though an old concept, it has been receiving a growing interest in the recent years due to the increasing amount of information we face, from disciplines as diverse as information science, cognitive psychology, database management or human-computer interaction [21].

Among personal information tools, two main approaches have been considered: searching versus semantic ordering. The first category contains systems such as MyLifeBits [10] or Stuff I've Seen [7]. They are improved query-based desktop search engines allowing querying all types of personal information in a unified way. Stuff I've Seen presents the query results as lists than can be sorted chronologically or by relevance. MyLifeBits also provides more flexible time-based visualizations of query results. Keys for retrieving personal information in these systems include the content of documents and other metadata than can be directly extracted (time, type of document, author). MyLifeBits also allows user-created collections of documents, opening the door to semantic annotations of personal information. Systems falling in the second category focus on the semantic aspect of personal information. Haystack [13] is a platform allowing the user to define annotations and collections over its own personal information to enable semantic browsing. Alternatively, SEMEX [29] tries to generate semantically meaningful associations automatically.

Several studies have been conducted to understand users' habits and needs in terms personal information. A particularly interesting one was conducted by Boardman [4]. Among other conclusions, he notes the preference of users for browsing over searching and the potential of integration between emails and files that often appear to have strong similarities. We leverage those conclusions in our approach of personal information management.

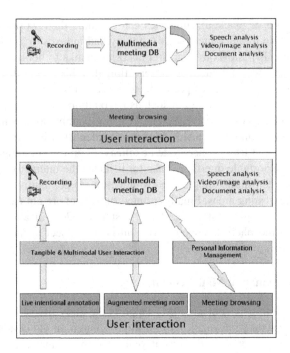

Fig. 1. Standard versus ego-centric meeting recoring, indexing and browsing

3 Ego-Centric Indexing and Browsing of Meetings

This section presents our ego-centric approach to index and browse meetings. Figure 1 represents our approach, at the applicative level, in perspective with the standard approach generally followed for recording, analyzing, indexing and browsing meetings. On the top part, the standard approach contains three sequential steps: (1) first synchronized audio/video recording of meeting; (2) in a post-production phase, analysis of multimedia meeting records is performed on raw-data to extract search-able indexes, for instance speech to text, facial expressions recognition, document analysis, etc.; (3) and finally, users can browse on meetings using the previously extracted annotations as cues to access the searched parts. In our approach, at the bottom, our goal is to enrich this process with personal annotations and with personal in-formation. These two aspects are reflected in the following two tasks of our ego-centric approach:

1. Personal live annotation of meetings using intentional tangible interaction techniques during the recording phase;
2. Browsing multimedia meeting archives via personal information structure and tangible shortcuts as specified by the user during the recording phase.

3.1 Indexing and Browsing Meetings Through Personal Tangible Anchors

Figure 2 illustrates how meetings and personal information can be intentionally linked thanks to tangible user interactions. It shows that meetings are limited in time, whereas personal information evolves during an entire person's life. Consequently, we could represent all the relationships between these applications from meeting time point of view, which is in general decomposed in pre-production, in-meeting and post-production phases.

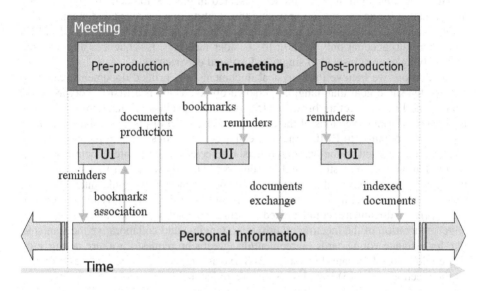

Fig. 2. The schema synthesizes the interactions binding personal information to meetings in function of time

Pre-production consists in preparing all the material the person aims to present during the meeting. In classical systems dealing with meeting, the participants prepare slideshows, printed documents, etc. At this stage, in our approach, tangible user interfaces can be used to create tangible shortcuts to the prepared documents, i.e. they allow further access to stored multimedia information during the meeting. Moreover, a participant could define a set of tangible annotators to use in the in-meeting phase, in order to bookmark interesting information according to freely defined categories that match their personal information structure.

At **in-meeting** phase, personal tangible shortcuts created in the previous step can be used to present documents to participants or share information. Furthermore, participants can bookmark some meeting part by intentionally putting an object, representing a personal category (e.g. project's name, person's name, etc.) extracted from their personal information structure, or a personal annotation (interesting, to-do, etc.), on the RFID reader. Tagged printed documents can also be exchanged between participants. This possibility has the great advantage of 1) presenting new documents not uploaded in the pre-production phase and 2) allowing to annotate paper in a human-natural way (pen strokes, hand gesture, etc.).

The **post-production** phase includes all the analysis applied to recorded data. In our model, the participant can not only replay (the summary of) the meeting, but also access to some specific parts thanks to his personal tangible bookmarks, categories or annotations. Moreover, "category" bookmarks used during the meeting allow to automatically cluster meeting's parts, organizing the information in respect of person needs and experience.

The main contribution of the model presented in this section consists in proposing an alternative for accessing and browsing the meeting in a user-centered manner. This solution can also solve another interesting problem, which has been identified in [22] and taken into account only in [20]: to consider meetings as sequences of correlated and evolving events, instead of single and isolated experiences.

In addition, we believe that physical documents, i.e. printed documents, can help achieve the same goal than tangible personal object. TDoc is a first prototype extension of the FriDoc meeting browser [15], providing exclusively tangible interaction. By means of printed versions of the documents discussed or presented during a meeting, identified thanks to RFIDs, the user can access the vocal content recorded during that meeting, related to the paragraph he/she selected in the printed document using simple Phidgets such as sliders and IR sensors [11]. This work is a first step towards the implication of physical documents, and more generally tangible interactions, at several stages of meeting recording, analysis and browsing. Similarly to tangible objects, printed documents could indeed be engaged in three different tasks: (1) Aid "live" annotation of the meeting by providing tangible and collaborative mechanisms, like for instance voting cards (agreement/disagreement, request for a turn taking, etc.) that could be used for later browsing; (2) Serve as triggers to access services provided by the meeting room; (3) Help browsing a recorded and fully annotated meeting.

3.2 Indexing and Browsing Meetings Through Personal Information Structure

As we have recalled in section 2.3, our personal information (PI) contains valuable signs of the different roles and interests we have in life. Therefore, an abstract structure put on top of our PI, gathering pieces of meta-information about us, could be helpful in a professional context, by proposing an egocentric vision of professional document archives or events, and particularly meetings.

The extraction of an abstract structure from raw PI data is not an easy task, though. Extending the previously mentioned conclusions of Boardman's study [4], our approach is based primarily on the emails, which form a rich subset of PI, obviously linked to other pieces of PI (agenda, documents, visited websites, etc.). Indeed, a standard mailbox typically contains clues about the activities, topics of interest, friends and colleagues of the owner, as well as relevant temporal features associated to activities or relationships. Therefore, it is a rich entry point into the personal information space. Three dimensions of personal information are particularly well represented in emails: the thematic, social and temporal dimensions. We extract features pertaining to each dimension from the raw email archive. Thematic features can be extracted from the email subjects and bodies using traditional text mining methods. Social features stem from the social network built from the email archive considering the co-occurrences of addresses in emails headers as social links between people. Temporal features consist in emails timestamps. On top of the features, we use clustering methods to help structuring the email archive according to the different dimensions.

Once a structure has been elicited from emails, the remaining personal information can be linked to it. For this purpose, multi-media information mining techniques shall be used on personal information in order to extract thematic, temporal and social features. Further, cross-media alignment techniques shall be used to link those to the elicited email archive structure. Obviously, professional documents (e.g. related to meetings) can be introduced in the process as well and thus become integrated into the whole structure.

Furthermore, the personal information structure acts as a lens for visualizing PI and browsing through it. Visual clusters, filtering mechanisms, as well as views related to different dimensions of PI can be implemented more easily thanks to the PI structure extracted from emails.

In the course of our research on personal information management, a system aiming to extract the first-level PI structure from emails is currently being implemented. The data on which it works consists of the mailbox of one individual containing nearly 6000 emails and 3500 addresses. The social and thematic dimensions of emails have already been explored to some extent: (1) a social network has been built using similarity measures between people's email addresses based on the frequency of exchanged emails between people, and this network can be visualized as a graph using the "spring" layout method; (2) exploiting the statistical similarity based on the co-occurrences of words in the subjects and contents of emails, an agglomerative hierarchical clustering has been performed, which aims at finding a thematic organization of emails. The result of this clustering has been fed into a treemap visualization. However, the study conducted so far tends to show that no dimension alone (whether thematic, social or temporal) can fully grasp the complexity of one's mailbox. Therefore, our plan is to combine and link several visualization techniques applied on each dimension to help the user browse through his personal email archive.

Plans for future works mainly include the reinforcement of the PI structure extraction from emails and the alignment of PI and professional information with this structure. As new dataset, the AMI meeting corpus, which notably includes emails, will be

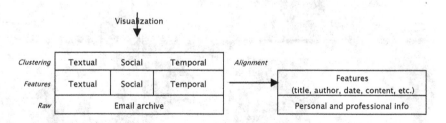

Fig. 3. (1) Extraction of the personal information structure from email, (2) alignment with the remaining personal and professional information at the feature level and finally (3) browsing in the personal and professional information through the email visual clusters

used in order to lay the foundations of an egocentric meeting browser, taking profit of the PI structure's metadata to guide the user towards the desired piece of information in meetings.

4   Towards an Ego-Centric Meeting Room

The Meeting Room in Fribourg has been built in the context of the NCCR on Interactive Multimodal Information Management (IM2) [15] as show on figure 4. This application aims at recording meeting with audio/video devices, and further at analyzing the captured data in order to create indexes to retrieve interesting parts of meeting recordings. In this context the University of Fribourg was in charge of analyzing the static documents and to align them, i.e. to synchronize them, with the other meeting media.

Roughly 30 hours of meeting have been recorded at the University of Fribourg thanks to a room equipped with 10 cameras (8 close-ups, one per participant, 2 overviews), 8 microphones, a video projector, a camera for the projection screen capture and several cameras for capturing documents on the table. Camera and microphone pairs' synchronization is guaranteed and they are all plugged to a single capture PC, thanks to three IVC-4300 cards. A meeting capture application pilots the capture. It has a user-friendly interface to start, pause and stop recording, to control post-processing operations such as compression (for streaming and archiving) and to control file transfers to a server. This application is part of a more general Organizer tool for specifying the cameras and microphones to be used, the participants' position, camera's frame-rate, etc. The Organizer tool also assists users in the preparation, management and archiving of a meeting. This includes services for registering meeting participants, gathering documents and related information.

At the time of writing, RFID readers have been integrated in our meeting room, one for each meeting participant. First of all, they enable participants to register to the meeting room, thus automatically entering the metadata related to the recordings, and also identifying the tagged documents. Secondly, thanks to the synchronization of

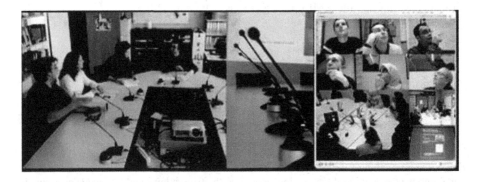

Fig. 4. Fribourg Smart Meeting Room environment

RFID readers with the audio/video recording, users can intentionally bookmark interesting meeting instants.

In our current environment, meetings data and personal information have not been fully linked. The AMI meeting corpus, which notably includes emails, is currently used in order to lay the foundations of our egocentric meeting browser, taking profit of the PI structure's metadata to guide the user towards the desired piece of information in meetings.

Further, we are currently working on ways to augment live meetings with multimodal interaction techniques such as voice or gesture. There are three aspects we plan to handle:

- Controlling meeting room services (printing, projection, etc.): how multimodal interaction can help interacting naturally with the meeting room to project documents, exchange documents, control a slideshow, etc.;
- Annotating the meeting records live (personal bookmarks): how tangible interaction combined with multimodal interaction can enable book-marking or annotation of moments of interest in a meeting for future replay or browsing (for instance using voice to label the bookmark);
- Augmenting human/human communication and collaboration capabilities: the goal is to build an ambient room able to reflect and enhance the productivity of a meeting, for instance by projecting on the table a visualization of the dialog structure or the speakers' time of intervention.

Towards this end, a prototype allowing the control of a multimedia presentation via voice commands, gestures and tangible objects has been developed; this prototype is built upon HephaisTK, a toolkit allowing rapid prototyping of multimodal interfaces. This toolkit is based on a multi-agents middleware, using meaning frames fed by a central blackboard as a preliminary fusion mechanism. HephaisTK allows developers to rapidly prototype multimodal human-machine dialogs using SMUIML (Synchronized Multimodal User Interaction Markup Language) [8]. HephaisTK and its SMUIML will be gradually extended and applied to handle the three applications described above.

5 Conclusion

This article presents an ego-centric and tangible approach to meeting recording, indexing and browsing. The approach proposed takes benefit (1) of the alignment of personal information with meeting archives to enable ego-centric browsing and (2) of tangible interactions during the meeting to add personal annotation in real time onto meeting data, linking meeting data with personal information.

The article presents a preliminary solution and implementation for managing personal information through emails mining and clustering, that we believe is the core of personal information. Based on this central information structure we expect to build a personal access to meetings' archives, thanks to emails/meetings alignment. The article further explains how tangible interaction, as well as printed documents, can be another way to bridge the gap between meetings and personal information, and presents in detail its involvement in pre- and post-production as well as during meetings.

The article finally presents Fribourg smart meeting room and how it implements this ego-centric vision, and the future plans to augment it with multimodal interactions.

Last but not least, the various stage of the vision presented in this article should be carefully evaluated in the future with users in order to measure not only the benefits gained following this ego-centric approach at the retrieval/browsing stage, but also how it modifies the meeting structure itself at the recording stage.

References

1. Abou Khaled, O., Lalanne, D., Ingold, R.: MEMODULES as tangible shortcuts to mutimedia information. Project specification (2005)
2. Back, M., Cohen, J., Gold, R., Harrison, S., Minneman, S.: Listen Reader: An Electronically Augmented Paper-Based Book. In: proceedings of CHI 01, pp. 23–29. ACM Press, New York
3. Bertini, E., Lalanne, D.: Total Recall Survey Report., Technical Report, Fribourg (2007)
4. Boardman, R., Sasse, M.A.: stuff goes into the computer and doesn't come out: A cross-tool study of personal information management. In: CHI 2004. Proceedings of the SIGCHI conference on Human factors in computing systems, pp. 583–590. ACM Press, New York, NY, USA (2004)
5. Chiu, P., Boreczky, J., Girgensohn, A., Kimber, D.: LiteMinutes: An Internet-Based System for Multimedia Meeting Minutes. In: proceedings of the 10th international conference on World Wide Web, Hong Kong, pp. 140–149 (2001)
6. Cutler, R., Rui, Y., Gupta, A., Cadiz, J.J., Tashev, I.: He, L.-w., Colburn, A., Zhang, Z., Liu, Z., Silverberg, S.: Distributed meetings: A meeting capture and broadcasting system. In: proceedings of Multimedia 02, Juan-Les-Pins, France, 503-512 (2002)
7. Dumais, S., Cutrell, E., Cadiz, J.J., Jancke, G., Sarin, R., Robbins, D.: Stuff I've seen: A system for personal information retrieval and re-use. In: SIGIR 2003. Proceedings of the 26th annual international ACM SIGIR conference on Research and development in information retrieval, pp. 72–79. ACM Press, New York (2003)
8. Dumas, B., Lalanne, D., Ingold, R.: Creating Multimodal Interfaces with HephaisTK, a Programming Toolkit using SMUIML Markup Language. In: Poster at the 4th Joint Workshop on Multimodal Interaction and Related Machine Learning Algorithms, Brno, Czech Republic (June 2007)
9. Erol, B., Li, Y.: An Overview of Technologies for E-Meeting and E-Lecture. In: proceedings of ICME 2005 (2005)
10. Gemmell, J., Bell, G., Lueder, R., Drucker, S., Wong, C.: Mylifebits: fulfilling the memex vision. In: MULTIMEDIA 2002. Proceedings of the 10th ACM international conference on Multimedia, pp. 235–238. ACM Press, New York (2002)
11. Greenberg, S., Fitchett, C.: Phidgets: Easy Development of Physical Interfaces through Physical Widgets. In: proceedings of the ACM UIST 2001, pp. 209–218. ACM Press, Orlando, USA (2001)
12. van den Hoven, E.: Graspable Cues for Everyday Recollecting, PhD Thesis, Department of Industrial Design, Eindhoven University of Technology, The Netherlands (2004), ISBN 90-386-1958-8
13. Karger, D., Bakshi, K., Huynh, D., Quan, D., Sinha, V.: Haystack: A general-purpose information management tool for end users based on semistructured data. In: CIDR, pp. 13–26 (2005)

14. Klemmer, S.R., Graham, J., Wolff, G.J., Landay, J.A.: Books with Voices: Paper Transcripts as a Tangible Interface to Oral Histories. In: proceedings of CHI 2003, pp. 89–96. ACM Press, Ft. Lauderdale, USA (2003)
15. Lalanne, D., Ingold, R., von Rotz, D., Behera, A., Mekhaldi, D., Popescu-Belis, A.: Using static documents as structured and thematic interfaces to multimedia meeting archives. In: Bengio, S., Bourlard, H. (eds.) MLMI 2004. LNCS, vol. 3361, pp. 87–100. Springer, Heidelberg (2005)
16. Lalanne, D., Lisowska, A., Bruno, E., Flynn, M., Georgescul, M., Guillemot, M., Janvier, B., Marchand-Maillet, S., Melichar, M., Moenne-Loccoz, N., Popescu-Belis, A., Rajman, M., Rigamonti, M., von Rotz, D., Wellner, P.: The IM2 Multimodal Meeting Browser Family, IM2 technical report (2005)
17. McGee, D.R., Cohen, P.R., Wu, L.: Something from nothing: Augmenting a paper-based work practice with multimodal interaction. In: proceedings of DARE 2000, Helsingor, Den-mark, pp. 71–80 (2000)
18. Nelson, L., Ichimura, S., Pedersen, E.R., Adams, L.: Palette: A Paper Interface for Giving Presentations. In: proceedings of CHI 1999, pp. 354–361. ACM Press, Pittsburgh, USA (1999)
19. Norrie, M.C., Signer, B.: Overlaying Paper Maps with Digital Information Services for Tourists. In: Proceedings of ENTER 2005. 12th International Conference on Information Technology and Travel & Tourism, Innsbruck, Austria, pp. 23–33 (2005)
20. Rigamonti, M., Lalanne, D., Evéquoz, F., Ingold, R.: Browsing Multimedia Archives Through Intra- and Multimodal Cross-Documents Links. In: Renals, S., Bengio, S. (eds.) MLMI 2005. LNCS, vol. 3869, pp. 114–125. Springer, Heidelberg (2006)
21. Teevan, J., Jones, W., Bederson, B.: Introduction. In communication ACM 49(1), 40–43 (2006)
22. Tucker, S., Whittaker, S.: Reviewing Multimedia Meeting Records: Current Approaches. In: International Workshop on Multimodal Multiparty Meeting Processing (ICMI 2005). Trento, Italy (2005)
23. Ullmer, B., Ishii, H.: MediaBlocks: Tangible Interfaces for Online Media. In: Proceedings of CHI 1999, pp. 31–32 (1999)
24. Ullmer, B., Ishii, H.: Emerging frameworks for tangible user interfaces. IBM Systems Journal 39(3-4), 915–931 (2000)
25. Ullmer, B.: Tangible Interfaces for Manipulating Aggregates of Digital Information. Doctoral dissertation, MIT Media Lab (2002)
26. Waibel, A., Schultz, T., Bett, M., Denecke, M., Malkin, R., Rogina, I., Stiefelhagen, R., Yang, J.: SMaRT: The Smart Meeting Room Task at ISL. In: proceedings of ICASSP 2003, vol. 4, pp. 752–755 (2003)
27. Wellner, P.: Interacting with Paper on the Digital Desk. Commun. ACM 36(7), 86–96 (1993)
28. Whittaker, S., Bellotti, V., Gwizdka, J.: Email in personal information management. Commun. ACM 49(1), 68–73 (2006)
29. Dong, X.L., Halevy, A.Y.: A platform for personal information management and integration. In: CIDR, pp. 119–130 (2005)

Integrating Semantics into Multimodal Interaction Patterns

Ronnie Taib[1,2] and Natalie Ruiz[1,2]

[1] ATP Research Laboratory, National ICT Australia
Locked Bag 9013, NSW 1435, Sydney, Australia
{ronnie.taib,natalie.ruiz}@nicta.com.au
[2] School of Computer Science and Engineering
The University of New South Wales, NSW 2052, Sydney, Australia

Abstract. A user experiment on multimodal interaction (speech, hand position and hand shapes) to study two major relationships: between the level of cognitive load experienced by users and the resulting multimodal interaction patterns; and how the semantics of the information being conveyed affected those patterns. We found that as cognitive load increases, users' multimodal productions tend to become semantically more complementary and less redundant across modalities. This validates cognitive load theory as a theoretical background for understanding the occurrence of particular kinds of multimodal productions. Moreover, results indicate a significant relationship between the temporal multimodal integration pattern (7 patterns in this experiment) and the semantics of the command being issued by the user (4 types of commands), shedding new light on previous research findings that assign a unique temporal integration pattern to any given subject regardless of the communication taking place.

1 Introduction

Multimodal interaction allows users to communicate more naturally and interact with complex information with more freedom of expression than traditional computer interfaces. The use of multiple modalities expands the communication channel between human and computer, and hence facilitates completion of complex tasks, compared to unimodal interfaces [8]. Over the past decade, research in the field has led to a number of theoretical advances in understanding the mechanisms governing multimodal interaction. The fine-grained analysis of multimodal interaction patterns (MIP) unveils broad classes of interaction preferences across users that could benefit automatic recognition of speech or gesture. Moreover, multimodal interaction research leverages recent progress in fields of both psychology and education, using cognitive load theory (CLT) in particular as theoretical background to explain the cognition processes behind the acquisition and selection of modalities during human communication.

The overall aim of our research is to provide more consistent, integrated and intuitive human-computer interaction by decreasing users' cognitive load

A. Popescu-Belis, S. Renals, and H. Bourlard (Eds.): MLMI 2007, LNCS 4892, pp. 96–107, 2007.

through the use of a multimodal paradigm. This paper presents the results of a user experiment addressing the effects of cognitive load on multimodal behaviour. Using speech and hand gesture as input modalities, we explored the effect of task complexity on temporal and semantic characteristics of MIP.

1.1 Related Work on Cognitive Load Theory and Analysis

Cognitive Load Theory (CLT) can be used in multimodal interaction research as a theoretical framework for hypotheses and interpretation of fine-grained multimodal behaviour under different conditions. The concept of cognitive load was developed within the field of educational psychology and refers to the amount of mental demand imposed by a particular task. It is associated with novel information and the limited capacity of working memory [11]. Since high levels of cognitive load can impede performance, efficiency and learning [11], CLT has been primarily used to help teachers design more efficient educational material, including multimedia presentations.

The close association between CLT and working memory can be understood through the perspective of Baddeley's theory of modal working memory [1]. This theory contends that certain areas of working memory are reserved exclusively for modal use e.g., the visuo-spatial sketchpad, for spatial and image representations, and the phonological loop for verbal, linguistic and audio representations [1]. In human computer interaction research, the benefits of multimodal interaction in complex tasks are well established [8], and there is evidence to suggest that users adapt their multimodal behaviour in complex situations to increase their performance [8].

Currently, cognitive load can be measured in a number of ways, for a variety of scenarios. Most commonly, it is measured by soliciting subjective load ratings after a task is completed [11]. Performance measures, such as scores, error-rates and time-to-completion measures have also been used. Subjects are rarely assessed in real-time and the probe-method, which solicits subjective ratings during task completion, interrupts the user's task flow and potentially adds to the cognitive load [11]. A new method for measuring cognitive load which is unobtrusive and provides results in real time is needed.

Given the known links between cognitive load and multimodality, we hypothesise that many aspects of users' multimodal interaction will change with increases in cognitive load. From spatial and temporal arrangement of individual modalities within multimodal productions; as well as the semantic and syntactic structures of such productions, and even the selection of preferred modality and changes in modality specific features, many features could be cues to increases in cognitive load. The detection of these trends in user interaction could lead to reliable indicators of increased load for the user with the added benefit of being unobtrusively captured and continually assessed automatically by the system.

In this research, we differentiate the term "designed cognitive load", which refers to the complexity of the task (intrinsic and extraneous), from the term "experienced cognitive load" which is the actual degree of demand felt by a particular subject. Individuals' experienced cognitive load, for the same task, may change from time to time. The designed (or expected) cognitive load of the task can provide a coarse indication for the degree of experienced (or actual) cognitive load, but the latter can only be interpreted relative to other tasks.

1.2   Related Work on Multimodal Interaction Patterns

Multimodal interaction (or integration) patterns refer to micro level relationships between the inputs comprising a multimodal production, for example in the temporal, spatial or semantic domain. Temporal relationships, in particular, are crucial for correctly interpreting individual productions, in both human-human and human-computer communication. Multimodal input fusion (MMIF) systems rely on such knowledge to validate or reject potential fusion of inputs. Most systems in the past have employed artificially or arbitrary defined values, even though some methods using syntactic or semantic combinations based on machine learning have been developed [3].

The qualitative and quantitative aspects of the temporal relationships have been studied in a bid to provide a better design of MMIF modules, but also to progress the fundamental understanding of human communication. Oviatt et al. analysed the preferred integration patterns when using pen and speech in the QUICKSET system; experiments involving speech and handwriting helped distinguish two groups of multimodal integrators: sequential and simultaneous. The first group produced multimodal inputs sequentially, one modality at a time, whereas the second group overlapped modalities, at least partially, in the time dimension. The authors also show that machine learning could be used to quickly classify the user's temporal pattern as soon as they begin to interact [5,10].

Further to simple temporal analysis, their study also determined the types of tasks more likely to induce multimodal interaction, as opposed to unimodal interaction. It was found that spatial location commands (e.g. modify, move) represented 86% of all multimodal inputs, against 11% for selection commands (e.g. zoom, label), and 3% for function commands (e.g. print, scroll). Finally, the study also reported that the order of semantic constituents in multimodal interactions from their corpus was different from the order in spoken English, but mainly due to the position of the locative constituents at the beginning in multimodal, and at the end in corresponding spoken inputs [10].

The study of MIP has been less prominent for other modality combinations, especially when input recognition is a major concern, for example in systems involving hand gesture. Bolt used a Polhemus sensor, based on electromagnetic field variations, to obtain the position and orientation of the hand from a fixed transmitter. The worn device is a cube of about 2 cm edge, and requires a chord, hence is fairly cumbersome and obtrusive while sensing only limited information [2]. Data gloves later allowed more complex gestural inputs due to finger position sensing but impede movement. Vision-based hand tracking and shape recognition became a reality a few years later, involving a range of algorithms and set-ups, e.g., single or stereo cameras. However, vision-based recognition rates are not yet satisfactory, so successfully combining such input with other modalities is currently seen as the most promising path to success.

Attempts at MIP analysis in gesture and speech systems were made, such as Hauptmann's framework to formalise and quantify speech and gesture interaction, with a view to automation. The work studied user preferences and intuitive behaviours when using speech and hand gesture to move a 3D virtual object [4]. Very detailed statistics were reported on the usage of words, lexicon, syntactic structure, hand and finger gestures, highlighting the type of multimodal interaction preferred by users and when it occurs. However, the results appear as an unstructured set of percentages, rather difficult to put in perspective or implement in a generic way.

Finally, a psychological approach has been used to provide a formal framework for the study of speech and gesture interaction. For example, McNeill, Quek et al. established the existence of low level relationships between gesture, speech and gaze, suggesting cross-modal segmentation based on specific features of those modalities [12]. More recently, an HMM-based implementation was used to improve disambiguation of speech with gestures [6].

1.3 Objectives and Hypotheses

This study was designed to identify the relationships between combined speech and gesture input productions and users' cognitive load. The two input modalities are very familiar to users and psychologically closely interrelated, both in terms of planning and execution.

Specifically, we hypothesise firstly that variations in redundant and complementary multimodal productions can reflect cognitive load changes experienced by the user. Redundant multimodal productions are ones that semantically double-up information over a number of modalities (first example in Table 1). Complementary productions are, conversely, those that convey different semantic information over a number of different modalities in the same multimodal production (second example in Table 1). Partially redundant productions occur when only part of the production (i.e. the function or the object) is expressed redundantly across modalities, while the rest of the production is unimodal or complementary. We expected many redundant productions when the cognitive load was low, and as cognitive load increased, complementary productions would be more prevalent, as users would begin to instigate strategies to maximise use of available working memory. To this end, all multimodal productions would be classified according to the degree of semantic redundancy in each one.

The second hypothesis relates to the temporal structure of multimodal productions. While there is evidence that users fall into one of two major groups of

Table 1. Redundant vs. complementary multimodal productions

Turn	Modality	Productions	Semantics
REDUNDANT - Ex 1:	Speech	"Select library"	<Fn:select><Obj: library>
Select the library on a map	Gesture	Point to library	<Fn:select><Obj: library>
COMPLEMENTARY - Ex 2:	Speech	"Event"	<Fn: event>
Mark an event at a location	Gesture	Point to library	<Obj: library>

integrator pattern (i.e. simultaneous vs. sequential integrators), we also hypothesised that other factors of the communication, such as the type of information being manipulated, e.g., a function or an object, would also affect the way the temporal structure of the production is planned and executed. Knowledge of these patterns, the types of users who employ them and the domains in which they are likely to occur could make a significant contribution to the algorithms and strategies implemented in adaptive multimodal input recognition and fusion systems. Temporal integration patterns between two modalities can fall into one of the nine integrator 'shapes', one sequential pattern and eight distinct simultaneous patterns [8,9,10]. However, we hypothesised that the temporal integration patterns of each user's production would be affected by the semantics of the information the user is trying to convey. For example, a user may exhibit different simultaneous patterns when zooming-in the map than when selecting a map entity. Hence, our null hypothesis is that the integrator pattern is chosen independently of the information semantic.

2 Experiment Design

2.1 Wizard of Oz Set-Up

The selected modalities for the experiment were speech, hand motion and a set of specific hand gestures, hence very prone to errors by automatic recognisers, so we opted for a Wizard of Oz (WOz) implementation. This technique has been identified as an essential tool for the design of multimodal interfaces, where novel interaction patterns were expected to appear [14]. Indeed, WOz allows intuitive interaction while removing the bias caused by input recognition errors and misinterpreted semantic fusion of multimodal signals.

Our wizard only performed manual recognition of speech and hand shapes, while hand motion detection was automated with an in-house, video-based recognition module. The hand tracking module was used by the subject to move a visual pointer on the system graphical interface, which was echoed on the wizard's machine. This allowed the wizard be aware of the actions such as button

Fig. 1. Subject interacting with the system

clicks or manual selections made by the subjects. Inputs directly generated by the subject, or via the wizard, were then processed by the application logic in order to update the interface and progress the task.

Subjects stood 2m away from a large, wall-sized screen displaying the application. The camcorder and video camera used for tracking were both located on the right hand side (all subjects reported to be right handed) as shown in Fig. 1.

2.2 Application Scenario: Tasks and Modalities

Eliciting multimodal interaction under various levels of cognitive load requires carefully crafted task design. The use of novel modalities may only become salient for novel application functionality, as opposed to well-known applications as suggested by Lisowska and Armstrong [7]. Moreover, it has been shown that spatial location tasks such as map interaction are more likely to generate combined multimodal interaction [10]. The study scenario realised the update of a geographical map with traffic condition information, using either natural speech, or manual gesture, or a combination of these. Available gestures comprised:

- Deictic pointing to map locations, items, and function buttons;
- Circling gestures for zoom functions; and
- Predefined hand shapes (fist, scissors...) for item tagging.

All tasks were designed to provide as much semantic equivalence as possible, meaning that atomic actions required could be completed using any single modality or combinations thereof. Subjects were shown multiple examples of interaction involving various blends of modalities and system functionality and could practise them during a 30-min training session. Table 2 provides examples of atomic actions and possible realisations using various modality combinations. Instructions for each new task were high-level situation descriptions and subjects were allowed freedom of inspection during their response actions. Typical sets of actions required for each situation was taught in training.

Table 2. System functionality and examples of inputs

System Functionality	Example of Interaction
Zooming in or out of a map	<Point at quadrant>; or Say: "Zoom in to the top right quadrant"
Selecting a location/item of interest	<Point at location>; or Say: "St Mary's Church"
Tagging a location of interest with an 'accident', 'incident' or 'event' marker	<Select location>and: Say: "Incident"; or Scissors shape
Notifying a recipient (item) of an accident, incident or an event	<Select accident>and Say: "notify"; or fist shape and <Select recipient>

Table 3. Levels of cognitive load

Level	Entities	Actions	Distractors	Time
1	6	3	2	∞
2	10	8	2	∞
3	12	13	4	∞
4	12	13	4	90 sec.

There were four levels of cognitive load, and three tasks of similar complexity for each level. The same map was used for each level to avoid differences in visual complexity, hence designed cognitive load. The designed cognitive load was controlled through the number of entities present on the map and actions required in order to complete the task. Table 3 summarises the various levels induced by changes in (i) number of distinct entities in the task description; (ii) number of distractors (entities not needed for task); (iii) minimum number of actions required for the task; and (iv) a time limit to complete the task.

2.3   Procedure

Twelve remunerated, randomly selected, native English-speaking participants (6 females, 6 males, aged 18-49) completed the study. As mentioned above, subjects were asked to perform a set of tasks under 3 different conditions: gesture-only, speech-only and multimodally. Each set consisted of 4 levels, with 3 tasks in each. The order of these conditions and the tasks within the levels was randomised to counter balance rank order effects. Video, hand position and UI interaction data were synchronised and recorded digitally. Users were also debriefed after each task level and were asked to provide a subjective ranking of the level of load relative to the other levels in that condition. The video data collected from the subjects was manually annotated: start and end time of speech and gesture were annotated with a precision of 1 frame (25fps). Gesture semantics were also tagged and speech orthographically transcribed.

3   Results

3.1   Cognitive Load Analysis Based on Multimodal Redundancy and Complementarity

Out of 12 subjects, only the data from 9 was usable, since two users had difficulty comprehending the tasks, such that they could not achieve the goals of the task, and one did not finish for external reasons. The data collected for the multimodal condition for Levels 1, 2 and 4 was annotated for 6 out of these 9 users. In total, 1119 modal inputs were annotated, forming 394 turns and 644 productions. However, smaller numbers were used for the analysis of individual levels. To measure the perceived level of cognitive load, users ranked the tasks in increasing levels of difficulty along a 9-point Likert scale, the average difficulty

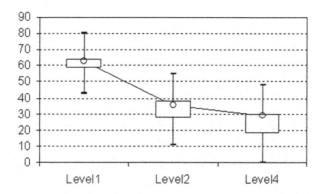

Fig. 2. Proportion of purely redundant turns by level

score for Levels 1, 2 and 4 across these 6 users was 2.2, 4.2 and 5 respectively. Level 3 data was not annotated due to lack of time. For each user, we classified the multimodal turns into three groups: purely redundant, purely complementary and partially redundant turns. Fig. 2 shows the mean percentage and range of purely redundant turns across users, for each level, over all multimodal turns. However, statistical analysis was carried out on 5 of the 6 users; one subject exhibited hardly any redundant behaviour, preferring to interact in a complementary manner.

We observed a steady decrease in redundancy as task difficulty increased. An ANOVA test between-users, across levels, shows there are significant differences between the means (F =3.88 (df=2); p<0.05). Subsequent t-tests show significant differences, 27.16% between Level 1 (62.91%) and Level 2 (35.74%) (p=0.03, <0.05, two-tailed) and 33.61% between Level 1 and Level 4 (29.29%) (p=0.01, <0.05, two-tailed). By the same token, we expected the rate of purely complementary productions to increase. In Level 1, the average percentage of purely complementary turns was 12.86%, increasing to 45.53% and 36.02% in Levels 2 and 4 respectively. Though not significantly different, there is a clear trend across users of an increased use of complementary multimodal productions in higher load tasks when comparing Level 1 and 2, and 1 and 4, corroborating with decreasing redundancy. There is also an increase in partially redundant productions between Levels 1 and 4, with averages of 24.24%, and 34.69% respectively, which can be interpreted as representative of the shift from purely redundant to partially redundant to purely complementary as shown in Fig. 3.

Overall, we explain the lack of difference between Levels 2 and 4 in the above features as a symptomatic of very similar subjective ratings of the difficulty levels (respectively 4.2 and 5), but at least 2 points higher than Level 1 (2). This was probably due to increased familiarity with the system by the time the 4th level was attempted, highlighting the semantic distinction between designed and experienced cognitive load. A higher designed load does not necessarily imply a higher experienced load. Further results and discussion of this study have been presented in [13].

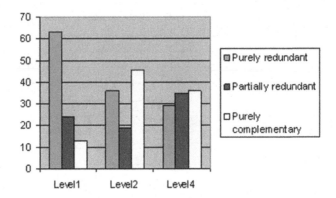

Fig. 3. Averages of modal input blends

3.2 Impact of Semantics on Multimodal Interaction Patterns

Annotation Schema. This paper focuses on bi-modal inputs only, ignoring unimodal inputs (no tri-modal input was performed by the subjects). For the same 6 subjects mentioned previously, these 177 bi-modal inputs constitute 29% of all annotated inputs. Our annotation schema is based on the bi-modal integrator patterns defined by Oviatt et al. [9,10], which comprises a sequential pattern (SEQ) and 9 possible simultaneous patterns (SIM). However, since three modalities were available in our system, we analysed the temporal relationship within any pair of inputs, disregarding the *type* of modality used. This reduces to 1 SEQ and 6 SIM patterns only as shown in Fig. 4. The patterns are based on the temporal features defined in [9,10], with a coordination between signals accuracy increased to 50ms (100ms in cited work).

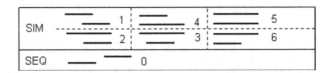

Fig. 4. Graphical representation of integrator patterns

The semantics of the performed commands were also annotated and regrouped into 4 major productions: select (any entity selection), zoom (zoom in or out of the current map), tag (tagging of any entity), end (end of task). We did not perform statistics on the notify production since it received only 6 multimodal inputs across users.

Results. In order to test the hypothesis that integrator patterns are affected by the semantics of the command, we carried out a Chi-square test of independence

Table 4. Observed frequencies of integrator patterns vs. production

Integrator pattern	Productions				Total
	Select	Zoom	Tag	End	
0	10	2	0	10	22
1	10	3	2	6	21
2	5	0	0	1	6
3	62	43	14	4	123
4	0	0	0	0	0
5	3	0	0	0	3
6	0	0	2	0	2
Total	90	48	18	21	177

using the observations from the experiment shown in Table 4. This data provides evidence of a significant relationship between the integrator pattern and the production ($\chi^2 = 67.01 < \chi^2 critical = 25, \alpha = 0.05$, df=15). The integrator pattern 4 had to be discarded in this analysis as no input of this type was collected during the experiment.

Similarly, we tested the independence of integrator patterns and subject identity. Again, the data provides evidence of a significant relationship between the integrator pattern and the subject identity ($\chi^2 = 77.59 < \chi^2 critical = 37.65, \alpha = 0.05$, df=25). The integrator pattern 4 was discounted in this analysis as no such input was collected during the experiment.

4   Discussion and Future Work

The experiment presented in this paper is part of our ongoing research on multimodal interaction behaviour. In particular, we found that the reduced level of redundancy, combined with the increased level of complementarity variations across modalities reflects changes in cognitive load. As the cognitive demand increases, users' multimodal productions tend to become complementary. We interpret that this is due to Baddeley's modal Model of Working Memory, which suggests there are areas of working memory that are dedicated exclusively for modal use. During interaction at high levels of load, productions are planned so as to maximise the usage of modal working memory. This could be achieved by channelling the required semantic chunks to areas allocated to different modalities, with the least amount of replication possible. This hypothesis results in increased purely complementary productions and a reduction in purely redundant productions, as cognitive load increases. The results of this study give initial evidence for this behavioural symptom of cognitive load management employed by users, and provide a potential multimodal cue candidate for revealing changes in cognitive load. The results raise some interesting questions about the use and suitability of multimodal features as indices of cognitive load. It is recognised that the number of subjects used in the results is relatively low, and further validation and evaluation of the results is necessary. The subjects who did complete the study rated the high load tasks only as moderately difficult, and this is attributed to a lack of understanding of the full requirements of the solution. The subjects did not receive feedback as to how well they 'scored' – hence had no indication as to their own performance. It is sufficient that these tasks were ranked higher than the rest, anecdotally we can report high levels of frustration.

We also analysed the relationship between temporal integration patterns and the semantics of the information conveyed by the user. The data collected during our experiment indicates that there is, in fact, a significant relationship between the integration pattern (7 classes in this experiment) and the semantics of the command (4 classes in this experiment). Recognising that users may slightly adapt their integration pattern according to the semantics of the message being conveyed extends the previous findings of Oviatt et al. They originally established a relationship between semantics and multimodal vs. unimodal interaction [10], then concentrated on user categorisation according to integration patterns [8,9]. However they do not seem to have explored the finer relationships between semantics and integrator patterns.

In parallel developments, Oviatt et al. demonstrated through longitudinal studies that users tend to entrench in a given pattern, allowing automatic selection through machine learning for example [5]. Our experimental data seems to be valid in view of these findings since it indicates that integration patterns are significantly linked to the identity of the subjects. This suggests further, however, that multimodal integration patterns are a personal trait of individual users, and prompts the question of the existence of broad classes of users, who interact with similar temporal pattern-semantic combinations. While simple machine learning techniques are effective to assign a unique integration pattern to a user [5], broader user classes may be required to train learning models to assign integration patterns according to the semantics being conveyed. Also, the specific format and affordances of the user interface, as well as the system functionality may impact the integration patterns used during interactions.

In the long term, the benefits of such a refinement include more accurate multimodal input fusion but may also help predict multimodal blends and patterns based on the dialogue state. At present though, we plan further user experiments to capture longitudinal data that can counter balance the learning effects (or discomfort [7]) caused by the use of both novel modalities and novel functionality.

Acknowledgements. The authors wish to thank the twelve study participants and two pilot participants for their involvement in the experiment.

References

1. Baddeley, A.D.: Working Memory. Science 255, 5044, 556–559 (1992)
2. Bolt, R.A.: Put-That-There: Voice and Gesture at the Graphics Interface. In: 7th annual conference on Computer Graphics and Interactive Techniques, Seattle, Washington, United States, pp. 262–270. ACM Press, New York (1980)

3. Gupta, A.K., Anastasakos, T.: Dynamic Time Windows for Multimodal Input Fusion. In: Proc. 8th International Conference on Spoken Language Processing (INTERSPEECH 2004 - ICSLP), Jeju, Korea, October 4-8, 2004, pp. 1009–1012 (2004)

4. Hauptmann, A.G.: Speech and Gestures for Graphic Image Manipulation. In: CHI 1989, SIGCHI Conference on Human Factors in Computing Systems: Wings for the Mind, pp. 241–245. ACM Press, New York (1989)

5. Huang, X., Oviatt, S., Lunsford, R.: Combining user modeling and machine learning to predict users' multimodal integration patterns. In: Renals, S., Bengio, S., Fiscus, J.G. (eds.) MLMI 2006. LNCS, vol. 4299, pp. 50–62. Springer, Heidelberg (2006)

6. Kettebekov, S.: Exploiting Prosodic Structuring of Coverbal Gesticulation. In: ICMI 2004: 6th international conference on Multimodal interfaces, State College, PA, USA, October 13-15, 2004, pp. 105–112. ACM Press, New York (2004)

7. Lisowska, A., Armstrong, S.: Multimodal input for meeting browsing and retrieval interfaces: preliminary findings. In: Renals, S., Bengio, S., Fiscus, J.G. (eds.) MLMI 2006. LNCS, vol. 4299, pp. 142–153. Springer, Heidelberg (2006)

8. Oviatt, S., Coulston, R., Lunsford, R.: When Do We Interact Multimodally? Cognitive Load and Multimodal Communication Pattern. In: ICMI 2004: 6th international conference on Multimodal interfaces, State College, PA, USA, October 13-15, 2004, pp. 129–136. ACM Press, New York (2004)

9. Oviatt, S., Coulston, R., Tomko, S., Xiao, B., Lunsford, R., Wesson, M., Carmichael, L.: Toward a Theory of Organized Multimodal Integration Patterns during Human-Computer Interaction. In: ICMI 2003, 5th international conference on Multimodal interfaces, Vancouver, British Columbia, Canada, November 05-07, 2003, pp. 44–51. ACM Press, New York (2003)

10. Oviatt, S., DeAngeli, A., Kuhn, K.: Integration and Synchronization of Input Modes During Multimodal Human-Computer Interaction. In: SIGCHI conference on Human factors in computing systems, Atlanta, Georgia, United States, March 22-27, 1997, pp. 415–422 (1997)

11. Paas, F., Tuovinen, J.E., Tabbers, H., Van Gerven, P.W.M.: Cognitive Load Measurement as a Means to Advance Cognitive Load Theory. Educational psychologist 38(1), 63–71 (2003)

12. Quek, F., McNeill, D., Bryll, R., Kirbas, C., Arlsan, H., McCullough, K.E., Furuyama, N., Gesture, A.R.: Speech, and gaze cues for discourse segmentation. In: IEEE conference on computer vision and pattern recognition (CVPR 2000), Hilton head island, South Carolina, USA, June 13-15, 2000, pp. 247–254 (2000)

13. Ruiz, N., Taib, R., Chen, F.: Examining the redundancy of multimodal input. In: Proc. 20th annual conference of the Australian computer-human interaction special interest group (OzCHI 2006), Sydney, Australia, November 20-24, 2006, pp. 389–392 (2006)

14. Salber, D., Coutaz, J.: Applying the Wizard of Oz technique to the Study of Multimodal Systems. In: Bass, L.J., Unger, C., Gornostaev, J. (eds.) EWHCI 1993. LNCS, vol. 753, pp. 219–230. Springer, Heidelberg (1993)

Towards an Objective Test for Meeting Browsers: The BET4TQB Pilot Experiment

Andrei Popescu-Belis[1], Philippe Baudrion[2], Mike Flynn[1], and Pierre Wellner[1]

[1] IDIAP Research Institute,
Centre du Parc, Av. des Prés-Beudin 20,
Case postale 592,
CH-1920 Martigny, Switzerland
{andrei.popescu-belis,mike.flynn,pierre.wellner}@idiap.ch
[2] University of Geneva, School of Translation and Interpretation,
40, bd. du Pont d'Arve,
CH-1211 Geneva 4, Switzerland
philippe.baudrion@adm.unige.ch

Abstract. This paper outlines first the BET method for task-based evaluation of meeting browsers. 'Observations of interest' in meetings are empirically determined by neutral observers and then processed and ordered by evaluators. The evaluation of the TQB annotation-driven meeting browser using the BET is then described. A series of subjects attempted to answer as many meeting-related questions as possible in a fixed amount of time, and their performance was measured in terms of precision and speed. The results indicate that the TQB interface is easy to understand with little prior learning and that its annotation-based search functionality is highly relevant, in particular keyword search over the meeting transcript. Two knowledge-poorer browsers appear to offer lower precision but higher speed. The BET task-based evaluation method thus appears to be a coherent measure of browser quality.

Keywords: Multimedia meeting browsers, task-based evaluation, human-computer interaction, human factors.

1 Introduction

As more and more meetings are being recorded and stored, the demand for applications which access this data to find relevant information increases as well. The goal of this paper is to outline the BET evaluation method for meeting search and browsing interfaces, and to argue that this method captures significant aspects of meeting browser quality, based on the analysis of first-time usage of several meeting browsers. The BET evaluation of the TQB interface aims first at finding the most useful features of the meeting browser, i.e. the ones that appear to be used in correlation with the highest BET scores. In addition, the experiment aims also at comparing these with those obtained for other browsers, and to assess the validity of the BET method itself.

A. Popescu-Belis, S. Renals, and H. Bourlard (Eds.): MLMI 2007, LNCS 4892, pp. 108–119, 2007.

The BET will be briefly explained and discussed in Section 2, followed by a comparison with other approaches in Section 3. The features of the TQB annotation-based meeting browser will be described in Section 4. The details of the main evaluation experiment reported here appear in Section 5, while results are discussed in Section 6. These results are compared to a similar experiment with two knowledge-poorer browsers in Section 7. Perspectives for further analyses appear in Section 8.

2   The Browser Evaluation Test (BET)

2.1   Designing a BET Evaluation

The BET is an extensive framework containing guidelines and tools that allow evaluators to construct a browser-independent evaluation task, and then to test the performances of a given browser on that task [1]. Each evaluation task is meeting-specific and consists of a set of *observations of interest* determined by a pool of *observers* who have watched closely the meeting recording, and have noted the most salient facts and events that occurred in the meeting (observers are not meeting participants). The observations are sampled, and possibly edited, to produce a final list for each meeting. The actual testing of a browser requires *subjects* to answer as many binary-choice *test questions* as possible in a fixed amount of time, by using the meeting browser to access the meeting. The binary-choice test questions are pairs of true/false statements constructed by the observers from their observations of interest, as explained below.

Using the BET requires therefore a one-time investment in collecting and possibly annotating the corpora, collecting and preparing the observations, and possibly running benchmark tests with baseline browsers such as media players. Subsequent browser tests take advantage of this one-time effort to run tests and to produce comparable scores. While the details of the testing protocol can vary according to the evaluators' goals, we believe that the list of observations will remain a valuable resource associated to these meetings, which should be extended to other meetings in the future.

From the very first BET experiments [1], two important parameters characterized the subjects' performance. The first one is *precision* (or accuracy), i.e. the proportion of correctly answered questions among all true/false statements that were seen, a number between 0 and 1. The second one, called *speed*, is the average number of questions that were processed per minute. These scores parallel somewhat the precision and recall scores used in information retrieval. None of them is sufficient alone to capture the overall quality of a meeting browser, as trivial strategies can maximize them independently, but not jointly. However, while it is certainly possible to compute the average of precision and speed, a more nuanced integrative score, which factors out the different strategies of the subjects (maximizing either precision or speed), must yet be found.

2.2   Collecting the Observations

BET questions are derived from observations of interest produced by a set of observers using dedicated interfaces. Observers can see the full recordings of every media source—audio, video and slides—for each meeting they work on. There is no time limit, but observers are asked to produce a minimal amount of observations, for instance 50 observations for a 50-minute meeting.

Each observer is instructed to produce observations about facts or events *that the meeting participants appeared to consider interesting*. The instruction is kept generic on purpose, in order not to influence observers towards a particular type of observation. Such a definition is compatible with many types of observations, even though it is possible to argue that some facts which do not seem important to participants could be important to an external observer, depending on their interest, which is a relative notion. We argue however that by selecting various subsets from the lists of observations produced using the BET, one can accommodate a wide range of evaluation objectives.

Observers create first a list of observations, which are automatically time-stamped by the BET observer interface with the media time. Observers are also asked to estimate the "locality" of each observation, i.e. whether it applies around the current media time or throughout the meeting. Observations should also be difficult to guess without access to the recording, and must be stated in a simple and concise manner. After they have completed their list, observers are asked to rate the importance of observations (on a five-point scale) and to create a false version of each of them. The result for each observation is a complementary pair of statements, one true and one false, both of which will be later presented to subjects during testing.

2.3   Validation, Editing, Grouping and Ordering

Once collected, observation pairs (a true and a false statement) are discussed by the BET experimenters and by browser designers. At this stage, some of the observations can be rejected for a number of reasons, which are carefully explicited to ensure that they are browser-neutral, and do not select observations that are better suited to a particular kind of browsing technique. These reasons are: (1) statements that are true at one moment but false at another moment of the meeting; (2) statements that are considered incomprehensible to native English speakers because of serious grammatical or typographical errors, or unclear formulation; (3) statements that are too easily guessable; (4) true and false statements that aren't parallel enough, or are not mutually exclusive; (5) statements based on "censored" material, i.e. on segments which participants had asked to be left out of the recording. Rejection of observations requires consensus among different experimenters, working on potentially very different browser designs. In addition to rejection, only very limited editing of statements is also allowed (for any of the above reasons) in order to avoid rejecting too many observations.

In many cases, different observers make similar observations—a proof of inter-observer agreement which is exemplified below (Section 2.4). Therefore, observations must be manually grouped by the experimenters, so that subjects are tested using a single representative from the group, in order to avoid redundancy. The representative observation of the group is manually selected by the experimenters based on the following criteria, which (again) avoid favouring one type of browser over another. The selected pair of true/false statements must (1) meet the validity criteria stated above; (2) be concise and crisply expressed; (3) express, if possible, only one factual point; (4) share the same keywords as the whole group; and (5) be difficult to guess.

The edited representative observations are finally ordered, first by size of group, because this represents the number of times the observation was made by independent observers, then (for groups of equal size) by median importance adjusted per observer, then by mean adjusted importance, and finally by media time. The ordering can be changed to suit the evaluators' goals, though in some cases the answer to one question could reveal the answer to following ones.

2.4   Resulting Test Material

Three meetings from the AMI Corpus [2] were selected for the observation collection procedure: IB4010, IS1008c, and ISSCO-Meeting_024. The meetings are in English, and involve four participants, native or non-native English speakers. In the first meeting, the managers of a movie club select the next movie to show; in the second one, a team discusses the design of a remote control; in the third one, a team discusses project management issues. Although the first two meetings are in reality enacted by researchers or students, the movie club meeting (IB4010) appears to be more natural than the remote control meeting (IS1008c), probably due to the familiarity of the participants with the topic. For each of these three meetings, BET observations were collected, edited and ordered, this resource being now publicly available at http://mmm.idiap.ch. In the evaluations below, the order based on importance was kept constant.

For these meetings, respectively 222, 133 and 217 raw observations were collected, from respectively 9, 6 and 6 observers, resulting in respectively 129, 58 and 158 final pairs of true/false observations. As initial observations are grouped according to their similarity, as explained above, the average size of the groups (1.72, 2.29 and 1.37 observations per group) provides a measure of inter-observer agreement. While these values are not very high with respect to the number of observers, it is more eloquent to consider only the agreement for the observations that were answered by at least half of the subjects in the experiments on TQB (i.e. 16 for IB4010 and 8 for IS1008c). As these were ranked by importance, the average number of observers having made these observations was around 5 for both meetings, i.e. 55% and 83% of the observers agreed upon them.

As an example, the first two pairs of true/false observations for IB4010 were: "The group decided to show The Big Lebowski" vs. "The group decided to show Saving Private Ryan", and "Date of next meeting confirmed as May 3rd" vs "Date of next meeting confirmed as May 5th". For the IS1008c meeting the first pair is: "According to the manufacturers, the casing has to be made out of wood" vs. "According to the manufacturers, the casing has to be made out of rubber".

3   Comparison to Other Approaches

The evaluation of interactive software, especially of multi-modal dialogue systems, is still an open problem [3,4,5]. A possible approach in the case of meeting browsers is based on the ISO/IEC standards for software evaluation, especially for task-based evaluation or for evaluation in use [6,7]. The three main aspects of quality that are evaluated in such approaches are often summarized as *effectiveness* (the extent to which the system helps the user to successfully accomplish a task), *efficiency* (the speed with which the task is accomplished) and *user-satisfaction* (measured using questionnaires). Depending on the nature of the system that is evaluated, these three broad quality characteristics can be substantially particularized and/or extended [8]. A well-known approach to dialogue system evaluation, PARADISE [9], predicts user satisfaction from task completion success and from a number of computable parameters related to dialogue cost. However, depending on the specificity of the modules of a dialogue system, each of them can also be evaluated separately using black-box methods [10].

The goal of the BET is to provide an evaluation framework that sets as few *a priori* constraints as possible on the task of meeting browsing and on the functionalities of a meeting browser. The BET differs from classic usability testing as the details of the task are not predetermined by designers or evaluators. The process of collecting observations determines only in an indirect manner what the users of a meeting browser would primarily look for in a meeting (information type and content), and therefore the BET can be applied independently of the specifications of a meeting browser. This approach tempers undue influence of each observer's own special interests, and avoids the introduction of experimenter bias regarding the relative importance of particular types of meeting events, e.g. related to a particular modality that a given browser might focus on. The precision and speed of the subjects using a specific browser to answer questions based on BET observations respectively reflect the effectiveness and efficiency of the meeting browser, if the subjects' abilities are factored out across a large pool of subjects. Finally, in the setting described here, the BET measures browser quality at first-time usage, and not occasional or long-term usage, which would require extensive training of the subjects.

4   The Transcript-Based Query and Browsing Interface

In the study reported here, the TQB interface [11] was submitted to the BET via a web browser. TQB was designed to provide access to the transcript of meetings and to their annotations, as well as to the meeting documents, as these language-related modalities are considered to be the main information vector for human interaction in meetings. Along with the transcript, the following annotations are stored in a database to which the TQB interface gives access: segmentation of individual channels into utterances, labelling of utterances with dialogue act tags (e.g. statement, question, command, or politeness mark), segmentation of the meeting into thematic episodes, labelling of episodes with salient keywords, and document-speech alignment using explicit references to documents [2,12]. For the BET evaluation, manual transcripts and annotations from the AMI Corpus are used, in order to focus the test on the quality of the interface and not on the quality of automatic annotation.

TQB allows users to search, within a given meeting, for the particular utterances that satisfy a set of constraints on the transcript and the annotations (Figure 1). TQB displays in one frame the annotation dimensions that are searchable for the selected meeting, with a menu of possible values for each of them, except for the transcript, which can be searched as free text. The results of a query, i.e. the utterances that match all the constraints, are displayed in another frame. These utterances can be used as a starting point to browse the meeting, by clicking on one of the retrieved utterances, which makes the transcript frame scroll automatically to its position. The enriched transcript and the meeting documents constitute the two principal frames occupying the center of the TQB interface. Browsing through these frames is enhanced by the possibility to listen to the recording of each utterance, and to display documents that are explicitly referred to at a given point of the conversation [11].

To increase the informativeness of the BET evaluation, the users' interactions with TQB are closely monitored. The logging mechanism has two components: the first one logs all the queries to the database with their timestamps, while the second one logs the actions performed by the user in the other frames. These include the state and position of the audio player and of the scrollbars (sampled every 30 seconds), and the user's mouse clicks.

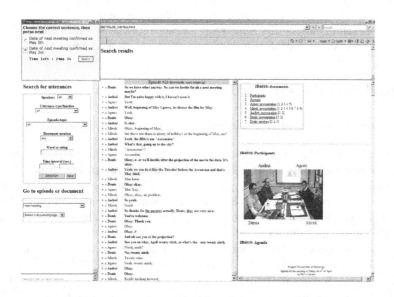

Fig. 1. View of the Transcript-based Query and Browsing Interface (TQB) with BET true/false observations displayed in the upper-left corner

5   BET Setup for TQB Evaluation

Two of the three meetings for which BET observations exist were used for the evaluation of TQB, namely IB4010 and IS1008c. The evaluation proceeds as follows. The subjects register with their email as a unique identifier and state their proficiency in English. The subjects then read the instructions for the experiment on their computer screen, explaining first the BET guidelines, and then the principles of the TQB interface, using a snapshot and 4-5 paragraphs of text. The subjects did not have the opportunity to work with TQB before being tested on the first meeting, so this was their very first occasion to explore the functions of TQB.

The BET master interface displays one by one the pairs of true/false statements corresponding to observations, following the order described above. Using TQB to browse the meeting, each subject must determine the true statement and (implicitly) the false one. When the choice is validated, the BET interface automatically displays the following pair of statements, and so on until the time allowed for the meeting is over. After a short break, the subject proceeds to the second meeting. The duration allowed for each meeting was half the duration of the meeting: 24'40" for IB4010, and 12'53" for IS1008c; the timing was managed by the BET master interface.

TQB was tested with 28 subjects, students at the University of Geneva, mainly from the School of Translation and Interpreting. Results from 4 other students were discarded for not completing the two meetings. The average proficiency on a 4-point scale (from 'beginner' to 'native') was 2.6, median value being 3 ('advanced'). Half of the subjects started with IB4010 and continued with IS1008c, and the other half did the reverse order, thus allowing for differentiated results depending on whether a meeting was seen first or second within the trial. Performance was measured using precision and speed, as defined above.

6   BET Results for TQB

6.1   Overall Scores and Variations

The overall precision, averaged for 28 subjects on two meetings, is 0.84 with a ±0.05 confidence interval at 95% level (confidence intervals will be regularly used below). The overall average speed is 0.63±0.09 questions per minute. These values do not vary significantly (less that 1%) when they are computed for the two subgroups of 14 subjects who saw the meetings in a different order (IB/IS or IS/IB): only the confidence intervals increase, by 20 to 50%.

The average speed and precision vary more markedly across the two meetings, though however these differences are not significant at the 95% confidence level: speed and precision are 0.67±0.10 and respectively 0.85±0.05 for IB4010, both higher than the respective values for IS1008c, 0.57±0.13 and 0.79±0.10. If the statistical significance was higher, one could conclude that IB4010 is easier than IS1008c from the BET perspective.

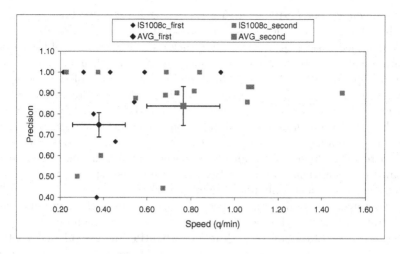

Fig. 2. IS1008c: scores of each subject and average score with 95% confidence intervals when the meeting is seen first (diamonds ◇) and when it is seen second (squares ☐). Speed appears to be significantly higher in the second case, but not precision.

The performance of each group (IB/IS vs. IS/IB) on the IS1008c meeting is shown in Figure 2: diamonds (◇) correspond to subjects seeing IS1008c as their first meeting, while squares (☐) represent subjects seeing IS1008c as their second meeting, after training on IB4010. The average values of precision and speed are higher when the meeting is seen second, certainly because subjects were able to get training with the TQB interface. The 95% confidence intervals are however strictly disjoint only for speed (0.38 ± 0.12 vs. 0.77 ± 0.17 questions per minute) but not for precision (0.75 ± 0.17 vs. 0.84 ± 0.09). Similarly, for IB4010, speed is significantly higher (exactly with 94% confidence) when the meeting is seen second than when it is seen first, while precision increases less significantly.

These results point to an important property of the TQB interface that is highlighted by the BET evaluation, namely its learnability. A single previous trial appears to be sufficient to improve scores, indicating that TQB is an easily learnable interface. These results seem to hold independently of the individual performances of the subjects (which might maximize either precision or speed), but an assessment using more meetings would enable us to extend the study of the learning curve beyond the first two ones.

6.2 Use of TQB Features by BET Subjects

The analysis of TQB features used during the experiments shows that queries to the transcript and annotation database are quite extensively used to browse meetings. Subjects submit on average 2.50 ± 0.54 queries for each BET question, with no significant differences between the two groups (IB/IS: 2.40 ± 0.92 vs. IS/IB: 2.59 ± 0.57). There is however a significant difference between the two meetings: the remote control one elicits twice as many queries per BET question as the movie club one (IS1008c: 4.16 ± 1.53 vs. IB4010: 2.12 ± 0.44), a fact that could be related to the differences in meeting difficulty alluded to above.

When subjects use TQB queries, they click on average in 35% of the cases on one or more utterances returned by the query, to visualize them in context within the transcript window—this figure provides thus a measure of the relevance of query results. Again, while the average is basically the same for the two groups, there is a difference between meetings: 39% for IB4010 vs. 27% for IS1008c. So, although more queries per BET question are used for IS1008c, clicking on the results is less frequent for this meeting, both facts being consistent with a higher perceived difficulty. Viewing the utterances within the meeting transcript appears to be sufficient to answer BET questions, as listening to the related audio is very infrequent, only about twice per meeting.

Another measure of the importance of TQB queries is the increasing correlation, from the first to the second meeting of a trial, of the number of queries and the precision of the answers. Pearson correlation (across subjects) between the precision scores and the average number of queries launched for each BET question is 0.49 overall (IS/IB group: 0.70 and IB/IS group: 0.37). For each meeting, the correlation increases after learning: for IB4010 it goes from 0.33 to 0.76, and for IS1008c from -0.39 to 0.28. Quite naturally, speed is however negatively correlated with the number of queries per BET question, overall at -0.32. Put simply, these figures show that using queries helps subjects to increase their precision, while at the same time slowing them down slightly.

Statistics over all the 550 queries produced by the 28 subjects indicate that most of the queries are keyword related: 43% look only for a specified word (or character string), while an additional 31% look for a specific word uttered by a particular speaker, and 7% for a word within a given topic episode. Some other constraints or combinations of constraints are used in 1–3% of the queries each: word(s) + dialogue-act, word(s) + person + dialogue-act, topic, person, word(s) + topic + person, etc. The fact that subjects use the query functionality mainly to do keyword search over the transcripts probably reflects the influence of popular Web search engines, and suggests that annotations other than transcript could better be used for automated meeting processing (e.g. for summarization) rather than directly for search by human users.

6.3   Question-Specific Scores

Moving further into the analysis of subjects' answers, it is possible to compute the above statistics separately for the correct answers, and for the wrong ones, and to compare the results. For instance, the average number of queries per BET question, computed only for the questions to which a subject answered correctly, is 2.41 ± 0.58, while the same average over the wrong answers is 2.01 ± 0.53. The difference is more visible for the more difficult meeting, IS1008c (3.38 ± 1.10 queries per correct answer vs. 2.37 ± 1.43 for wrong answers) than for IB4010 (2.09 ± 0.47 vs. 1.88 ± 0.74).

The detailed analysis of scores indicates that these vary considerably with each question. As the order of the questions was kept constant, there are less and less available answers per question as one moves forward through the list. For instance, all 28 subjects answered the first eight questions for IB4010, but only the very first question for IS1008c was answered by all subjects. When IS1008c was seen first, two subjects spent all their time on the first question only, and only 8 subjects (out of 14) managed to answer the first five questions; when IS1008c was seen second, 13 subjects managed to answer the first five questions. As an example, average values for precision and speed are shown in Figure 3 for the first six questions of IS1008c: while scores generally increase after learning, there is considerable variation across questions, e.g. improvement of precision is not the same for all questions, while speed sometimes even degrades in the second round (for the 4th and 6th questions).

These results indicate that performances should be analyzed separately for each question, as their nature requires different competencies and browser functionalities. To take an example, it appears that the most clicked utterance among all those retrieved through TQB queries is the following one, from IB4010: "Uh Goodfellas, I didn't see it". This utterance was clicked 18 times, out of which 16 were in relation to the fifth BET question for IB4010: "No one had seen Goodfellas" vs. "Everyone had seen Goodfellas". Quite obviously, in this case, finding this utterance provides implicitly the answer to the BET question through an immediate inference, as the second statement cannot be true.

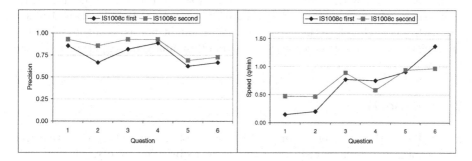

Fig. 3. IS1008c: precision and speed for the first six BET questions, when the meeting is seen first (diamonds ◇) and when it is seen second (squares □). Performances generally increase in the second case, but there is considerable variation across questions.

7 BET Results for Other Browsers

In another series of experiments [13], conducted by the IDIAP Research Institute and the University of Sheffield, four meeting browsers or "conditions" were tested with the BET, in a slightly different setting than the one described above. Usable data was obtained from 39 subjects: each subject performed a calibration task (answering questions using a very simple browser), and one of the following browsers: base (15 subjects), speedup (12 subjects), and overlap (12 subjects). Unlike TQB, none of these meeting browsers relied on manual annotation of the data or on human transcripts. The ISSCO-Meeting_024 was used for calibration, and the other two meetings (IB4010 and IS1008c) were used alternatively in the different conditions.

The calibration condition presented a large slide view, 5 video views, the audio, a timeline, and slide thumbnails. The base condition played audio and included a timeline, scrollable speaker segmentations, a scrollable slide tray, and headshots with no live video. The speedup condition was exactly like the base condition except that it allowed accelerated playback with a user-controlled speed between 1.5 and 3 times normal speed. The overlap condition duplicated the speedup condition by offering simultaneously the first half of meeting on the left audio channel of the subject's headphone, and the second half of the meeting on the right channel, requiring the subjects to focus on one channel at the time.

Raw performance scores for both meetings were as follows for the three conditions (see [13, Section 3] for more details). For the base condition, average precision and speed were respectively 0.77 and 1.2 questions per minute; for the speedup condition, 0.83 and 0.9 questions per minute; and for the overlap condition, 0.74 and 1.0 questions per minute. The average precision is generally below the values obtained by TQB (0.84 ± 0.05 for TQB), while speed is always higher (0.63 ± 0.09 for TQB). These results are quite surprising, as TQB provides access to the transcript, which should considerably improve its information extraction capabilities. In addition, although TQB subjects were not native English speakers unlike those of the other two browsers, data from TQB shows that proficiency is in fact better correlated with precision (at 0.65 level) and much less with speed, therefore the proficiency factor might not explain the difference in precision. Other factors must thus be found, by analyzing experimental logs, to account for these differences.

8   Perspectives

The results of the Browser Evaluation Test method presented here show that the BET captures a number of properties related to browser quality, which match our *a priori* intuitions and therefore contribute to validate the BET evaluation method itself. These results must be further confirmed through future analyses, in particular question-specific ones, and possibly through experiments with more browsers and subjects. Future analyses could better model, for instance, the notion of 'strategy', i.e. a subject's bias towards maximizing either precision or speed, in order to construct a more global performance score, as an "average" of precision and speed.

The BET method offers a generic, task-based solution to the problem of evaluating very different meeting browsers, as it sets few constraints on their functionalities. The set of BET observations created for three meetings will constitute a valuable resource for future evaluations, along with the scores obtained in the experiments presented here, which will provide an initial baseline to which future interfaces can be compared.

Acknowledgments. This work has been supported by the Swiss National Science Foundation, through the IM2 National Center of Competence in Research, and by the European IST Programme, through the AMIDA Integrated Project FP6-0033812. The authors would like to thank Gerwin Van Doorn, who implemented the speedup and overlap browsers at IDIAP.

References

1. Wellner, P., Flynn, M., Tucker, S., Whittaker, S.: A meeting browser evaluation test. In: CHI 2005 (Conference on Human Factors in Computing Systems), Portland, OR, pp. 2021–2024 (2005)
2. Carletta, J., Ashby, S., Bourban, S., Flynn, M., Guillemot, M., Hain, T., Kadlec, J., Karaiskos, V., Kraaij, W., Kronenthal, M., Lathoud, G., Lincoln, M., Lisowska, A., McCowan, I., Post, W., Reidsma, D., Wellner, P.: The AMI Meeting Corpus: A pre-announcement. In: Renals, S., Bengio, S. (eds.) MLMI 2005. LNCS, vol. 3869, pp. 28–39. Springer, Heidelberg (2006)
3. Gibbon, D., Mertins, I., Moore, R.K. (eds.): Handbook of Multimodal and Spoken Dialogue Systems: Resources, Terminology and Product Evaluation. Kluwer Academic Publishers, Dordrecht (2000)
4. Devillers, L., Maynard, H., Paroubek, P.: Méthodologies d'évaluation des systèmes de dialogue parlé: réflexions et expériences autour de la compréhension. Traitement Automatique des Langues 43(2), 155–184 (2002)
5. Dybkjær, L., Bernsen, N.O., Minker, W.: Evaluation and usability of multimodal spoken language dialogue systems. Speech Communication 43(1-2), 33–54 (2004)
6. ISO/IEC: ISO/IEC TR 9126-4:2004 (E) – Software Engineering – Product Quality – Part 3: Quality in Use Metrics. International Organization for Standardization / International Electrotechnical Commission, Geneva (2004)
7. Bevan, N.: International standards for HCI and usability. International Journal of Human-Computer Studies 55, 533–552 (2001)
8. Möller, S.: A new ITU-T recommendation on the evaluation of telephone-based spoken dialogue systems. In: LREC 2004 (4th International Conference on Language Resources and Evaluation), Lisbon, pp. 1607–1610 (2004)
9. Walker, M.A., Litman, D.J., Kamm, C.A., Abella, A.: PARADISE: A framework for evaluating spoken dialogue agents. In: ACL/EACL 1997 (35th Annual Meeting of the Association for Computational Linguistics), Madrid (1997) pp. 271–280 (1997)
10. Traum, D.R., Robinson, S., Stephan, J.: Evaluation of multi-party virtual reality dialogue interaction. In: LREC 2004 (4th International Conference on Language Resources and Evaluation), Lisbon, pp. 1699–1702 (2004)
11. Popescu-Belis, A., Georgescul, M.: TQB: Accessing multimodal data using a transcript-based query and browsing interface. In: LREC 2006 (5th International Conference on Language Resources and Evaluation), Genova, pp. 1560–1565 (2006)
12. Popescu-Belis, A., Clark, A., Georgescul, M., Zufferey, S., Lalanne, D.: Shallow dialogue processing using machine learning algorithms (or not). In: Bengio, S., Bourlard, H. (eds.) MLMI 2004. LNCS, vol. 3361, pp. 277–290. Springer, Heidelberg (2005)
13. Cremers, A., Post, W., Elling, E., van Dijk, B., van der Wal, B., Carletta, J., Flynn, M., Wellner, P., Tucker, S.: Meeting browser evaluation report. Technical report, AMI Project Deliverable D6.4 (December 2006)

Face Recognition in Smart Rooms

Hazım Kemal Ekenel, Mika Fischer, and Rainer Stiefelhagen

interACT Research, Computer Science Department, Universität Karlsruhe (TH)
Am Fasanengarten 5, Karlsruhe 76131, Germany
{ekenel,mika.fischer,stiefel}@ira.uka.de
http://isl.ira.uka.de/

Abstract. In this paper, we present a detailed analysis of the face recognition problem in smart room environment. We first examine the well-known face recognition algorithms in order to observe how they perform on the images collected under such environments. Afterwards, we investigate two aspects of doing face recognition in a smart room. These are: utilizing the images captured by multiple fixed cameras located in the room and handling possible registration errors due to the low resolution of the aquired face images. In addition, we also provide comparisons between frame-based and video-based face recognition and analyze the effect of frame weighting. Experimental results obtained on the CHIL database, which has been collected from different smart rooms, show that benefiting from multi-view video data and handling registration errors reduce the false identification rates significantly.

1 Introduction

Face recognition has attracted significant research efforts that are mainly fueled by security applications. Recently, face recognition for smart interactions has become another application area of significant interest [1]. There have been many papers published on the use of face recognition technology in human-robot interactions [2], smart cars [3], human-computer interfaces [4] as well as image and video retrieval applications [5], [6], [7].

One of the most interesting smart interaction applications is face recognition in smart rooms. Sample application areas can be a smart store that can recognize its regular customers while they are entering the store; a smart home, where family members can be identified while they are entering the rooms of the house and their location can be determined in order to automatically route incoming phone calls; a smart lecture or meeting room, where the participants can be identified automatically and their behaviours can be analyzed throughout the meeting or the lecture. This group of applications requires identification of people without any cooperation, and under uncontrolled conditions, without any constraints on head-pose, illumination, use of accessories, etc. Moreover, according to the distance between the camera and the subject the face resolution varies, and generally the face resolution is low. In these respects, face recognition in smart rooms is a very difficult task. The only factor that can help to improve

A. Popescu-Belis, S. Renals, and H. Bourlard (Eds.): MLMI 2007, LNCS 4892, pp. 120–131, 2007.

Fig. 1. Sample views of the smartrooms

the face recognition performance in smart rooms is the video data of the individuals from multiple views, provided by several cameras that are mounted in the smart room. Sample images from different smart rooms are shown in Figure 1.

Taking these facts into consideration, in this paper we present a detailed analysis of the face recognition problem in a smart room environment. We first compare the well-known face recognition algorithms in order to observe how they perform on the images collected in such environments. Afterwards, we investigate two typical aspects of doing face recognition in a smart room. These are: utilizing the images captured by multiple fixed cameras located in the room and handling possible registration errors due to the low resolution of the aquired face images. We propose a camera-weighting scheme in order to be able to give higher weights to the cameras that have a better view of the person. To be able to handle registration errors, we generate additional registered samples from the manually labelled training images by moving the manual eye label locations in the neighborhood and doing registration with respect to the newly obtained eye coordinates. Note that, even with manual labelling, due to the low resolution of the face images, there can be slight errors in the eye center coordinates. In addition, we also provide comparisons between frame-based and video-based face recognition and analyze the effect of frame weighting. We conduct the experiments on a data corpus that has been collected at different smart rooms. The experimental results indicate that utilizing video data and generating additional samples reduces the false identification rates significantly. Camera and frame weighting have been found to improve the performance further.

The organization of the paper is as follows. In Section 2, local appearance-based face recognition using the discrete cosine transform is explained briefly. A baseline face recognition system is described in Section 3. In Section 4, experimental results are presented and discussed. Finally, in Section 5, conclusions are given.

2   Local Appearance-Based Face Recognition Using Discrete Cosine Transform

Local appearance-based face recognition was proposed as a fast and generic approach [8], [9] and does not require detection of any salient local regions, such as eyes, as in the modular or component based approaches [10], [11]. The underlying ideas for preferring a local appearance-based approach over a holistic appearance-based approach are as follows: (i) In a holistic appearance-based face recognition approach, a change in a local region can affect the entire feature representation, whereas in local appearance-based face recognition it affects only the features that are extracted from the corresponding block while the features that are extracted from the other blocks remain unaffected. This property provides robustness against both local registration imperfections and expression variations, (ii) a local appearance-based algorithm can facilitate weighting of local regions. It can put more weight to the regions which are found to be more discriminant.

In order to represent the local regions, the discrete cosine transform (DCT) is used. Its compact representation ability is superior to that of the other widely used input-independent transforms like the Walsh-Hadamard transform. Although the Karhunen-Loeve transform (KLT) is known to be the optimal transform in terms of information packing, its data dependent nature makes it infeasible for some practical tasks. Furthermore, DCT closely approximates the compact representation ability of the KLT, which makes it very useful for representation both in terms of information packing and in terms of computational complexity.

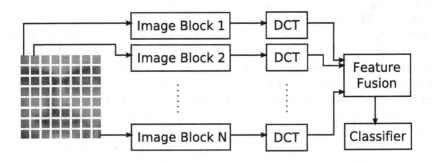

Fig. 2. System architecture of the face recognition system

Feature extraction using the local appearance-based face representation can be summarized as follows: A detected and normalized face image is divided into blocks of 8×8 pixels size. Then the DCT is applied on each block. The obtained DCT coefficients are ordered using zig-zag scanning. From the ordered coefficients, M are selected according to a feature selection strategy, and then normalized to unit norm, resulting in an M-dimensional local feature vector. These extracted local features are then concatenated to represent the entire face image (Figure 2). For details of the algorithm please see [8], [9].

3   Baseline Face Recognition System

In order to provide an identity estimate, the face recognition system processes multi-view, multi-frame visual information. The system components are: image registration, feature extraction, score normalization, fusion over camera-views and fusion over image sequence.

The baseline system receives an input image and the eye-coordinates of the face in the input image. The face image is cropped and registered according to the eye coordinates. The local appearance-based face recognition that is mentioned in Section 2 is used for feature extraction. The first DCT coefficient is removed since it only represents the average value of the image block. The first M coefficients are selected from the remaining ones. To remove the effect of intensity level variations among the corresponding blocks of the face images, the extracted coefficients are normalized to unit norm.

Classification is performed by comparing the extracted feature vectors of the test image with the ones in the training database. Each camera view is compared with all the others. Distance values of the 10-best matches obtained from each frame are normalized using the Min-Max rule, which is defined as:

$$ns = 1 - \frac{s - \min(S)}{\max(S) - \min(S)} \tag{1}$$

where, s corresponds to the distance value of the test image to one of the ten closest training images in the database, and S corresponds to a vector that contains the distance values of the test image to the ten closest training images. The division is subtracted from one, since the lower the distance is, the higher the probability that the test image belongs to that identity class. This way, we obtain a confidence score that is normalized to the value range of $[0, 1]$, closest match having the score '1', and the furthest match having the score '0'. These scores are then normalized by dividing them by the sum of the confidence scores. The obtained confidence scores are summed over camera-views and over the image sequence. The identity of the face image is assigned as the person who has the highest accumulated score at the end of a sequence.

4   Experiments

The experiments have been conducted on a database that has been collected by the CHIL consortium for the CLEAR 2006 evaluations [12]. The recordings are

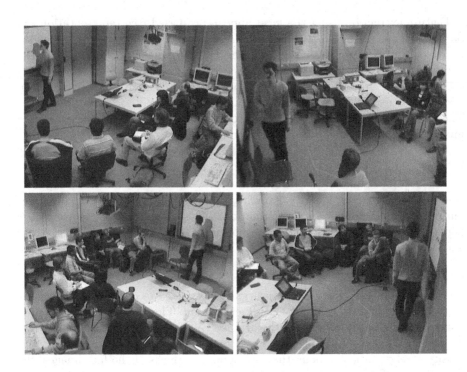

Fig. 3. Views of the UKA smartroom taken at the same instant from the four cameras

from lecture-like seminars and interactive small working group seminars that have been held at different CHIL sites: AIT, Athens, Greece, IBM, New York, USA, ITC-IRST, Trento, Italy, UKA, Karlsruhe, Germany and UPC, Barcelona, Spain. Each smartroom has four cameras, one at each corner, which are labeled from 1 to 4. Sample images from the recordings can be seen in Figures 1 and 3. The evaluation data for the visual identification task consists of short video sequences taken from the database. The recording conditions are uncontrolled and lead to low resolution faces ranging between 10 to 50 pixels resolution, depending on the camera view and the position of the presenter/participant. The presenter to be recognized moves around the projection screen without facing the cameras. Shadows and the beam of the projector result in largely varying face illumination conditions. There are 26 subjects in the database. Two different training and four different testing durations are used in the experiments as presented in Table 1. The training sets contain one sequence for each subject, whereas the number of sequences for each subject in the testing set is varying. Identity estimates are provided at the end of each test segment using the available video data. In addition frame-based recognition — where an identity estimate is provided for every single frame — is also performed and the corresponding results are also presented.

In the database, eye center labels are available for every 200 ms. We only used the frames where both of the eyes are visible and are labelled at the same time. In total we processed 26494 images for the experiments, where 8689 of them belong

Table 1. Duration of Training and Testing Segments

Train/Test ID	Segment duration (sec)	No. of segments
Train A	15	26
Train B	30	26
Test 1	1	613
Test 2	5	411
Test 3	10	289
Test 4	20	178

Table 2. False Identification Results of the Baseline System

	A1	A5	A10	A20	B1	B5	B10	B20
Frame-based	50.1	50.9	50.7	50.4	43.5	43.6	43.7	43.7
Video-based	35.1	25.9	24.9	19.3	32.3	22.3	22.1	17.1

to the training set and the remaining 17805 belong to the testing set. The face images are aligned according to the labelled eye-center coordinates and scaled to 64×64 pixels resolution. The aligned images are then divided into 8×8 pixels resolution non-overlapping blocks making 64 local image blocks. From each image block five-dimensional DCT-based feature vectors are extracted and they are concatenated to construct the final 320-dimensional feature vector. The classification is performed using a nearest neighbor classifier. The L1 norm is selected as the distance metric, since it has been observed that it consistently gives the best correct recognition rates when DCT-based feature vectors are used. The distance values are converted to the matching scores and then the normalized matching scores are combined in order to provide the identity estimate. The identity candidate that has the highest score is assigned as the identity of the person.

The baseline results of both the frame-based and video-based identification are presented in Table 2. In this experiment, all the camera views are compared with each other and the cameras are weighted equally. No frame weighting is performed and no additional samples are used. Each column shows the results for a different training-testing duration combination. The letter indicates whether the training is from set A or B which corresponds to 15 and 30 second training durations, respectively. The number indicates the duration of the testing segment in seconds. For frame-based identification all the frames in the training-testing duration combination are used. For example in the combination 'A5', all the frames in the Train A set and all the frames in 5 seconds duration testing segments are used. Two main observations can be derived from the table. The first one is that using the video data improves the results significantly compared to the single frame classification and the second one is that as the duration of training or testing increases the false identification rate decreases.

In the following experiments the baseline parameters — comparing all camera views with each other, no camera weighting, no frame weighting, no additional samples — will be kept and whenever a parameter is changed it will be indicated in the section.

4.1   Comparison of the Well-Known Face Recognition Algorithms

In the first part of the experiments, well-known face recognition algorithms have been tested on the smart room data. The experiments are conducted frame-based. That is, an identity estimate is provided for each frame. We used all the frames from the Train B set for training and all the frames in the 20 second segments for testing. We have compared our local appearance-based face recognition (LAFR) algorithm with Eigenfaces [13], [14], linear discriminant analysis (LDA) [15] and Bayesian face recognition [16] algorithms. In the Eigenfaces and Bayesian face recognition algorithms we kept the first 320 eigenvectors, in order to have the same dimensional feature vector that we used for the LAFR approach. For Bayesian face recognition we used 1000 intra-personal and extra-personal samples. For LDA, we used the LDA+PCA algorithm provided in the CSU face identification evaluation system [17]. This version of LDA uses a soft distance measure proposed by Zhao et al. [15]. We both used the L1 and MAHCOS [14] distance metrics in the Eigenfaces algorithm. The false identification rates are given in Table 3. As can be seen the local appearance-based face recognition approach outperforms the other well-known face recognition algorithms. The most interesting result that can be observed in this table is the very high false identification rate obtained by Bayesian face recognition which has been known to be one of the best performing algorithms in the FERET evaluations [18] and which has inspired many other algorithms that utilize intra-personal and extra-personal variations. The main reason for the bad performance on the smart room database is the multiple sources of variations that exist in the database. Varying pose and illumination changes, registration errors and low resolution make the intra-personal and extra-personal variations almost identical, therefore the approach loses its discriminative capability.

Table 3. Performance Comparison of Well-known Face Recognition Algorithms

Recognizer	FI rate (%)
LAFR	43.6
Eigenfaces L1	48.6
Eigenfaces MAHCOS	59.5
LDA	49.6
Bayesian	87.4

Table 4. False Identification Results of Camera-wise Classification

Recognizer	FI rate (%)
LAFR	37.8
PCA L1	45.8
PCA MAHCOS	60.8
Fisherfaces	46.5
Bayesian	82.5

Table 5. False Identification Results of Camera-wise and All Camera Classification for LAFR

	A1	A5	A10	A20	B1	B5	B10	B20
All cameras	50.1	50.9	50.7	50.4	43.5	43.6	43.7	43.7
Camera-wise	46.7	46.9	46.7	46.4	39.7	38.1	38.2	37.8

4.2 Camera-Wise vs. All Camera Classification

In the second part of the experiments, we compared camera-wise and all camera classification. In camera-wise classification, each camera-view is handled separately. That is, the testing image acquired by a camera is only compared with the training images acquired at each site by the camera with the same label. For example, if the testing image was acquired by a camera with label 1, we only compare it with training images also acquired by a camera with label 1. On the other hand, in all camera classification the testing image acquired by a camera is compared with the training images acquired by all the cameras. Camera-wise classification has many advantages. First of all, it speeds up the system significantly. That is, if we have N images from each camera for training, and if we have R images from each camera for testing, and if we have C cameras that do recording, $(C \cdot N) \cdot (C \cdot R)$ similarity calculations are performed between all the training and testing images. However, when we do camera-wise image comparison, then we only need to do $C \cdot (N \cdot R)$ comparisons between the training and testing images. Apparently, this reduces the amount of required computation by $1/C$. In addition to the improvement in the system's speed, it also provides a kind of view-based approach that separates the comparison of different views, which was shown to perform better than doing matching between all the face images without taking into consideration their view angles [10].

Table 4 shows the false identification results of the well-known face recognition algorithms. Again, all the frames from the Train B set and all the frames in the 20 second segments are used for training and testing, respectively. This time camera-wise classification is done instead of comparing all the camera views with each

Table 6. Effect of Camera Weighting

	A1	A5	A10	A20	B1	B5	B10	B20
Video-based	34.4	25.1	22.5	19.3	31.3	22.3	21.1	15.9

other. Compared to the results in Table 3, it can be noticed that the results have been improved for each recognizer except PCA MAHCOS.

In Table 5, camera-wise and all camera classification results are presented for the LAFR algorithm for different training-testing duration combinations. Both of the classifications are performed frame-based. As can be seen at each training-testing duration combination the results improved with camera-wise classification.

4.3 Camera Weighting

In the third part of the experiments, the effect of camera weighting is analyzed. The camera weighting is performed with respect to the distance between the eyes. The higher inter-eye distance implies either a high resolution face image or a lower resolution face image with a close to frontal pose. On the other hand, a small inter-eye distance implies either low resolution face image or a higher resolution face image with a close to profile head pose. Since we would like to weight the cameras that have better view of the subject more and since higher resolution or close to frontal face images are more desirable for face recognition, we did the weighting by taking into consideration the inter-eye distance. We put more weights to the camera views with high inter-eye distances, by using weights proportional to the inter-eye distance. The obtained results can be seen in Table 6. Compared to the results at the second row of Table 2 a slight decrease in the false identification rates can be observed.

4.4 Additional Samples

In the fourth part of the experiments, we analyze the contribution of additional training sample generation to the face recognition performance on the smart room data. Note that, even with manual labelling, due to the low resolution of the face images, there can be slight errors in the eye center coordinates. To be able to handle registration errors, we generate additional registered samples from the manually labelled training images. In order to do this, we move the left and right eye center labels in their 4-neighborhood, $(x + 1, y), (x - 1, y), (x, y - 1), (x, y + 1)$. This gives 5 locations for each eye and 25 combinations of eye positions (including the original eye coordinates). The face image is then registratered using each of these 25 eye coordinates. This way, we generated 24 additional training samples per original training sample. Table 7 shows the results. Both the frame-based and video-based results improved significantly, around 10% absolute decrease is achieved in the false identification rates.

24 additional training samples implies 24 times more processing time that must be spent in a nearest neighbor classification scheme which is not desirable. To

Table 7. Effect of Using Additional Samples

	A1	A5	A10	A20	B1	B5	B10	B20
Frame-based	38.7	39.0	38.8	38.3	31.9	32.1	32.1	31.8
Video-based	28.2	17.5	17.2	11.9	22.8	13.2	11.9	9.1

Table 8. Effect of Using Additional Samples with Clustering

	A1	A5	A10	A20	B1	B5	B10	B20
Frame-based	40.5	41.9	41.9	41.7	35.0	34.3	34.2	34.1
Video-based	25.3	15.7	17.9	13.6	20.7	12.2	12.6	9.1

reduce the number of training samples we used k-means clustering. We chose k to be the number of original samples and used the resulting cluster centers as representatives. This way the processing time for classifying new images remains the same. The resulting false identification rates are shown in Table 8. The results are very close to the ones that were obtained without clustering. Even, at some cases the false identification rates decrease. These indicate that there is no need to sacrifice from the processing time in order to obtain better results using the additional samples.

4.5   Frame Weighting

In the fifth part of the experiments, we investigated the effect of frame weighting. It has been observed that the distance between the closest and the second closest training samples is generally smaller in the case of a false classification than in the case of a correct classification [19]. It has been found that the distribution of these distances resembles an exponential distribution:

$$\varepsilon(x; \lambda) = 0.1\lambda e^{-\lambda x} \quad \text{with } \lambda = 0.05 \qquad (2)$$

The weights are then computed as the cumulative distribution function:

$$\mathcal{E}(x; \lambda) = 1 - e^{-\lambda x} \qquad (3)$$

Note that this distribution is extracted completely on a different database and is not specific to the mentioned smart room scenario [19]. We weighted each frame using this formula. The results are given in Table 9. Again an improvement over the baseline system is achieved.

4.6   Combining All the Parameters

In the last experiment, we combined all the parameters we have analyzed so far. We used additional samples with clustering, camera weighting and frame

Table 9. Effect of Frame Weighting

	A1	A5	A10	A20	B1	B5	B10	B20
Video-based	34.9	25.4	22.1	17.6	31.5	19.3	18.3	14.8

Table 10. Effect of Combining all the Parameters

	A1	A5	A10	A20	B1	B5	B10	B20
Video-based	26.3	17.0	16.1	11.9	21.5	12.9	11.2	9.1

weighting for this experiment. Interestingly, the improvements observed in the previous experiments do not sum up in the combined experiment. We noticed that the largest impact comes from using additional samples with clustering.

5   Conclusions

In this paper we provided a detailed analysis of face recognition in smart rooms. We first compared the well-known face recognition algorithms in order to observe how they perform on the images collected under such environments. We found the local appearance-based face recognition algorithm to be superior to the other well-known face recognition algorithms. We also observed that the Bayesian face recognition approach, which is based on intra- and extra-personal variations, does not work well on this kind of uncontrolled data. Afterwards, we investigated two typical aspects of doing face recognition in a smart room. The first one is utilizing the video data captured from multiple fixed cameras located in the room. The obtained results show that benefiting from video data provided by multiple cameras decreases the false identification rates significantly compared to the frame-based results. The second aspect is handling possible registration errors due to the low resolution of the aquired face images. We generated additional registered samples from the manually labelled training images by moving the manual eye label locations in the neighborhood and did registration with respect to the newly obtained eye coordinate pairs. We also clustered the newly generated additional samples in order to have the same number of representative training samples as we had original training samples. In both cases — without and with clustering — the false identification rates decreased significantly, which indicates that registration errors are one of the most important problems in low resolution face recognition. In addition, we also analyzed the effect of camera and frame weighting. Camera and frame weighting have been found to improve the performance further.

Acknowledgement

This work is sponsored by the European Union under the integrated project CHIL—Computers in the Human Interaction Loop, contract number 506909.

References

1. Pentland, A., Choudhury, T.: Face recognition for smart environments. Computer 33(2), 50–55 (2000)
2. Nickel, K., Ekenel, H.K., Voit, M., Stiefelhagen, R.: Audio-Visual Perception of Humans for a Humanoid Robot. In: 2nd Intl. Workshop on Human-Centered Robotic Systems, Munich, Germany (2006)
3. Erzin, E., Yemez, Y., Tekalp, A.M., Ercil, A., Erdogan, H., Abut, H.: Multimodal Person Recognition for Human-Vehicle Interaction. IEEE Multimedia 13(2), 18–31 (2006)
4. Stiefelhagen, R., Bernardin, K., Ekenel, H.K., McDonough, J., Nickel, K., Voit, M., Wölfel, M.: Audio-Visual Perception of a Lecturer in a Smart Seminar Room. Signal Processing 86(12) (2006)
5. Berg, T., Berg, A.C., Edwards, J., Forsyth, D.: Who is in the picture. In: Neural Information Processing Systems(NIPS) (2004)
6. Arandjelovic, O., Zisserman, A.: Automatic face recognition for film character retrieval in feature-length films. In: CVPR 2005, Washington, DC, USA, pp. 860–867 (2005)
7. Sivic, J., Everingham, M., Zisserman, A.: Person spotting: Video shot retrieval for face sets. In: Proc. Intl. Conf. on Image and Video Retrieval, Springer, Heidelberg (July 2005)
8. Ekenel, H.K., Stiefelhagen, R.: Local appearance-based face recognition using discrete cosine transform. In: 13th European Signal Processing Conference, Antalya, Turkey (2005)
9. Ekenel, H.K., Stiefelhagen, R.: Analysis of Local Appearance-based Face Recognition: Effects of Feature Selection and Feature Normalization. In: CVPR Biometrics Workshop, New York, USA (June 2006)
10. Pentland, A., Moghaddam, B., Starner, T.: View-based and modular eigenspaces for face recognition. In: CVPR 1994 (1994)
11. Heisele, B., Ho, P., Poggio, T.: Face recognition with support vector machines: Global versus component-based approach. In: ICCV 2001, pp. 688–694 (2001)
12. Stiefelhagen, R., Bernardin, K., Bowers, R., Garofolo, J., Mostefa, D., Soundararajan, P.: The CLEAR 2006 Evaluation. In: Stiefelhagen, R., Garofolo, J. (eds.) CLEAR 2006. LNCS, vol. 4122, pp. 1–45. Springer, Heidelberg (2007)
13. Turk, M., Pentland, A.: Eigenfaces for recognition. Journal of Cognitive Science, 71–86 (1991)
14. Draper, B.A., Yambor, W.S., Beveridge, J.R.: Analyzing pca-based face recognition algorithms: Eigenvector selection and distance measures. In: Empirical Evaluation Methods in Computer Vision, Singapore (2002)
15. Zhao, W., Chellappa, R., Phillips, P.J.: Subspace linear discriminant analysis for face recognition, Technical Report, UMD (1999)
16. Moghaddam, B., Jebara, T., Pentland, A.: Bayesian Face Recognition. Pattern Recognition 33(11), 1771–1782 (2000)
17. The, C.S.U.: Face Identification Evaluation System:
http://www.cs.colostate.edu/evalfacerec/
18. Phillips, P.J., Moon, H., Rauss, P.J., Rizvi, S.: The FERET evaluation methodology for face recognition algorithms. IEEE Trans. on PAMI 22(10) (October 2000)
19. Stallkamp, J.: Video-based Face Recognition Using Local Appearance-based Models, Thesis report, Universität Karlsruhe (TH) (November 2006)

Gaussian Process Latent Variable Models for Human Pose Estimation

Carl Henrik Ek[1], Philip H.S. Torr[1], and Neil D. Lawrence[2]

[1] Oxford Brookes University, Department of Computing, United Kingdom
cek@brookes.ac.uk, philiptorr@brookes.ac.uk
http://cms.brookes.ac.uk/research/visiongroup/
[2] University of Manchester, School of Computer Science, United Kingdom
neill@cs.man.ac.uk
http://www.cs.man.ac.uk/~neill/

Abstract. We describe a method for recovering 3D human body pose from silhouettes. Our model is based on learning a latent space using the Gaussian Process Latent Variable Model (GP-LVM) [1] encapsulating both pose and silhouette features Our method is generative, this allows us to model the ambiguities of a silhouette representation in a principled way. We learn a dynamical model over the latent space which allows us to disambiguate between ambiguous silhouettes by temporal consistency. The model has only two free parameters and has several advantages over both regression approaches and other generative methods. In addition to the application shown in this paper the suggested model is easily extended to multiple observation spaces without constraints on type.

1 Introduction

We consider the problem of estimating 3D articulated human pose from monocular silhouettes. Silhouettes are commonly used for pose estimation [2,3,4,5,6] as they contain strong cues for pose while at the same time being invariant to texture and lighting. Pose estimation from silhouettes is difficult because of inherent ambiguities leading to a one to many mapping from silhouette to pose. These ambiguities can be split into two types, (i) mis-labeling or limb flips and (ii) out of plane rotations. The first type appears for in plane rotations when lack of occlusion cues makes it hard to differentiate between limbs. The out-of-plane ambiguities appear when the subject is facing the view plane: the perspective distortions do not give strong enough cues to disambiguate limbs position out-of-plane. Algorithms with silhouette inputs need to handle these ambiguities.

There are two lines of work on pose estimation from silhouettes, (i) methods modeling the silhouette as a generative process from pose [4,6,7], (ii) methods based on regression from image observations to pose [2,5,8,9]. Generative methods model the space of silhouettes as a function of pose. This will correctly reflect the structure of the problem as each silhouette could have been generated by several different poses but each pose can only generate one single silhouette. The

A. Popescu-Belis, S. Renals, and H. Bourlard (Eds.): MLMI 2007, LNCS 4892, pp. 132–143, 2007.

problem arises when trying to infer pose from silhouette as, due to the multi-modality no inverse functional mapping from silhouette to pose exists. Finding each mode in pose space for a given silhouette is often very complicated due to the high-dimensionality of the pose space, even approximative methods like particle filters will be very expensive as a very large number of particles are needed to explore the high-dimensional pose space.

Regression based techniques try to model the pose space as a function of silhouettes. However due to the multi-modality no such functional exists. To overcome this problem it has been suggested to divide the silhouette space into subspaces for which functionals exist [2,5]. The pose space can then be described as a mixture of these single regressors. The structure of such a mixture has to reflect the multi-modality such that a *one-to-one* mapping exist between silhouette and pose for each subspace. In [2] the mixture centers region of support are decided by clustering in pose space. This is based on the assumption that ambiguities will occur between poses that have a significantly different joint angle configuration.

In [5] clustering is initially done in silhouette feature space then each cluster is split into several sub-clusters based on their corresponding poses. A clustering approach will not resolve all the ambiguities as it is based on the assumption that ambiguous silhouettes are "clearly" separated in pose space. This is only true for a small subset of ambiguities as especially the out-of-plane type occurs for continuous ranges in pose space. The final number of regressors will need to be decided based on some heuristic assumption about the occurrence of ambiguities or an error measure. There is a trade-off between generalization and training error as the minimal error would be given if each pose were to be represented by a separate regressor, but this would remove all generalizing capabilities of the model.

In this paper we take a learning based approach where we model both silhouette observations, joint angles and their dynamics as generative models from a shared low dimensional latent representations using the GP-LVM [1]. In line with other work on pose estimation [2,3,4,5,6] we have chosen to represent each image by its silhouette. As in [2,5] each silhouette is represented using shape context histograms [10]. We subsample each contour with one pixel spacing, acquiring about $100 - 150$ histograms for each image. To reduced the dimensionality of the descriptor and remove the the the effects of ordering we vector quantize the histograms using K-means clustering as described in [11], resulting in a 100D silhouette descriptor.

As described above a generative process will correctly handle the multi-modality between silhouette and pose. As we are learning a low-dimensional representation of pose we are not forced to fall-back on approximative methods for solving the inverse of this generative mapping. Our latent representation reflects the dynamics of of the data and can therefore predict poses over time in simple manner. The model requires no manual initialization when predicting sequential data but automatically initializes from training data.

2   Gaussian Processes

Gaussian Processes (GP) [12] are generalizations of Gaussian distributions defined over infinite index sets. Thereby a GP can be used to specify distribution over functions. It is completely defined by its mean function $\mu(\mathbf{x}_i)$, which is often taken to be zero, and its covariance function $k(\mathbf{x}_i, \mathbf{x}_j)$. The covariance function k characterizes the nature of the functions that can be sampled from the process. One widely used covariance function is

$$k(\mathbf{x}_i, \mathbf{x}_j) = \theta_1 e^{-\frac{\theta_2}{2}||\mathbf{x}_i - \mathbf{x}_j||^2} + \theta_3 + \beta^{-1}\delta_{ij}, \tag{1}$$

where the parameters are given by $\Phi = \{\theta_1, \theta_2, \theta_3, \beta\}$ and δ_{ij} is Kronecker's delta function. This covariance function combines an RBF function, a bias and a white-noise term. The parameters Φ of the covariance function k will be referred to as the hyper-parameters of the GP.

2.1   Prediction

By definition of a GP any finite number of variables specified by the process will have a joint Gaussian distribution [12]. For regression $y_i = f(\mathbf{x}_i) + \epsilon$, with noise $\epsilon \sim N(0, \beta^{-1})$, where $y_i \in \Re$ and $\mathbf{x}_i \in \Re^q$ placing a GP prior with zero mean and covariance function $k(x_i, x_j)^1$ over f, leads to the joint distribution,

$$\begin{bmatrix} \mathbf{y} \\ y_* \end{bmatrix} \sim N \left(\mathbf{0}, \begin{bmatrix} \mathbf{K} & \mathbf{K}_* \\ \mathbf{K}_*^\mathbf{T} & k(\mathbf{x}_*, \mathbf{x}_*) \end{bmatrix} \right) \tag{2}$$

of a set of observed data $\{\mathbf{x}_i, y_i\}_{i=1}^{N}$ and an unseen point $\mathbf{x}_*$, where $K_{ij} = k(\mathbf{x}_i, \mathbf{x}_j)$. Conditioning on the observed data leads to a posterior distribution over functions. From this posterior we obtain the predictive equations of a GP for an unseen point $\mathbf{x}_*$,

$$\overline{y}_* = k(\mathbf{x}_*, \mathbf{X})\mathbf{K}^{-1}\mathbf{Y} \tag{3}$$

$$\sigma_*^2 = k(\mathbf{x}_*, \mathbf{x}_*) - k(\mathbf{x}_*, \mathbf{X})^T\mathbf{K}^{-1}k(\mathbf{x}_*, \mathbf{X}), \tag{4}$$

where $\mathbf{X} = [\mathbf{x}_1, \ldots, \mathbf{x}_N]^T$, $\mathbf{Y} = [y_1, \ldots, y_N]^T$, $\overline{y}_*$ is the mean prediction and σ_*^2 is the variance.

2.2   GP Training

By maximizing the marginal likelihood over functions f,

$$p(\mathbf{Y}|\mathbf{X}, \Phi) = \int p(\mathbf{Y}|f, \mathbf{X}, \Phi)p(f|\mathbf{X}, \Phi)df \tag{5}$$

$$p(f|\mathbf{X}, \Phi) = N(\mathbf{0}, \mathbf{K}),$$

---

[1] Including a white-noise term with variance β^{-1}.

the hyper-parameters Φ of the GP can be learned from the observed data. This is referred to as training in the GP framework. It might seem undesirable to optimize over the hyper-parameters as the model might over-fit the data[2] Inspection of the logarithm of equation (5),

$$\log p(\mathbf{Y}|\mathbf{X}) = \underbrace{-\frac{1}{2}\mathrm{tr}\left(\mathbf{Y}^T\mathbf{K}^{-1}\mathbf{Y}\right)}_{data-fit} \underbrace{-\frac{1}{2}\log|\mathbf{K}|}_{complexity} -\frac{N}{2}\log 2\pi, \quad (6)$$

shows two "competing terms", the data-fit and the complexity term. The complexity term measures and penalizes the complexity of the model, while the data-fit term measures how well the model fits the data. This "competition" encourages the GP model not to over-fit the data.

3   GP-LVM

Lawrence [13] proposed an algorithm for dimensionality reduction using Gaussian Processes called the Gaussian Process Latent Variable Model (GP-LVM). The GP-LVM is a generative model where each observed data point, $\mathbf{y}_i \in \Re^D$, is generated through a noisy process from a latent variable $\mathbf{x}_i \in \Re^q$,

$$\mathbf{y}_i = f(\mathbf{x}_i) + \epsilon, \quad (7)$$

where $\epsilon \sim N(\mathbf{0}, \beta^{-1}\mathbf{I})$. Placing a zero mean GP-prior on the generative function f the marginal likelihood $P(\mathbf{Y}|\mathbf{X}, \Phi)$ can be formulated by integration over f,

$$P(\mathbf{Y}|\mathbf{X}, \Phi) = \prod_{j=1}^{D} \frac{1}{(2\pi)^{\frac{N}{2}}|\mathbf{K}|^{\frac{1}{2}}} e^{-\frac{1}{2}\mathbf{y}_{:,j}^T\mathbf{K}^{-1}\mathbf{y}_{:,j}}, \quad (8)$$

where $\mathbf{y}_{:,j}$ is the jth column from the data matrix, $\mathbf{Y}$. The GP-LVM maximizes the marginal likelihood (8) with respect to both the latent points $\mathbf{X}$ and the hyper-parameters Φ of the covariance function,

$$\{\hat{\mathbf{X}}, \hat{\Phi}\} = \mathrm{argmax}_{\mathbf{X},\Phi} P(\mathbf{Y}|\mathbf{X}, \Phi). \quad (9)$$

In general[3] there is no closed form solution for (9) and we must turn to gradient based optimization to make progress. The only parameter of the GP-LVM that can not be found through maximum likelihood is the dimensionality of the latent space, q, which must be set by hand.

---

[2] By setting the noise variance β^{-1} to zero the function f will pass exactly through the observed data $\mathbf{Y}$.

[3] An exception is when the linear kernel is used, in which case the optimization becomes an eigenvalue problem [1].

3.1   Back Constrained GP-LVM

Using a smooth covariance function the GP-LVM will specify a smooth mapping from the latent space $\mathbf{X}$ to the observation space $\mathbf{Y}$, this means that points close in the latent space will be close in the observed space. Having a smooth generative mapping does not imply that an inverse functional mapping exists.

Recently Lawrence and Quiñonero Candela [14] proposed an extension to the GP-LVM where the model is constrained by representing each latent point as a smooth parametric mapping from its corresponding observed data point, $\mathbf{x}_i = g(\mathbf{y}_i, \mathbf{W})$, where $\mathbf{W}$ is the mapping parameter set. This constrains points that are close in the observed space to also be close in the latent space. The mapping from observed data $\mathbf{Y}$ to $\mathbf{X}$ will be referred to as back-constraint. Including a back-constraint in the GP-LVM model changes the maximization in equation (9) from optimization with respect to the latent variables $\mathbf{X}$ to optimizing the parameters of the back-constraining mapping,

$$\{\hat{\mathbf{W}}, \hat{\Phi}\} = \text{argmax}_{\mathbf{W}, \Phi} P(\mathbf{Y}|\mathbf{W}, \Phi). \tag{10}$$

3.2   GP Dynamics

For embedding sequential data Wang *et. al.* [15] proposed an extension to the GP-LVM to find a latent space that would reflect the ordering of the observed data. This is done by specifying a predictive function over the sequence in latent space,

$$\mathbf{x}_t = h(\mathbf{x}_{t-1}) + \epsilon_{dyn}, \tag{11}$$

where $\epsilon_{dyn} \sim N(\mathbf{0}, \beta_{dyn}^{-1}\mathbf{I})$. A GP prior can then be placed over the function $h(\mathbf{x})$. Marginalizing this mapping results in a distribution over the latent points which, through combination with the marginalized likelihood for the GP-LVM, specifies a new objective function,

$$\{\hat{\mathbf{X}}, \hat{\Phi}_Y, \hat{\Phi}_{dyn}\} = \text{argmax}_{X, \Phi_Y, \Phi_{dyn}}$$
$$P(\mathbf{Y}|\mathbf{X}, \Phi_Y)P(\mathbf{X}|\Phi_{dyn}). \tag{12}$$

4   GP-LVM for Pose Estimation

The aim of our model is to learn a shared latent representation $\mathbf{X} = [\mathbf{x}_1, \ldots, \mathbf{x}_N]^T$ that relates corresponding pairs of feature $\mathbf{Y} = [\mathbf{y}_1, \ldots, \mathbf{y}_N]^T$ and pose $\mathbf{Z} = [\mathbf{z}_1, \ldots, \mathbf{z}_N]^T$. In [16] a shared latent structure between two joint angle spaces, one corresponding to a humanoid robot and the other corresponding to a human is learned. This is done by modifying the GP-LVM to learn separate sets of Gaussian Processes to each of the different observation spaces from a shared latent space. The latent representation is found by maximizing the joint marginal likelihood of the two observation spaces,

$$P(\mathbf{Y}, \mathbf{Z}|\mathbf{X}, \Phi_s) = P(\mathbf{Y}|\mathbf{X}, \Phi_Y)P(\mathbf{Z}|\mathbf{X}, \Phi_Z), \tag{13}$$

where $\Phi_s = \{\Phi_Y, \Phi_Z\}$. We want to learn a latent structure that preserves local distances from the pose space. This can be achieved by incorporating a back-constraint from the pose space onto the latent space by representing the latent points as a function of the pose. Incorporating a back-constraint from pose implies we are trying to enforce a *one-to-one* mapping between the pose space and the latent space. This is desirable as it will force the mapping from latent to feature to be *many-to-one* which means that we have contained the multi-modality of the system to this mapping.

To back-constrain the latent space, we represent the latent points by a regression over a kernel induced feature space that allows for non-linearities,

$$\mathbf{x}_i = \sum_{j=1}^{N} w_j \phi(\mathbf{z}_i, \mathbf{z}_j) \tag{14}$$

$$\phi(\mathbf{z}_i, \mathbf{z}_j) = e^{-\frac{\gamma}{2}(\mathbf{z}_i - \mathbf{z}_j)^T (\mathbf{z}_i - \mathbf{z}_j)} \tag{15}$$

This leads to a modified objective where the positions of the latent variables are optimized indirectly by maximizing $P(\mathbf{Y}, \mathbf{Z}|\mathbf{W}, \Phi_Y, \Phi_Z) = P(\mathbf{Y}|\mathbf{W}, \Phi_Y)P(\mathbf{Z}|\mathbf{W}, \Phi_Z)$ with respect to the parameters of the back-constraint. The latent representation is shared by the feature space and the pose space. By back-constraining the pose space we are encouraging the mapping between the latent space and pose space to be *one-to-one*, thereby forcing the GP-LVM from latent space to silhouette features to be *many-to-one* to reflect the ambiguities in the silhouette features.

4.1   Dynamical Model

Many of the pose ambiguities from our silhouette representation can be resolved by considering sequential data. Including a model that can predict poses over time allows us to resolve ambiguous silhouettes by temporal consistency. However, by learning a latent representation we can do even better. We can incorporate the dynamic model when learning the latent representation, forcing the latent representation to respect the data's dynamics. This can be done by specifying a GP over the latent space as in [15], incorporating this within our back-constrained, shared latent space representation leads to the following objective,

$$P(\mathbf{Y}, \mathbf{Z}, \mathbf{W}|\Phi) = P(\mathbf{Y}, \mathbf{Z}|\mathbf{W}, \Phi_s)P(\mathbf{W}|\Phi_{dyn}),$$

where $\Phi = \{\Phi_s, \Phi_{dyn}\}$. Sequences of pose usually form locally smooth but globally complex trajectories through joint angle space. This makes it difficult to fit a dynamic model when pose is represented as joint angles. Learning a dynamical model jointly with the latent representation is beneficial as the non-linear mapping from latent space to pose space allows for a significantly different structure for the latent representation and the joint angle representation. We use this property to smoothly[4] arrange the latent representation according to the dynamics

---

[4] Due to the smooth covariance function in the dynamic GP.

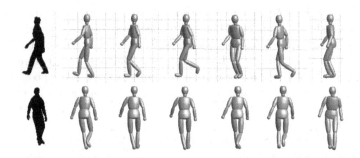

Fig. 1. Single Image Pose Estimation: Input silhouette followed by output poses associated with modes on the latent space ordered according to decreasing likelihood

of the data. Along with the dynamics our model also contains a back-constraint. The back-constraint will encourage that the local smoothness of the joint angle trajectories is preserved. Our experiments show that the structure of the latent space changes significantly when incorporating the dynamics, this difference is evidence of the complex nature of trajectories through joint angle space.

4.2   Model Summary

Training of the model implies that we are finding a shared latent representation $\mathbf{X}$ of both the observation spaces $\mathbf{Y}$ (the silhouette features) and $\mathbf{Z}$ (the pose angles), learning two sets of GP regressors from the latent space, $\mathbf{X}$, to reconstruct each of the observation spaces. Additionally the latent space incorporates a set of GPs which give temporal predictions. We are also learning a parametric mapping from the pose space, $\mathbf{Z}$, to the latent space, $\mathbf{X}$, thereby enforcing the latent space to preserve the local similarities of the pose space. All the remaining parameters of the model, except two, are found through maximum likelihood. The two remaining parameters are: (i) the width γ of the kernel that specifies the back-constraining mapping (15). This parameter was estimated by viewing the scatter matrix of the kernel response to the pose training data (in all our experiments it is set $\gamma = 10^{-3}$). (ii) The dimensionality of the latent space q which we set to 4 for all our experiments. All other parameters of the model (*i.e.* the parameters of each of the three covariance functions, the parameters of the back-constraint and the coordinates of latent representation) are learned from training data.

5   Pose Inference

Given a trained model which jointly represents the pose angles, $\mathbf{Z}$, and the silhouette features, $\mathbf{Y}$, in terms of a shared sub-space, $\mathbf{X}$, we wish to infer the most likely sequence of pose angles given a set of silhouette features. We will first describe inference for a single frame and then show how inference is done for a sequence of frames.

5.1   Single Image Pose Estimation

Inference in this model is a two stage process. In the first stage the position on the latent space that is most likely to have generated the observed features is found. This is done by maximizing the predictive likelihood for the GP that maps the latent space to the given silhouette features,

$$\hat{\mathbf{x}} = \text{argmax}_{\mathbf{x}_*} p(\mathbf{y}_* | \mathbf{x}_*, \mathbf{Y}, \mathbf{X}, \Phi_Y). \tag{16}$$

Having forced the multi modality to be handled by the GP from latent to feature we expect (16) to have several maxima for an ambiguous silhouette. Equation (16) needs to be maximized using gradient based methods which require initialization. For each initialization for $\mathbf{x}_*$ we will find one (not necessarily unique) maximum. To recover multiple solutions we need multiple initializations. We chose the initializations of $\mathbf{x}_*$ from the latent points, $\mathbf{X}$, that correspond to the training data, choosing the 20 most likely points to have generated $\mathbf{y}_*$.

As explained in the previous section we back-constrain the latent space with a smooth mapping from pose space with the aim of enforcing a *one-to-one* correspondence between the latent space and the pose space. This means that given a latent representation the pose can simply be found by mapping each of the optimized latent points to pose space using the mean prediction from each latent point (3) of the GP as the most likely pose for each of the modes,

$$\hat{\mathbf{z}} = k(\hat{\mathbf{x}}, \mathbf{X})^T \mathbf{K}^{-1} \mathbf{Z}, \tag{17}$$

and accounting for the width of the distribution around each mode using the variance (4).

5.2   Sequence Estimation

A single feature descriptor is likely to correspond to several different poses, however a sequence of feature descriptors are less likely to be ambiguous. Learning a GP to predict latent points over time we can formulate the joint likelihood for a sequence of features and their latent coordinates using the dynamical model. We can maximize this joint likelihood to find the most likely latent coordinates for the observed sequence of silhouette features.

$$\hat{\mathbf{X}} = \text{argmax}_{\mathbf{X}_*} p(\mathbf{Y}_*, \mathbf{X}_* | \mathbf{Y}, \mathbf{X}, \Phi_Y, \Phi_{dyn}) \tag{18}$$

Having found the corresponding latent points the most likely poses can, as in the case of the single frame estimation, be found through the mean prediction of the GP from the latent space to the pose space,

$$\hat{\mathbf{Z}} = k(\hat{\mathbf{X}}, \mathbf{X})^T \mathbf{K}^{-1} \mathbf{Z} \tag{19}$$

5.3   Sequence Initialization

As with the initialization for a single image we want to initialize each frames latent point with a point from the latent space from the training data $\mathbf{X}$. The most likely sequence through the training data for an unseen sequence $\mathbf{Y}_*$ can be found by interpreting the sequence as a hidden Markov model (HMM) where the latent states of the HMM correspond to the training points. The likelihood for each observation is specified by the GP point likelihood (16) associated with each each latent point, and the transitions are given by the dynamical GP that predicts over time in the latent space. The most probable path $\mathbf{X}_{init} = \operatorname{argmax}_{\mathbf{x}^{(1)},\ldots\mathbf{x}^{(n)}}$ $p(\mathbf{x}^{(1)},\ldots\mathbf{x}^{(n)}|\mathbf{y}_*^{(1)},\ldots\mathbf{y}_*^{(n)})$ through this lattice can be found using the Viterbi algorithm [17]. The optimization of the sequence objective in (18) can then be initialized with $\mathbf{X}_{init}$.

6   Results

We will consider the data presented in [2]. This dataset contains 1927 training poses and 418 poses for testing from human motion capture data. Each pose is parametrized by a 54 dimensional joint vector. From each pose vector an image has been generated using the computer graphic package Poser from Curious Labs.

6.1   Single Image

In Figure 1 results for a single pose estimate are shown. Each estimate is initialized using the 20 most likely points from the training data. The top row shows an ambiguous silhouette of the mis-labeling or limp flip type, followed by the model estimates sorted according to likelihood. We expect a mis-labeling ambiguity to correspond to a discrete set of poses with significantly different joint angle configurations. Only the three first estimates have a good correspondence to the silhouette, corresponding to significantly different joint configurations of both legs and arms as expected.

The bottom row show a silhouette corresponding to an out-of-plane ambiguity. This silhouette contains very little information about the position of the limbs moving out-of-plane, we can see that this is well reflected by the model, suggesting several different configuration of the legs and the arms. The least likely of the shown estimates is a heading ambiguity which is also a plausible estimate to the silhouette.

6.2   Sequence

In Figure 2 every $20th$ frame for a circular walk sequence is shown. Our model does well for most of the frames but misestimates one stride in a turn. This bad estimate is due to lack of training data, each turn in the training data from this position is taken with the opposite leg compared to the test data. Therefore the dynamical model does not agree with the observations and the estimated pose is a suboptimal minimum. The estimate waits until the stride with the "unexpected" leg is finished and then latches on in the correct stride. Outside this turn our model correctly estimates the true pose.

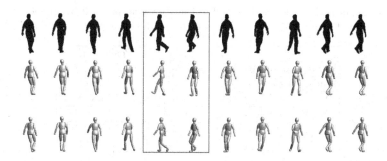

Fig. 2. Every 20*th* frame from a circular walk sequence, Top Row: Input Silhouette, Middle Row: Model Pose Estimate, Bottom Row: Ground Truth. The box indicates bad estimates by our model.

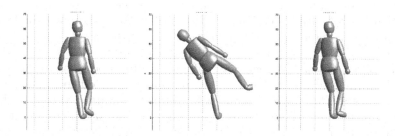

Fig. 3. Angle error: The image on the left is the true pose, the middle image has an angle error of 1.7°, the image on the right has an angle error of 4.1°. An angle error higher up in the joint-hierarchy will effect the positions for all joints further down. As the errors for the middle image are higher up in the hierarchy this will effect each limb connected further down the chain from this joint thereby resulting in a significantly different limp positions.

6.3   Quantitative Results

Results on Human Pose estimation are normally reported by the means square error between the estimated pose and ground truth [2,18]. A mean square error treats all dimension of the joint angle space with equal importance and do not reflect the hierarchical structure of the human physiology, Figure 3. Table 1 shows mean RMS angle and joint position error for our model, a set of regression algorithms and the mean pose in the training data over the test sequence. We can see that our single estimate is worse than RVM regression. This is expected as the multi-modal prediction in our model will either predict the correct pose or we find an ambiguous pose in these cases a regression based methods would predict the mean pose which will result in a smaller error. Neither angle or joint position error can correctly reflect visual similarity for sequences as humans have strong and complex priors with regards to motion.

Table 1. Mean RMS Angle and Joint Position Error normalized by the height of the model. Note that the only the GP-LVM Sequence method is using temporal information.

	Angle Error	Joint Position Error
Mean Training Pose	$8.3°$	$37.3 \cdot 10^{-2}$
Linear Regression	$7.7°$	$33.5 \cdot 10^{-2}$
RVM	$5.9°$	$15.8 \cdot 10^{-2}$
GP-LVM Single	$6.5°$	$17.2 \cdot 10^{-2}$
GP-LVM Sequence	$5.3°$	$15.0 \cdot 10^{-2}$

7   Discussion

The model presented in this paper learns a shared low-dimensional representation of a single observation space and a target domain. For the task of estimating pose from silhouette the information in the observation space in not sufficient to determine pose why we in this paper have used temporal consistency to disambiguate between multiple solutions. Another strategy is to incorporate additional observation spaces that together will better represent the target domain. A common example for the application presented is to use information from additional views, another example is combining information from both visual and audio cues. This leads to the problem of how to "merge" the different observation spaces into a single representation. In [8] features are extracted from multiple views and concatenated into a larger feature vector from which pose is inferred. In the presented model additional observations can simply be added and a shared low-dimensional representation can be learned of all the observation spaces and the target domain. An additional advantage with the model is that inference can be done if any non-zero subset of the observations are given, presenting additional observations will constrain the problem further.

8   Conclusion and Future Work

We have presented a method for human pose estimation from silhouettes using Gaussian Process Latent Variable Models. Our model represent both image observations and pose parameters in a shared latent space. The structure of the latent space is constrained to produce smooth trajectories over time by incorporating a GP to predict over the latent space. The model only has two free parameters and requires no manual initialization.

In future work we would like to extend the model to learn two separate sets of dynamics, one dynamic for the human relative motion (e.g. stride) and one model for the motion of the root of the body. This should hopefully solve the problems in our estimation and also reduced the amount of training data needed.

Acknowledgment

This work was supported by EPSRC, the IST Programme of the European Community, under the PASCAL Network of Excellence, IST-2002-506778 and

Sharp Laboratories Europe. We would like to thank Ankur Agarwal, Guido Sanguinetti and Nathaniel J. King.

References

1. Lawrence, N.D.: Probabilistic non-linear principal component analysis with gaussian process latent variable models. Journal of Machine Learning Research 6, 1783–1816 (2005)
2. Agarwal, A., Triggs, B.: Recovering 3d human pose from monocular images. IEEE Trans. Pattern Anal. Mach. Intell. 28(1), 44–58 (2006)
3. Grauman, K., Shakhnarovich, G., Darrell, T.: Inferring 3d structure with a statistical image-based shape model. In: ICCV 2003, pp. 641–648 (2003)
4. Kehl, R., Bray, M., Gool, L.J.V.: Full body tracking from multiple views using stochastic sampling. In: CVPR(2), pp. 129–136 (2005)
5. Sminchisescu, C., Kanaujia, A., Li, Z., Metaxas, D.N.: Discriminative density propagation for 3d human motion estimation. In: CVPR (1), pp. 390–397 (2005)
6. Sminchisescu, C., Telea, A.: Human pose estimation from silhouettes - a consistent approach using distance level sets. In: WSCG, pp. 413–420 (2002)
7. Sidenbladh, H., Black, M.J., Fleet, D.J.: Stochastic tracking of 3d human figures using 2d image motion. In: Vernon, D. (ed.) ECCV 2000. LNCS, vol. 1843, pp. 702–718. Springer, Heidelberg (2000)
8. de Campos, T.E., Murray, D.W.: Regression-based hand pose estimation from multiple cameras. In: CVPR(1), pp. 782–789 (2006)
9. Sun, Y., Bray, M., Thayananthan, A., Yuan, B., Torr, P.: Regression-based human motion capture from voxel data. In: BMVC (2006)
10. Belongie, S., Malik, J., Puzicha, J.: Shape context: A new descriptor for shape matching and object recognition. In: NIPS, pp. 831–837 (2000)
11. Mori, G., Belongie, S.J., Malik, J.: Efficient shape matching using shape contexts. IEEE Trans. Pattern Anal. Mach. Intell. 27(11), 1832–1837 (2005)
12. Rasmussen, C.E., Williams, C.K.: Gaussian Processes for Machine Learning. MIT Press, Cambridge (2006)
13. Lawrence, N.D.: Gaussian process latent variable models for visualisation of high dimensional data. In: NIPS (2003)
14. Lawrence, N.D., Candela, J.Q.: Local distance preservation in the gp-lvm through back constraints. In: ICML, pp. 513–520 (2006)
15. Wang, J., Fleet, D.J., Hertzmann, A.: Gaussian process dynamical models. In: NIPS (2005)
16. Shon, A.P., Grochow, K., Hertzmann, A., Rao, R.P.N.: Learning shared latent structure for image synthesis and robotic imitation. In: NIPS (2005)
17. Viterbi, A.J.: Error bounds for convolutional codes and an asymptotical optimum decoding algorithm. IEEE Transactions on Information Theory (1967)
18. Shakhnarovich, G., Viola, P.A., Darrell, T.: Fast pose estimation with parameter-sensitive hashing. In: ICCV, pp. 750–759 (2003)

Automatic Labeling Inconsistencies Detection and Correction for Sentence Unit Segmentation in Conversational Speech

Sébastien Cuendet[1], Dilek Hakkani-Tür[1], and Elizabeth Shriberg[1,2]

[1]International Computer Science Institute (ICSI)
1947 Center Street
Berkeley, CA 94704, USA
[2]Speech Technology and Research Laboratory, SRI International
333 Ravenswood Avenue
Menlo Park, CA 94025, USA
{cuendet,dilek,ees}@icsi.berkeley.edu

Abstract. In conversational speech, irregularities in the speech such as overlaps and disruptions make it difficult to decide what is a sentence. Thus, despite very precise guidelines on how to label conversational speech with dialog acts (DA), labeling inconsistencies are likely to appear. In this work, we present various methods to detect labeling inconsistencies in the ICSI meeting corpus. We show that by automatically detecting and removing the inconsistent examples from the training data, we significantly improve the sentence segmentation accuracy. We then manually analyze 200 of noisy examples detected by the system and observe that only 13% of them are labeling inconsitencies, while the rest are errors done by the classifier. The errors naturally cluster into 5 main classes for each of which we give hints on how the system can be improved to avoid these mistakes.

Keywords: Automatic relabeling, error correction, boosting, sentence segmentation, noisy data.

1 Introduction

Sentence segmentation from speech is part of a process that aims at enriching the unstructured stream of words output by automatic speech recognizers (ASR). The role of sentence segmentation is to find the sentence units in the stream of words output by the ASR. It is of particular importance for speech related applications, as most of the further processing steps, such as parsing, machine translation, information extraction, assume the presence of sentence boundaries [1,2].

Sentence segmentation can be seen as a binary classification problem, in which every word boundary has to be labeled as a sentence boundary or as a non-sentence boundary. In the usual learning task, when provided with data, one has to manually label a consequent amount of them to perform automatic learning.

A. Popescu-Belis, S. Renals, and H. Bourlard (Eds.): MLMI 2007, LNCS 4892, pp. 144–155, 2007.
© Springer-Verlag Berlin Heidelberg 2007

Speaker 1: `So is this OK with you?` (**question**)
Speaker 2: `Yes` (**statement**) `but I do-` (**disruption**)
Speaker 1: `Come on` (**floor grabber**)
          `I want this very much` (**statement**)
Speaker 2: `Uh huh` (**backchannel**)
Speaker 1: `And I want ...` (**statement**)

Fig. 1. Example of a dialog along with dialog acts

`but the phrase is not part of the sentence. and neither is`
`the sentence part of the phrase.`

Fig. 2. Example of a dialog along with dialog acts

In this work, we focus on sentence segmentation for conversational speech from the ICSI meeting corpus, which has been manually labeled with 5 dialog acts: statement, question, backchannel, floor-grabber/holder, incomplete. Backchannels are short phrases such as *yeah* or *uh huh* to indicate that the listener is actually following the speaker. Floor grabbers indicate that the person wants to start talking; similarly floor holders indicate that the speaker has not yet finished. Disruptions (also called incompletes) stand for statements that remain uncompleted for some reason. Figure 1 shows an example of a dialog along with dialog acts.

For the sentence segmentation, we merge all DAs into one class, the "sentence" class, and the goal of the classification is to find the correct locations for the beginning and end of each sentence unit. It is therefore crucial that the DAs have been consistently labeled beforehand. Consistent labeling is however not always guaranteed, since labels are attributed by humans who make mistakes because of the difficulty of DAs labeling in conversational speech. Indeed, conversational speech comprises incomplete and gramatically incorrect sentences which make some candidate boundaries likely to be labeled as sentence boundary as well as non-sentence boundary[1]. Therefore, additionally to the inter-labeler inconsistencies due to a possible different interpretation among the labelers, the complexity of the task leads to inconsistencies. Figure 2 shows a case where the labeler has labeled the word boundary after the word *sentence* as the end of a statement, but another labeler might as well have not inserted anything and just considered the whole example as one statement. Such inconsistencies in the labeling might confuse the classifier and decrease the sentence segmentation accuracy.

In this paper, we study four approaches to automatically detect these ambiguous or wrongly labeled examples. The first approach is based on a committee decision, the second one is based on the confidence attributed by the classifier to each instance, and the two last methods use the weights and edges measures of the learning algorithm used, AdaBoost. We show that the sentence segmen-

---

[1] More details about the labeling can be found in the guidelines that were given to the labelers in [3].

tation accuracy significantly increases when we remove the noisy examples from the training data, whereas relabeling them does not increase the performance much.

The rest of the paper is structured as follows: in the next section, we describe Boosting, the learning algorithm that we used and review the work done in automatic noise detection. In Section 3, we describe our four approaches to detect noisy examples. The results are presented in Section 4, and discussed in Section 5.

2   The Boosting Algorithm and Related Work

To perform the binary classification task of sentence segmentation, we use the AdaBoost.MH[2] algorithm introduced by Schapire and Singer [4], since it has been shown to be among the best classifiers for the sentence segmentation task [5]. Boosting is an iterative procedure that builds a new weak learner h_t at each iteration. Every instance of the training data set is assigned a weight. These weights are initialized uniformly and updated on each iteration so that the algorithm focuses on the instances that were wrongly classified on the previous iteration. At the end of the learning process, the weak learners used on each iteration t are linearly combined to form the classification function:

$$f(x, l) = \sum_{t=1}^{T} \alpha_t h_t(x, l)$$

with α_t the weight of the weak learner h_t and T the number of iterations of the algorithm, x the example to classify and l the label, with $l \in \mathcal{L}$. The label l with highest score $f(x, l)$ is attributed to x. More details on Boosting can be found in [6].

Noisy data has always been a problem in the field of statistical learning. Noise can arise from various sources, such as imprecision or error in the measurement, and labeling errors. Multiple approaches have been tried to identify the noisy instances. A method based on a committee of classifier has been successfully introduced for spoken language understanding in [7]. In [8], E. Eskin presents a technique to detect anomalies and applies it to network intrusion detection. The main idea is to consider two sets of data A and B with corresponding distributions D_A and D_B, one for the regular instances (A) and one for the anomalies (B). At the beginning, all instances belong to A. Each instance is then removed from A and added to B and D_A and D_B are recomputed. The difference between the log likelihood before and after the exclusion of the instance decides if the instance should be moved to B or kept in A. This approach can be used with any statistical classifier that gives an estimation of the distributions D_A and D_B. Other approaches specific to Boosting have also been tried. In [9], the authors suggest to use the weights over the instances at the end of the training

---

[2] As is commonly done in the literature, we abusively use the term "Boosting" in this paper to designate the AdaBoost.MH algorithm.

in Boosting to detect the mislabeled instances in part-of-speech tagging. The assumption is that instances that have been wrongly annotated are hard to classify, and thus have a high weight at the end of the training phase. A similar approach is used in [10], where the instances are selected according to their edge value instead of their weight. A detailed presentation of the weight and edge measures is provided in the next section. An interesting approach is presented in [11], where the weights of the attributes as well as the weights of the instances are used to detect the noisy data. The approach is evaluated on endowment insurance records and to our knowledge has not been used on other test sets, which makes it difficult to compare to other methods.

While all the methods introduced above use various measures of Boosting, Oza slightly modifies the Boosting algorithm in order to make it more robust to noise [12]. The main idea in this algorithm is to average the new distribution of the instance weights with the distributions of the previous iterations. Averaging the weights has a regularizing effect which leads to a highest training error bound, but a better generalization error bound. Dealing with the noisy examples is thus done implicitly by the classifier, while all other methods require a post-processing of the noisy instances, such as removing them from the training set or relabeling them.

3   Approach

In this section, we present four methods to detect the noisy examples in the training set. Once the noisy examples have been detected, we can either remove them from the training set or automatically relabel them. Relabeling is especially trivial in the case of binary classification, since if an example does not belong to the sentence boundary class, it belongs to the non-sentence boundary class, and inversely.

In the following description, we assume a data set D of training instances, with $|D| = N$. Each example x_i in D is represented by a set of features and belongs to a class $y_i \in \mathcal{Y}$ that has been assigned by human labelers and to which we refer as the true class. $\mathcal{Y} = \{s, n\}$ is the set of possible classes with s the class of examples which are sentence boundaries and n the class of examples that are non-sentence boundaries. The Boosting algorithm described in the previous section is used to output a probability $p(s|x_i)$ for each example x_i to belong to the class s. If the $p(s|x_i)$ is larger or equal to a threshold T, x_i is attributed the class s, i.e. declared as a sentence boundary, otherwise it gets class n, i.e. declared as a non-sentence boundary.

3.1   Committee-Based Method

The training set D is split into k mutually exclusive data sets d_j of size N/k each. A classifier c_j is trained on each of the reduced data sets d_j. The k classifiers c_j are then used to evaluate each example in D. Therefore, for each instance of the original data set D, we now have k votes. An example x_i is defined as noisy when all k classifiers c_j agree on a class y_i', and y_i' is different from the true class y_i.

We describe 3 variants of this method. One of them is to exclude an example x_i if k' classifiers agree on a class $y_i' \neq y_i$, where $k' < k$. This is a less strong excluding condition and is thus more likely to remove non-noisy examples. Another variant is to remove only the examples whose true class is n whereas the k classifiers have agreed on class s (false positives). The motivation behind this variant is that labelers are more likely to forget to label an instance as a sentence boundary, while it truly is a sentence boundary, than to add sentence boundaries where there is no reason to have sentence boundaries. The last variant is to use a variable threshold T. Optimizing the threshold for each of the k classifier would however be computationally too expensive and we therefore only use $T \in \{0.3, 0.5\}$.

3.2   Confidence-Based Method

The complete training set D is first used to train a classification model M_1. The model M_1 is then used to estimate the class of each example x_i in D. The noisy examples are those that have a true class y_i but are assigned a class y_i' by the classification model M_1, where $y_i \neq y_i'$ and the probability $p(y_i'|x_i)$ assigned to the class y_i' for example x_i is larger than a threshold Z optimized on the held-out set. In Section 4, we present 3 variants of the experiment: one where all the detected noisy examples are excluded, and two where the false negatives (resp. false positives) are relabeled. Note that the confidence-based method is a special case of the committee-based method, where $k = 1$ and the true class in the detecting phase is determined with an optimized threshold.

3.3   Boosting Weights Method

This method is based on the observations done in [9] and uses the weights attributed by Boosting to each training instance. We use a simplified version (since sentence segmentation has only two classes) of the original weight update function described in [4]:

$$W_{t+1}(i) = \frac{W_t(i)exp(-h_t(x_i) \cdot Y[i])}{Z_t} \tag{1}$$

$$Z_t = \sum_{i=1}^{N} W_t(i)exp(-h_t(x_i) \cdot Y[i])$$

where $W_t(i)$ is the weight of instance x_i at iteration t, and

$$Y[i] = \begin{cases} +1, \text{ if } y_i = s \\ -1, \text{ if } y_i = n. \end{cases} \tag{2}$$

Thus, if the current rule h_t classifies the example x_i incorrectly, the next weight $W_{t+1}(i)$ will increase, otherwise it will decrease. To decide which examples are noisy, we sort all examples according to their weight at the end of the training and declare the top X examples as being noisy.

The parameters are the number X of excluded examples, as well as the number of iterations used to train the classifier. If we train for too many iterations, we take the risk of having the examples that are noisy having their weights decreased because the final rules created by Boosting after several iteration eventually classify them correctly. On the other hand, if we train for too few iterations, the weights of the noisy examples might not have had the chance to increase enough compared to those of the regular examples.

3.4   Boosting Edges Method

The definition of edges has first been introduced by Breiman [13], and used to detect noisy data in [10]. The edge value edge$_i$ of an instance x_i at iteration t is the *total weight assigned to x_i by all h_t that misclassified x_i up to iteration t*:

$$\text{edge}_i = \sum_{t=1}^{T} h_t(x_i)I_t(x_i) \tag{3}$$

$I_t(x_i)$ is the following indicator function, where $[h_m(x_i)]$ is the class assigned by h_m to x_i:

$$I_m(x_i) = \begin{cases} 0, \text{ if } [h_m(x_i)] = y_i \\ 1, \text{ if } [h_m(x_i)] \neq y_i \end{cases} \tag{4}$$

Note that the edge values are always positive since the chosen class is by definition the one that has a positive weight in a two-class Boosting problem.

In [10], Wheway suggests to declare as noisy the 5% instances with highest edge value after 10-20 iterations. She however does not evaluate her suggestion and although we think this approach is reasonable, we will experimentally show that the percentage of instances declared as noisy, as well as the number of iterations after which the edge values are computed, are both parameters that have to be optimized.

4   Experiments and Results

Data Sentence segmentation is performed on conversational speech, which comes from the ICSI meeting corpus (MRDA) [14]. This corpus contains 73 meetings which are grouped in three main types (according to the speakers, the conversations type, etc.). We use the same split of training, test and held-out set as specified in [15], i.e. 51 meetings for the training set, 11 meetings for the test set and 11 meetings for the held-out set. More details about the data are shown in Table 1. We use the manual transcriptions of the meetings and feed the classifier with both lexical and prosodic features, for a total of 34 features. The prosodic features are various measures of the pitch, energy and pause duration across the boundary of interest. The lexical features are unigrams, bigrams, and trigrams formed with the words surrounding the word boundary of interest. More details on the features can be found in [5].

Table 1. Data characteristics of the MRDA corpus. Sizes and sets are given in number of words.

Training set size (words)	538,956
Test set size	101,510
Held-out set size	110,851
Vocabulary size	11,034
Average utterance length	6.54

Metrics. To measure the performance of the sentence segmentation, we use the F-measure and the NIST-SU error. The F-measure is the harmonic mean of the recall and precision measures of the sentence boundaries hypothesized by the classifier to the ones assigned by human labelers. The NIST-SU error rate is the ratio of the number of wrong hypotheses made by the classifier to the number of reference sentence boundaries. So if no boundaries are marked by sentence segmentation, it is 100%, but it can exceed 100%; the maximum error rate is the ratio of number of words to the number of correct boundaries.

4.1   Results

We now report the results for each of the methods introduced in Section 3. The baseline was obtained by training the classifier on the entire training set and evaluating the classification accuracy on the test set, with parameters optimized on the held-out set. The baseling settings yielded a 81.7% F-Measure and a 35.6% NIST-SU error rate. For each of the methods described above, we present results obtained by optimizing the parameters on the held-out set.

For the committee-based method, we tried values 8, 9 and 10 for k, values 0.3 and 0.5 for the Boosting threshold, and for each of the settings, we tried to exclude only the noisy examples that the labelers labeled with class n. The results of the 2 best settings on the held-out set for each value of k are shown in Table 2.

For the high confidence disagreement method, the optimal value for the threshold Z on the held-out set was 0.6. Table 3 shows the results when we excluded noisy examples or relabeled a subset of them.

For the weights and the edges experiment, we trained Boosting for M iterations, with $M \in \mathcal{M} = \{10, 20, 50, 100, 200, 300, 400, 500, 1000\}$ iterations. For each $M \in \mathcal{M}$ iteration, we removed the X examples with the top weight (resp. edge) score, with $X \in [1000, 2000, ..., 10000]$, and report the results in Tables 4 and 5 for the number of excluded examples X that yielded the best result on the held-out set. Note that when several examples had the same score and they were at the border of the X top examples, we excluded all examples that had the exact same value as the Xth example.

All presented methods outperformed the baseline with optimized parameters. The overall improvement can look small, but an F-Measure above 81.94% and a NIST error under 35.30% are both statistically significant improvements according to a Z-test with 95% confidence range. The overall best performance was

Table 2. Results for the committee-based method. The first column shows the number of votes required to tag an example as noisy, the second shows the threshold that distinguishes between the two classes n and s; the 3 last columns show the result for the setting of the 3 first columns: number of examples excluded and accuracy according to the F-Measure and the NIST error.

k	Ths.	# Noisy (% of total)	F-Meas.	NIST err.
Baseline		0	81.71	35.59
10	0.5	10,786	82.10 (2.0)	35.07
10	0.3	9,363	82.11 (1.7)	34.99
9	0.5	21,572	81.66 (4.0)	35.01
9	0.3	18,726	82.09 (3.5)	34.90
8	0.5	22,465	81.96 (4.2)	34.63
8	0.3	22,016	82.35 (4.1)	34.32

Table 3. Results for the confidence-based method. The first row shows the standard case described in the text, while for the results in rows 2 and 3, only the examples with true class s (resp. n) were kept, while examples from the other class were relabeled.

Processing	# Noisy (% of total)	F-Meas.	NIST err.
Baseline	0	81.71	35.59
Exclude all noisy	22,951 (4.3)	82.00	34.6
Relabel false positive	-	82.00	34.7
Relabel false negative	-	82.10	34.7

Table 4. Results for the weights methods. The first column shows the number of iterations after which the weight values are measured, the second column indicates the number of examples tagged as noisy and the last two columns report the sentence segmentation accuracy according to the F-Measure and the NIST error.

Iterations	# Noisy (% of total)	F-Meas.	NIST err.
Baseline	0	81.71	35.59
10	1,000	81.34 (0.2)	35.48
20	1,000	81.82 (0.2)	35.45
50	1,000	81.66 (0.2)	35.66
100	6,000	81.76 (1.1)	35.27
200	6,000	81.94 (1.1)	35.24
300	7,000	81.74 (1.3)	35.15
400	8,000	81.95 (1.5)	35.02
500	8,000	81.97 (1.5)	34.87
1000	10,000	81.83 (1.9)	34.87

obtained by the committee-based method with $k = 8$ and improved the baseline of 0.7% absolute for the F-Measure and 1.3% absolute for the NIST error. In some settings, the F-Measure was lower than for the baseline, as opposed to the NIST error which was always better than the baseline, which means that in any

Table 5. Results for the edges methods. The first column shows the number of iterations after which the edge values are measured, the second column indicates the number of examples tagged as noisy and the last two columns report the sentence segmentation accuracy according to the F-Measure and the NIST error.

Iterations	# Noisy (% of total)	F-Meas.	NIST err.
Baseline	0	81.71	35.59
10	2,000 (0.4)	81.52	35.31
20	1,000 (0.2)	81.84	35.31
50	3,000 (0.6)	82.00	34.95
100	4,000 (0.7)	82.18	34.45
200	5,000 (0.9)	82.03	34.68
300	6,000 (1.1)	81.94	35.06
400	8,000 (1.5)	81.67	35.16
500	10,000(1.9)	81.83	35.04
1000	6,000 (1.1)	82.00	34.72

of these settings, the number of wrong word boundary predictions done by the new classifier was lower than in the baseline.

The optimal parameters used for the edges method was different than those in [10], where the author suggests to stop the training after 10-20 iterations and to exclude the top 5% examples. Our optimal solutions used 100 iterations and excluded less than 2% of the examples.

Removing vs. Relabeling Examples. The methods presented above determine which examples were considered as noisy but not how to handle them. Once we have detected noisy examples, we can either remove them from the training set, or we can try to automatically relabel them. Since the sentence segmentation problem is a binary classification problem, relabeling is straightforward: noisy examples originally labeled with class s are changed to class n and vice versa. However, in all of our experiments, we observed that although better than the baseline, the performance obtained after relabeling the noisy examples was lower or equal to the one obtained by simply excluding them. One explanation for this is that the noisy relabeled examples do not bring much new knowledge to the classifier, while examples that were correctly labeled and detected as noisy add noise to the data when they are relabeled.

5   Discussion

In the previous section, we have shown that the sentence segmentation accuracy improves when we exclude the noisy examples. While this is already a valuable result, we believe there is more knowledge to extract from the noisy examples. In the rest of this section, we examine the noisy examples for the committee-based method with $k = 10$ and the exclusion of examples from the two classes.

In this setting, the system detected 10, 786 noisy examples; 23% of them are instances where the system introduced an additional sentence boundary, while the remaining 77% are sentence boundaries that the system missed.

Among all the examples whose true class is a sentence boundary, we observe that only 0.32% of the backchannels are noisy, 15.87% of the incompletes, 10.66% of the statements, 12.99% of the floor-grabbers/holders and 11.95% for the questions. This confirms the intuition that disruptions are the most difficult cases to label.

One possible source of errors in this work could be human mistakes in assigning original dialog act boundaries. To explore this possibility, and further understand errors made by the system, a researcher familiar with the original dialog act annotation project hand-analyzed 200 randomly drawn errors using transcripts only, but with information about reference human punctuation labels, including disfluency markers and markers for incomplete sentences. Speech from other talkers was also interspersed in the transcripts, and the length of pauses was supplied. Of the original 200 examples, 10% were found difficult to understand from transcripts alone; the analysis thus refers to the remaining 178 samples. Of this set, only 13% were errors in the original human boundary labels, with a nearly even split between missed boundaries and false alarms. Because the analysis looks only at errors to begin with, this rate of human labeling error is tolerable (although to estimate it properly would also require determining the rate of felicitous correct machine decisions due to erroneous human labels). The remaining of the 178 cases were deemed to have correct human boundary labels.

The analysis becomes more interesting as we look at the remaining errors, all attributable to the system. Percentages are given as the percentage of the 178 original cases referred to above. Over half (54%) of the remaining errors fell into one of five groups. The first group, at 15%, had either a false start or incomplete sentence preceding the boundary of interest. In a two-way classification of boundaries there is no good way to group such cases, since to the left of the disruption they reflect no boundary, but to the right of the disruption they begin a new sentence and thereby suggest a boundary. To handle such cases explicitly, one would need to train specific models for this third boundary type. The second group, at 14%, comprises boundaries directly following filled pauses or discourse markers. Considering floor-grabbers/holders as full sentence boundaries, as explained in Section 1, is certainly the cause of this second class of errors. Since these boundaries are not per se sentence boundaries, one way of dealing with them would be to simply consider them as non-sentence boundaries or to treat them as a separate class. The third class of errors, which really should not be counted as errors at all, are ambiguous examples in which a human would have trouble assigning boundaries. An example of such a case is shown in Figure 3, where the word boundary after the word *document* was labeled as a statement, but considering it as a non-sentence boundary would clearly be correct too. Fourth, boundaries after questions accounted for 9% of errors. It is likely that the model suffers here both because question prosody often leaves

```
... that's relative to the structure of the x. m. l. document. (0.0)
not to the structure of what you're representing ...
```

Fig. 3. Example tagged as noisy by the committee-based method. The parenthesis indicate the length of the pause between the 2 words document and not.

pitch high, unlike the majority of boundaries occurring for statements that show final falls. The language model may also have trouble with questions, which can end in verbs or other syntactic classes that are unusual for sentence ends in statements. Finally, about 5% of cases occurred at subsentential locations that in text should contain a colon or semicolon. Such errors can be viewed similarly to the errors made for disfluency boundaries: the subsentential boundaries really belong somewhere in between the no-boundary and boundary classes. The remaining machine errors (at 33% of the 178 samples) had no obvious cause. Within this set, missed boundaries were three times as likely as false alarms. We can balance this ratio by setting the threshold to less than 0.5 and thus globally detect more sentence boundaries, and this 3 to 1 ratio is thus not a general rule and depends on the optimal threshold.

6   Conclusion

We presented four automatic methods to detect labeling error for automatic sentence segmentation. Although tested only on the ICSI meeting corpus, the methods can be applied to other conversational speech data, such as broadcast conversation and telephone conversations. We showed that the sentence segmentation accuracy improved when the noisy examples were first excluded from the training set, with either of the four methods. Relabeling the noisy instead of excluding them did not further improve the performance. We analyze 200 noisy examples: 13% were found to be labeling errors, 54% were errors done by the system that we could explain and that could be clustered into 5 main classes, and the rest of them were errors done by the system for which there was no clear explanation.

Further work will consist of using the knowledge extracted from the noisy examples to improve the sentence segmentation accuracy. Significant improvement can especially be obtained by focusing particularly on incompletes and questions detection.

Acknowledgments. We want to thank Mathew Magimai Doss for many helpful discussions. This work was partly supported by the Swiss National Science Foundation through the research network IM2 and Defense Advanced Research Projects Agency (DARPA) GALE (HR0011-06-C-0023). Any opinions, findings, and conclusions or recommendations expressed in this material are those of the authors and do not necessarily reflect the views of the funders.

References

1. Mrozinski, J., Whittaker, E.W.D., Chatain, P., Furui, S.: Automatic sentence segmentation of speech for automatic summarization. In: Proc. ICASSP, Philadelphia, PA (2005)
2. Makhoul, J., Baron, A., Bulyko, I., Nguyen, L., Ramshaw, L., Stallard, D., Schwartz, R., Xiang, B.: The effects of speech recognition and punctuation on information extraction performance. In: Proc. of Interspeech, Lisbon (2005)
3. Shriberg, E., Dhillon, R., Bhagat, S., Ang, J., Carvey, H.: The ICSI meeting recorder dialog act (MRDA) corpus. In: Proc. SigDial Workshop, Boston, MA (2004)
4. Schapire, R.E., Singer, Y.: BoosTexter: A boosting-based system for text categorization. Machine Learning 39(2/3), 135–168 (2000)
5. Zimmermann, M., Hakkani-Tür, D., Fung, J., Mirghafori, N., Shriberg, E., Liu, Y.: The ICSI+ multi-lingual sentence segmentation system. In: Proc. ICSLP, Pittsburgh, PA (2006)
6. Schapire, R.: The boosting approach to machine learning: An overview. In: MSRI Workshop on Nonlinear Estimation and Classification, Berkeley, CA (2001)
7. Tur, G., Rahim, M., Hakkani-Tür, D.: Active labeling for spoken language understanding. In: Proceedings of EUROSPEECH, Geneva, Switzerland (2003)
8. Eskin, E.: Anomaly detection over noisy data using learned probability distributions. In: Proc. 17th International Conf. on Machine Learning, pp. 255–262. Morgan Kaufmann, San Francisco (2000)
9. Abney, S., Schapire, R., Singer, Y.: Boosting applied to tagging and pp attachment. In: Proceedings of the Joint SIGDAT Conference on Empirical Methods in Natural Language Processing and Very Large Corpora (1999)
10. Wheway, V.: Using boosting to detect noisy data. In: Mizoguchi, R., Slaney, J.K. (eds.) PRICAI 2000. LNCS (LNAI), vol. 1886, pp. 123–132. Springer, Heidelberg (2000)
11. Liu, X.-D., Shi, C.-Y., Gu, X.-D.: A boosting method to detect noisy data. In: Proc. of the Fourth International Conference on Machine Learning and Cybernetics, Guangzhou, China (2005)
12. Oza, N.C.: Aveboost2: Boosting for noisy data. In: Fifth International Workshop on Multiple Classifier Systems, Cagliari, Italy, June 2004, pp. 31–40. Springer, Heidelberg (2004)
13. Breiman, L.: Arcing the edge. Technical report, Statistics Department, UC Berkeley (1997)
14. Janin, A., Ang, J., Bhagat, S., Dhillon, R., Edwards, J., Macias-Guarasa, J., Morgan, N., Peskin, B., Shriberg, E., Stolcke, A., Wooters, C., Wrede, B.: The ICSI meeting project: Resources and research. In: Proceedings of ICASSP, Montreal (2004)
15. Ang, J., Liu, Y., Shriberg, E.: Automatic dialog act segmentation and classification in multiparty meetings. In: Proc. ICASSP, Philadelphia, PA (2005)

Term-Weighting for Summarization of Multi-party Spoken Dialogues

Gabriel Murray and Steve Renals

University of Edinburgh, Edinburgh, Scotland
gabriel.murray@ed.ac.uk, s.renals@ed.ac.uk
http://www.cstr.ed.ac.uk

Abstract. This paper explores the issue of term-weighting in the genre of spontaneous, multi-party spoken dialogues, with the intent of using such term-weights in the creation of extractive meeting summaries. The field of text information retrieval has yielded many term-weighting techniques to import for our purposes; this paper implements and compares several of these, namely *tf.idf*, Residual IDF and *Gain*. We propose that term-weighting for multi-party dialogues can exploit patterns in word usage among participant speakers, and introduce the *su.idf* metric as one attempt to do so. Results for all metrics are reported on both manual and automatic speech recognition (ASR) transcripts, and on both the ICSI and AMI meeting corpora.

1 Introduction

The primary focus of this research is to create extractive summaries of meeting speech, in order to present users with concise and informative overviews of the content of meetings. Such extractive summaries, when incorporated into a meeting browser, can act as efficient tools for navigating meeting records as a whole. This paper focuses on one fundamental component of the extractive summarization pipeline: the way that terms are weighted within a given meeting, and the bearing that various term-weighting schemes have on extraction performance.

Choosing and implementing a term weighting method is often the first step in building an automatic summarization system. Though the unit of extraction may be the sentence or the dialogue act, those units need to be weighted by the importance of their constituent words. Popular text summarization techniques such as Maximal Marginal Relevance (MMR) and Latent Semantic Analysis (LSA) begin by representing sentences as vectors of term weights. There is a wide variety of term weighting schemes available, from simple binary weights of word presence/absence to more complex weighting schemes such as *tf.idf* and *tf.ridf*. Several of these are described in the following section.

A central question of this paper is whether term-weighting techniques developed for information retrieval (IR) and summarization tasks on text are well-suited for our domain of multiparty spontaneous spoken dialogues, or whether the patterns of word usage in such dialogues can be exploited in order to yield superior term-weighting for our task. To this end, we devise and implement a novel

A. Popescu-Belis, S. Renals, and H. Bourlard (Eds.): MLMI 2007, LNCS 4892, pp. 156–167, 2007.

term-weighting approach for multi-party speech called *su.idf*, based on differing word frequencies among speakers in a meeting. This metric is compared with 3 popular term-weighting schemes - *tf.idf*, *ridf* and *Gain* - and the metrics are evaluated via an extractive summarization task on both AMI and ICSI corpora.

2 Previous Term Weighting Work

Term weighting methods form an essential part of any IR system. Terms that characterize a given document well and discriminate the document from the remainder of the document collection should be weighted highly [1]. The most popular term weighting schemes have therefore combined *collection frequency* metrics with *term frequency* metrics. The latter component measures the term's prevalence in the document at hand while the former component analyzes the term usage across many documents.

The most common method of calculating collection frequency is called the *inverse document frequency* (IDF) [2]. The IDF for term t is given by

$$IDF(t) = -\log(\frac{D_w}{D})$$

or equivalently,

$$IDF(t) = \log D - \log D_w$$

where D is the total number of documents in the collection and D_w is the number of documents containing the term t. A term will therefore have a high IDF score if it is rare across the set of documents.

For the *term frequency* component, the simplest method is a binary term weight: 0 if the term is not present and 1 if it is. More commonly, the number of term occurrences in the document is used. Thus the term frequency TF is given by

$$TF(t, d) = \frac{N(t)}{\sum_{k=1}^{T} N(k)}$$

where $N(t)$ is the number of times the term t occurs in the given document and $\sum_{k=1}^{T} N_k$ is the total word count for the document, thereby normalizing the term count by document length.

The classic method for combining these components is simply *tf.idf* [1], wherein a term is scored highly if it occurs many times within a given document but rarely across the set of all documents. This term weighting scheme *tf.idf* increases our ability to discriminate between the documents in the collection. While there are variants to the TF and IDF components given above [1], the motivating intuitions are the same. Another example of combining these tree types of data (collection frequency, term frequency and document length) is given by Robertson et al [3] and is called the Combined Weight. For a term t(i) and document d(j), the Combined Weight is described as:

$$CW(t, d) = \frac{IDF(t) \cdot TF(t, d) \cdot (K + 1)}{K \cdot ((1 - b) + (b \cdot (NDL(d)))) + TF(t, d)}$$

where K is a tuning constant regulating the impact of term frequency, b is a tuning constant regulating the impact of document length, and NDL is the normalized document length.

When relevance information is available, i.e. a subset of documents has been determined to be relevant to a user query, additional proven metrics are available for term relevance weighting and/or query expansion [3]. One example is the RSJ metric given in [4]:

$$RSJ(t,q) = \log \frac{\left(\frac{r}{R-r}\right)}{\left(\frac{n-r}{N-n-R+r}\right)}$$

where R is the number of documents known to be relevant to the query q and r is the number of relevant documents containing term t. The following variation is sometimes used instead, partly to avoid infinite weights under certain conditions:

$$RW(t,q) = \log(\frac{((r+0.5)(N-n-R+r+0.5))}{((n-r+0.5)(R-r+0.5))})$$

It is often the case, however, that there is little or no relevance information available when doing term weighting. Work by Croft and Harper [5] has shown that IDF is an approximation of the RSJ relevance weighting scheme when complete relevance information is unavailable. Robertson [6] further discusses the relationship between IDF and relevance weighting and places the IDF scheme on strong theoretical ground.

One extension of IDF called *ridf* [7] has proven effective for automatic summarization [8] and named entity recognition [9]. In *ridf*, the usual IDF component is substituted by the difference between the IDF of a term and its expected IDF according to the poisson model. The *ridf* score can be calculated by the formula

$$expIDF = -\log(1 - e^{(-f_w/D)})$$

$$ridf = IDF - expIDF$$

where f_w is the frequency of the word across all documents D.

Papineni [10] also provides an extension to IDF. Arguing that the IDF of a word is not synonymous with the *importance* of a word, but is rather an optimal weight for document self-retrieval, Papineni proposes a term-weighting metric *Gain* which is meant to measure importance or information gain of the term in the document:

$$Gain = \frac{D(t)}{D}(\frac{D(t)}{D} - 1 - \log \frac{D(t)}{D})$$

Very common and very rare words have low gain; this is in contrast with IDF, which will tend to give high scores to uncommon words. *ridf* also favors medium-frequency words [8]. As Papineni points out [10], the effective performance of metrics such as *ridf* and *Gain* seems to corroborate Luhn's observation that medium-frequency words have the optimal "resolving power" [11].

Mori et al [12] introduce a term weighting metric for automatic summarization called Information Gain Ratio (IGR). The underlying idea of IGR is that documents are clustered according to similarity, and further grouped into sub-clusters. If the information gain of a word increases after clusters are partitioned into sub-groups, then it can be said that the word contributes to that sub-cluster and should thus be rated highly.

Finally, Song et al [13] introduce a term weighting scheme for automatic summarization that is based on lexical chains. Building lexical chains in the manner of Barzilay [14], they weight chains according to how many word relations are in the chain, and weight each word in a chain according to how connected it is in the chain. On DUC 2001 data, they reported outperforming *tf* and *tf.idf* weighting schemes.

3 Term-Weighting for Meeting Speech

A common theme of most of the term-weighting metrics described in the previous section is that the distribution of words across a collection of documents is key to determining an ideal weight for the words. In general, words that are unique to a given document or cluster of documents should be weighted more highly than words that occur evenly throughout the entire document collection. For multiparty spoken dialogue, we have another potential source of variation in lexical usage: the speakers themselves. We introduce a new term weighting score for multi-party spoken dialogues by also considering how term usage varies across speakers in a given meeting. The intuition is that keywords will not be used by all speakers with the same frequency. Whereas IDF compares a given meeting to a set of all meetings, we can also compare a given speaker to a set of other speakers in the meeting. For each of the four speakers in a meeting, we calculate a surprisal score for each word that speaker uttered, which is the negative log probability of the term occuring amongst the other three speakers. The surprisal score for each word w uttered by speaker s is

$$surp(s, w) = -\log \left(\frac{\sum_{s' \neq s} tf(w, s')}{\sum_{r \neq s} N(r)} \right)$$

where $tf(w,s')$ is the term frequency of word w for speaker s' and $N(r)$ is the total number of words spoken by each speaker r. For each term, we total its speaker surprisal scores and divide by the total number of speakers to find the overall surprisal score $surp(w)$. Thus the surprisal score for a word is given by

$$surp(w) = \frac{1}{S} \sum_s surp(s, w)$$

This surprisal score, the first component of the term-weighting metric, is then multiplied by $\frac{s(w)}{S}$, where $s(w)$ is the number of speakers who speak that word and S is the total number of speakers in the meeting. The third component of the metric is the inverse document frequency, or *idf*. The equation for *idf* is

$$idf(w) = -\log(\frac{D_w}{D})$$

where D is the total number of documents and D_w is the number of documents containing the term w. Putting these three components together, our term weighting metric is

$$su.idf = surp(w) \cdot \frac{s(w)}{S} \cdot \sqrt{idf}$$

One motivation for this novel term weighting scheme is that many important words in such meeting corpora are not necessarily rare across all documents, e.g. *cost, design* and *colour*. They are also not necessarily the most frequent content words in the meetings. They would therefore not score highly on either component of . Though we retain inverse document frequency for our new metric, the square root of *idf* is used to lower its overall influence within the metric, so that a term will not necessarily be weighted low if it is fairly common or weighted high simply because it is rare. Results on the development and test sets show a significant improvement by using the square root of *idf* rather than *idf* itself.

The hypothesis is that more informative words will be used with varying frequencies between the four meeting participants, whereas less informative words will be used fairly consistently by all. The component $\frac{s(w)}{S}$ is included for two reason. First, because individuals normally have idiosyncrasies in their speaking vocabularies, e.g. one meeting participant might use a type of filled pause not used by the others or otherwise frequently employ a word that is particular to their idiolect. And second, a word that is used by multiple speakers but with much different frequency should be more important than a word that is spoken by only one person.

There are several reasons for hypothesizing that use of informative words will vary between meeting participants. One is that meeting participants tend to have unique, specialized roles relevant to the discussion. In the AMI corpus, these roles are explicitly labelled, e.g. "marketing expert." With a given role comes a vocabulary associated with that role, e.g. "budget" and "cost" would be associated with a finance expert and "scroll" and "button" would be associated with an interface designer. Second, even when the roles are not so clearly defined, different participants have different areas of interest and different areas of expertise, and we expect that their vocabularies reflect these differences.

4   Experimental Setup

In addition to *tf.idf* and *su.idf*, we also implemented Residual IDF (*ridf*) and *Gain* for comparison. A hybrid approach combining the rankings of *tf.idf* and *su.idf* was implemented in the hope that the two methods would be complementary, perhaps locating different types of informative terms. For all collection frequency measures, we used a collection of documents from the AMI, ICSI, Broadcast News and MICASE corpora. All term-weighting methods were run on both manual and ASR transcripts.

4.1 Data Description

We tested our term-weighting methods on the AMI and ICSI meeting corpora, which differ from one another in several important ways. The AMI meeting corpus [15] is a corpus of both scenario and non-scenario meetings, though for these experiments we used only scenario meetings. In these scenario meetings, four participants take part in each meeting and play roles within a fictional company. The scenario given to them is that they are part of a company called Real Reactions, which designs remote controls. Their assignment is to design and market a new remote control, and the members play the roles of project manager (the meeting leader), industrial designer, user-interface designer, and marketing expert. Through a series of four meetings, the team must bring the product from inception to market. The participants are also given real-time information from the company during the meetings, such as information about user preferences and design studies, as well as updates about the time remaining in each meeting. While the scenario given to them is artificial, the speech and the actions are completely spontaneous and natural.

The AMI test set consists of 19 meetings, or 4 sequences of 4 meetings each and 1 sequence of 3 meetings.

The second corpus used herein is the ICSI meeting corpus [16], a corpus of 75 natural, i.e. non-scenario, meetings, approximately one hour each in length. The ICSI test set consists of 6 meetings.

ASR for both corpora was kindly provided by the AMI-ASR group. The word-error rate (WER) for the AMI corpus is 43% while the WER for the ICSI corpus is 29.5%.

For both corpora, multiple human annotations were carried out for evaluation purposes. A human-authored abstract is created for each meeting, summarizing the most important aspects of the meeting in terms of decision, actions and goals of the meeting. Multiple human annotators then work through the meeting transcript and link dialogue acts to sentences in the human abstract when they find that a given dialogue act supports an abstract sentence. The result is a many-to-many mapping between dialogue acts and sentences in the abstract, so that a given dialogue act can be linked to more than one abstract sentence, and vice-verse.

4.2 Evaluation Protocol

For our evaluation, each term-weighting approach was used to create a brief summary of each test set meeting, and the resulting summaries were then evaluated. In each case we summed term-scores over dialogue acts to create scores for the dialogue acts, which are the summary extraction unit. Dialogue acts are ranked from most informative to least informative, and are extracted until a length of 700 words is reached. These summaries are then evaluated using the *weighted precision* metric originally introduced by Murray et al [17]. This metric is based on the multiple human annotations of dialogue act importance described above. Because each annotator creates a many-to-many mapping between dialogue acts and sentences within the human abstract, we can score each summary dialogue act according to how often each annotator linked it, and score the summary overall based on the constitutent dialogue act scores.

5   Results

The following sections detail the results on both the AMI and ICSI corpora.

5.1   AMI Results

On manual transcripts, the best approaches were *su.idf*, the hybrid approach combining *su.idf* and *tf.idf*, and *ridf*, all of which were significantly better than *tf.idf* (p>0.95) and not significantly different from one another, according to paired t-tests.

On ASR transcripts, *Gain* performed much better than it had on manual transcripts, with higher weighted precision results than the other approaches. *su.idf* also performed better on ASR, with its weighted precision increasing from 0.63 to 0.65. The hybrid approach slipped 2 points, while the *tf.idf* weighted

Table 1. Weighted Precision Results for AMI Test Set Meetings

Meet	sidf	sasr	tfidf	tfasr	com	comasr	ridf	ridfasr	gain	gainasr
ES2004a	0.50	0.51	0.50	0.59	0.55	0.55	0.59	0.64	0.63	0.63
ES2004b	0.59	0.67	0.58	0.55	0.59	0.60	0.67	0.65	0.64	0.69
ES2004c	0.66	0.63	0.69	0.64	0.76	0.67	0.76	0.71	0.59	0.67
ES2004d	0.69	0.75	0.85	0.77	0.99	0.78	0.77	0.79	0.78	0.85
ES2014a	0.67	0.70	0.68	0.71	0.67	0.71	0.70	0.76	0.65	0.73
ES2014b	0.76	0.81	0.74	0.70	0.86	0.72	0.79	0.75	0.77	0.83
ES2014c	0.74	0.78	0.69	0.67	0.88	0.77	0.83	0.80	0.69	0.71
ES2014d	0.51	0.40	0.44	0.40	0.48	0.44	0.43	0.36	0.33	0.43
IS1009a	0.85	0.73	0.68	0.72	0.69	0.73	0.74	0.78	0.74	0.73
IS1009b	0.65	0.83	0.50	0.68	0.57	0.70	0.65	0.73	0.57	0.78
IS1009c	0.50	0.52	0.34	0.36	0.46	0.42	0.44	0.45	0.44	0.56
IS1009d	0.74	0.60	0.73	0.58	0.81	0.71	0.75	0.69	0.58	0.50
TS3003a	0.53	0.50	0.48	0.48	0.54	0.52	0.57	0.60	0.63	0.61
TS3003b	0.63	0.73	0.64	0.59	0.57	0.55	0.68	0.67	0.70	0.72
TS3003c	0.89	0.93	0.89	0.87	0.86	0.90	0.80	0.79	0.80	0.92
TS3003d	0.46	0.54	0.41	0.51	0.46	0.54	0.59	0.56	0.63	0.63
TS3007a	0.37	0.54	0.35	0.54	0.37	0.52	0.50	0.57	0.51	0.57
TS3007b	0.62	0.61	0.57	0.54	0.66	0.56	0.67	0.62	0.59	0.59
TS3007c	0.70	0.64	0.61	0.48	0.64	0.60	0.55	0.57	0.60	0.73
AVERAGE	**0.63**	**0.65**	**0.60**	**0.60**	**0.65**	**0.63**	**0.66**	**0.66**	**0.62**	**0.68**

sidf=*su.idf* on manual, **sasr**=*su.idf* on ASR, **tfidf**=*tf.idf* on manual, **tfasr**=*tf.idf* on ASR, **com**=combined *su.idf* and *tf.idf* on manual, **comasr**=combined *su.idf* and *tf.idf* on ASR, **ridf**=residual IDF on manual, **ridfasr**=residual IDF on ASR, **gain**=*Gain* on manual, **gainasr**=*Gain* on ASR.

Table 2. Word Error Rates for Extracted (**Summ-WER**) and Non-Extracted Portions (**NonSumm-WER**) of Meetings

Meet	Summ-WER	NonSumm-WER
ES2004a	47.1	56.0
ES2004b	35.5	45.9
ES2004c	34.8	48.1
ES2004d	43.6	54.8
ES2014a	43.5	56.9
ES2014b	37.4	53.9
ES2014c	43.9	54.5
ES2014d	39.6	53.6
IS1009a	42.6	50.0
IS1009b	43.9	48.5
IS1009c	59.2	57.6
IS1009d	46.3	46.5
TS3003a	26.7	45.2
TS3003b	25.2	30.3
TS3003c	22.7	34.8
TS3003d	27.9	38.2
TS3007a	33.6	44.3
TS3007b	27.1	38.3
TS3007c	31.8	42.7
AVERAGE	**37.49**	**47.37**

precision scores stayed much the same. *Gain, su.idf* and *ridf* all performed significantly better than *tf.idf*. Table 1 gives results on both manual and ASR.

It was particularly surprising that some of the term-weighting approaches performed better on ASR than on manual transcripts. Previous research [18,19] has shown that informative portions of speech data tend to have lower word-error rates, but it is nonetheless unexpected that weighted precision would actually *improve* on errorful ASR transcripts. *Gain* and *su.idf* were particularly resilient to the errorful transcripts on this test set. Table 2 shows the word-error rates for the extracted and non-extracted portions of meetings using the *su.idf* summarizer. The WER for the extracted portions is nearly 10 points lower than for the non-extracted portions of meetings, at 37.49% versus 47.37%. The WER for the corpus as a whole is around 43.0%.

To get a better idea of how *su.idf* and *tf.idf* differ in the way they score and rank terms, and in particular why the performance gap increases on ASR, we plotted term-score against term-rank for both metrics on one of the AMI test set meetings, TS3003b. On manual transcripts, performance according to weighted precision was comparable for this meeting. However, on ASR transcripts weighted precision for *su.idf* increased by 10 points while the scores for *tf.idf* decreased by 5 points. As Figure 1 shows, the relationship between term-score and term-rank varies greatly depending on the metric. *tf.idf* tends to score only a few words highly, so that there is a sudden drop-off in scores for words that are ranked only slightly lower. In contrast, *su.idf* tends to score a larger number of words highly and the descent of scores is less steep as the rank decreases. This trend is found across meetings, and the difference between the approaches is particularly pronounced on ASR.

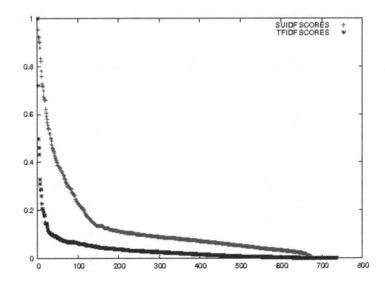

Fig. 1. Term Rank Plotted Against Term Score, ASR Transcripts

5.2   ICSI Results

On both manual and ASR transcripts there were fewer differences between term-weighting approaches than were found on the AMI test set. On manual transcripts, the highest scoring approaches were *Gain* and the hybrid of *su.idf* and *tf.idf*; however, there were no significant differences between approaches as a whole. As can be seen in Table 3, the weighted precision scores in general are much lower than on the AMI meetings.

On ASR, the highest scoring method was the hybrid approach, followed by *ridf*. There were again no significant differences between the various methods. Interestingly, however, all approaches tended to do better on ASR than on manual transcripts, as evidenced previously on the AMI test set above. Surprisingly, the only approach that showed decreasing weighted precision scores on ASR was

Table 3. Weighted Precision Results for ICSI Test Set Meetings

Meet	sidf	sasr	tfidf	tasr	com	comasr	ridf	ridfasr	gain	gainasr
Bed004	0.28	0.35	0.32	0.37	0.33	0.41	0.33	0.35	0.35	0.38
Bed009	0.53	0.45	0.44	0.38	0.45	0.42	0.38	0.38	0.39	0.39
Bed016	0.38	0.47	0.52	0.53	0.46	0.56	0.59	0.62	0.50	0.46
Bmr005	0.44	0.44	0.44	0.49	0.52	0.44	0.53	0.54	0.53	0.55
Bmr019	0.37	0.33	0.25	0.31	0.34	0.41	0.30	0.32	0.35	0.40
Bro018	0.36	0.36	0.39	0.32	0.41	0.36	0.36	0.32	0.39	0.29
AVERAGE	**0.39**	**0.40**	**0.39**	**0.40**	**0.42**	**0.43**	**0.42**	**0.42**	**0.42**	**0.41**

sidf=*su.idf* on manual, **sasr**=*su.idf* on ASR, **tfidf**=*tf.idf* on manual, **tfasr**=*tf.idf* on ASR, **com**=combined *su.idf* and *tf.idf* on manual, **comasr**=combined *su.idf* and *tf.idf* on ASR, **ridf**=residual IDF on manual, **ridfasr**=residual IDF on ASR, **gain**=*Gain* on manual, **gainasr**=*Gain* on ASR.

Gain, which slipped by a point. This is in contrast to the AMI results, where *Gain* did significantly better on ASR transcripts than on manual.

6 Discussion

There are several interesting aspects of the results reported above. Perhaps the most surprising is that some of the metrics, especially *su.idf* and *Gain*, are particularly resilient to ASR errors, and we found a general trend that weighted precision actually increased on ASR.

We also found that most of our metrics easily outperformed the classic *tf.idf* term-weighting scheme, with *su.idf*, the hybrid approach and *ridf* consistently performing the best. While *su.idf* outperformed *tf.idf* on the AMI corpus meetings, there was no statistical difference between the two approaches on the ICSI meetings. However, it was still advantageous to calculate *su.idf* on those meetings, as the hybrid approach was superior. Part of the reason for the difference in performance of those two metrics on AMI versus ICSI meetings may be due to the structure and set-up of the meetings themselves. As described above, the AMI meetings are scenario meetings with well-defined roles such as *project manager* and *marketing expert*, whilst roles in the ICSI corpus are much less clearly defined. Because roles are associated with certain vocabularies (e.g. the marketing expert being more likely to say "trend" or "survey" than the others), perhaps it would be expected that *su.idf* would perform better on those meetings than on meetings where roles are more opaque and the structure of the meetings is more loosely defined. Having said that, there were no significant differences between *any* of the term-weighting approaches on the ICSI meetings, and the results on a smaller test-set may simply be less reliable.

One clear result is that *tf.idf* is not as sensitive to term importance as the other metrics. It seems telling then that it is also the only metric that weights a term highly for occurring frequently within the given document. It is perhaps too blunt, favoring a few terms by scoring them highly and scoring the others dramatically lower, leading to a severely limited view of importance within the meeting. A strength of *su.idf* is that a term need not be very frequent within a document nor very rare across documents in order to receive a high score.

Our evaluation has relied on weighted precision of summaries that were created using each term-weighting scheme. We currently limit the evaluation to precision because the summaries are very brief and subsequently all recall scores are quite small. In the future we may wish to expand our evaluation to weighted precision, recall and f-measure, perhaps using longer automatic summaries. The weighted precision metric also, as currently formulated, does not have a theoretical maximum due to the fact that annotators may link each dialogue act as many times as they wish. One solution would be to use only the number of annotators who link each dialogue act, rather than the number of links they give to each dialogue act, thus providing a maximum score across summaries. However, doing so would cause us to lose a substantial amount of information in the form of annotator link counts.

7   Future Work

While exploiting differences in term usage among speakers has been promising, we believe there are additional speech-based features to exploit for term-weighting. One example is that informative terms used in meeting speech should tend to cluster into portions of the meeting roughly correlating to topic structure, whereas less informative words will be spread throughout the meeting. In addition, measures of prosodic prominence such as energy and F0 variance may be informative for locating more important words within the meeting.

8   Conclusion

We have presented an evaluation of term-weighting metrics for spontaneous, multi-party spoken dialogues. Three of the metrics, *tf.idf*, *ridf* and *Gain*, were imported from text IR to test for suitability with our data. A novel approach called *su.idf* was implemented, relying on the differing patterns of word usage among meeting participants. It was found to perform very competitively, both on its own and as part of a hybrid approach using combined rankings with *tf.idf*. In addition to the encouraging results for *su.idf*, we have provided evidence that *ridf* and *Gain* outperform *tf.idf* on our speech data.

Acknowledgements

Thanks to Thomas Hain and the AMI-ASR group for speech recognition output. This work was partly supported by the European Union 6th FWP IST Integrated Project AMI (Augmented Multi-party Interaction, FP6-506811, publication AMI-245).

References

1. Salton, G., Buckley, C.: Term-weighting approaches in automatic text retrieval. Information Processing and Management 24, 513–523 (1988)
2. Jones, K.S.: A statistical interpretation of term specificity and its application in retrievall. Journal of Documentation 28, 11–21 (1972)
3. Robertson, S., Jones, K.S.: Simple, proven approaches to text retrieval. University of Cambridge Computer Laboratory Technical Report TR-356 356 (1994)
4. Robertson, S., Jones, K.S.: Relevance weighting of search terms. Journal of the American Society for Information Science 35, 129–146 (1976)
5. Croft, W., Harper, D.: Using probabilistic models of information retrieval without relevance information. Journal of Documentation 35, 285–295 (1979)
6. Robertson, S.: Understanding inverse document frequency: On theoretical arguments for idf. Journal of Documentation 60, 503–520 (2004)
7. Church, K., Gale, W.: Inverse document frequency IDF: A measure of deviation from poisson. In: Proc. of the Third Workshop on Very Large Corpora, pp. 121–130 (1995)

8. Orasan, C., Pekar, V., Hasler, L.: A comparison of summarisation methods based on term specificity estimation. In: Proc. of LREC 2004, Lisbon, Portugal, pp. 1037–1041 (2007)
9. Rennie, J., Jaakkola, T.: Using term informativeness for named entity recognition. In: Proc. of SIGIR 2005, Salvador, Brazil, pp. 353–360 (2005)
10. Papineni, K.: Why inverse document frequency? In: Proc. of NAACL 2001, pp. 1–8 (2001)
11. Salton, G., McGill, M.J.: Introduction to Modern Information Retrieval. McGraw-Hill, NY, USA (1983)
12. Mori, T.: Information gain ratio as term weight: The case of summarization of ir results. In: Proc. of COLING 2002, Taipei, Taiwan, pp. 688–694 (2002)
13. Song, Y., Han, K., Rim, H.: A term weighting method based on lexical chain for automatic summarization. In: Gelbukh, A. (ed.) CICLing 2004. LNCS, vol. 2945, pp. 636–639. Springer, Heidelberg (2004)
14. Barzilay, R., Elhadad, M.: Using lexical chains for summarisation. In: Proc. of ACL 1997, Madrid, Spain, pp. 10–18 (1997)
15. Carletta, J., Ashby, S., Bourban, S., Flynn, M., Guillemot, M., Hain, T., Kadlec, J., Karaiskos, V., Kraaij, W., Kronenthal, M., Lathoud, G., Lincoln, M., Lisowska, A., McCowan, I., Post, W., Reidsma, D., Wellner, P.: The AMI meeting corpus: A pre-announcement. In: Renals, S., Bengio, S. (eds.) MLMI 2005. LNCS, vol. 3869, pp. 28–39. Springer, Heidelberg (2005)
16. Janin, A., Baron, D., Edwards, J., Ellis, D., Gelbart, D., Morgan, N., Peskin, B., Pfau, T., Shriberg, E., Stolcke, A., Wooters, C.: The ICSI meeting corpus. In: Proc. of IEEE ICASSP 2003, Hong Kong, China, pp. 364–367. IEEE Computer Society Press, Los Alamitos (2003)
17. Murray, G., Renals, S., Moore, J., Carletta, J.: Incorporating speaker and discourse features into speech summarization. In: Proc. of the HLT-NAACL 2006, New York City, USA, pp. 367–374 (2006)
18. Valenza, R., Robinson, T., Hickey, M., Tucker, R.: Summarization of spoken audio through information extraction. In: Proc. of the ESCA Workshop on Accessing Information in Spoken Audio, Cambridge UK, pp. 111–116 (1999)
19. Murray, G., Renals, S., Carletta, J.: Extractive summarization of meeting recordings. In: Proc. of Interspeech 2005, Lisbon, Portugal, pp. 593–596 (2005)

Automatic Decision Detection in Meeting Speech

Pei-Yun Hsueh and Johanna D. Moore

School of Informatics,
2 Buccleuh Place, Edinburgh EH8 9WL, United Kingdom

Abstract. Decision making is an important aspect of meetings in organisational settings, and archives of meeting recordings constitute a valuable source of information about the decisions made. However, standard utilities such as playback and keyword search are not sufficient for locating decision points from meeting archives. In this paper, we present the AMI DecisionDetector, a system that automatically detects and highlights where the decision-related conversations are. In this paper, we apply the models developed in our previous work [1], which detects decision-related dialogue acts (DAs) from parts of the transcripts that have been manually annotated as extract-worthy, to the task of detecting decision-related DAs and topic segments directly from complete transcripts. Results show that we need to combine features extracted from multiple knowledge sources (e.g., lexical, prosodic, DA-related, and topical class) in order to yield the model with the highest precision. We have provided a quantitative account of the feature class effects. As our ultimate goal is to operate AMI DecisionDetector in a fully automatic fashion, we also investigate the impacts of using automatically generated features, for example, the 5-class DA features obtained in [2].

keywords: Spoken language understanding, meeting tracking and analysis, argumentation modelling0.

1 Introduction

Recent advances in multimedia technologies have led to huge archives of audio and video recordings of meetings. Reviewing decisions is an aspect central to our organizational life [3,4]. For example, it would be helpful for a new engineer assigned to a project to review the major decisions that have been made in previous meetings by watching the recordings. However, while it is straightforward to archive a meeting, finding out what decisions have been made from the recording is still a challenging task. Unless all decisions are recorded in meeting minutes or annotated in the audio-video recordings, it is difficult to locate the decision points using existing browsing and playback utilities alone. Moreover, a recent study [5] has shown that even when a standard keyword search utility is provided, it is still difficult to recover information about the argumentative process in the discussion (e.g., decision points).

A. Popescu-Belis, S. Renals, and H. Bourlard (Eds.): MLMI 2007, LNCS 4892, pp. 168–179, 2007.
© Springer-Verlag Berlin Heidelberg 2007

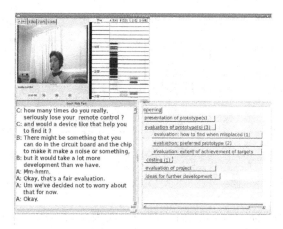

Fig. 1. Example application that demonstrates the use of decision-related topic segment information. The bottom right component shows a list of topic segments in an example meeting. The topic segments shaded in red are those that contain at least one decisions. The number shown in the parenthesis following each topic segment label indicates the number of decisions reached within the topic segment.

Banerjee and Rudnicky [6] have demonstrated that it is easier to recover information for user queries if the meeting record includes discourse-level annotations, such as topic segmentation, speaker role, and meeting state[1]. To assist users in revisiting decisions within meeting archives, our goal is thus to automatically annotate decision-related information on the dialogue acts and discussion segments where decisions are made. As the development of such an automatic decision detection component is critical to the re-use of meeting archives [7], it is expected to lend support to the development of other downstream applications, such as computer-assisted meeting tracking and understanding (e.g., assisting in the fulfilment of the decisions made in meetings) and group decision support systems (e.g., constructing group memory) [8,9].

Previous research has developed descriptive models of meeting discussions. Some of them focus on modelling the dynamics [10], while the others focus on modelling the content [11,4]. While automatically extracting these argument models remains a challenging task, researchers have begun to make progress towards this goal [12,13,14,1,15,16].

In this paper, we present the AMI DecisionDetector, which performs automatic decision detection in meeting speech and provides visual aids for users wishing to review decisions. In particular, we are interested in locating decision-related information at two levels of granularity: topic segments and dialogue acts. First, the system detects decision-related topic segments in which meeting participants have reached at least one decision. As shown in Figure 1, this allows

---

[1] Meeting states include discussion, presentation and briefing.

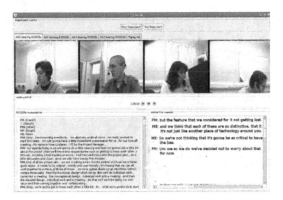

Fig. 2. Example application that demonstrates the use of decision-related DA information. The bottom right component shows a set of decision-related DA extracts that are representive of the design decision of "how to find (the remote) when misplaced".

users to get an overview of the decisions made in previous meetings by browsing the topics of the decision-related segments (e.g., those shaded in red in Figure 1).

Second, the system detects decision-related dialogue acts (DAs) by looking for DAs which are extract-worthy and reflective of the content of the decision discussions. As shown in Figure 2, this allows users to obtain details about the decisions they are particularly interested in by reviewing the relevant decision-related DAs. For example, if a user wants to know more about the design decision of "how to find (the remote) when misplaced", they can interpret the decision as "not to worry about designing a function to find the remote when misplaced" by looking at the extract shown in the bottom right component of Figure 2.

2   Data

In this study, we use a set of 50 scenario-driven meetings (approximately 37,400 DAs) that have been segmented into dialogue acts and annotated with decision information in the AMI meeting corpus [17]. These meetings are driven by a scenario, wherein four participants play the role of Project Manager, Marketing Expert, Industrial Designer, and User Interface Designer in a design team in a series of four meetings. Participants participated in only one series of 4 meetings. The corpus includes manual transcripts for all meetings as well as individual sound files recorded by close-talking microphones for each participant and cross-talking sound files recorded by an 8-element circular microphone array.

The meeting recordings have been annotated at several levels, including dialogue acts (DAs) and topics. The DA annotation scheme for the AMI corpus consists of 15 dialogue act types, which can be organised into five major groups:

- Information (31.9%): giving and eliciting information. E.g., "Suggestion".
- Action (9.8%): making or eliciting suggestions or offers. E.g., "Elicit-suggestion".

- Commenting on the discussion (22.6%): making or eliciting assessments and comments about understanding. E.g., "Assessment".
- Segmentation (31.8%): not contributing to the content but allowing segmentation of the discourse, E.g., "Backchannel", "Stall", and "Fragment".
- Other (3.9%): a remainder class for utterances which convey an intention, but do not fit into the four previous categories.

Topic segmentation and labels have also been annotated in the AMI meeting corpus. Annotators had the freedom to mark a topic as subordinated (down to two levels) wherever appropriate. In this work, we have flattened the structure into a hierarchy of two layers: top-level (TOP) and subtopic level (SUB). As the AMI meetings are scenario-driven, annotators are expected to find that most topics recur. Therefore, they are given a standard set of descriptions that can be used as labels for each identified topic segment. In particular, the annotators explicitly identify those parts of the meeting that refer to the meeting process (e.g., opening, closing, agenda/equipment issues), or are simply irrelevant (e.g., chitchat). To capture the common characteristics of these off-topic discussion segments, we have collapsed these segments into one category: functional segments (FUNC). The AMI scenario meetings takes, on average, 30 minutes (around 800 DAs) and contain eight top-level topic segments and seven subtopic segments per meeting.

2.1   Decision-Related Dialogue Acts

It is difficult to determine whether a DA contains information relevant to decision without knowing what decisions have been made in the meeting. Therefore, in this study decision-related DAs are annotated in a two-phase process: First, annotators are asked to browse through the meeting record and write an abstractive summary about the decisions that have been made in the meeting. In this phase, another group of three annotators are also asked to produce extractive summaries by selecting a subset (around 10%) of DAs which form a summary of this meeting. Annotators are instructed to produce these summaries for an absent project manager.

Finally, this group of annotators are asked judge whether the DAs in the extractive summary support any of the sentences in the abstractive decision summary; if a DA is related to any sentence in the decision section of the abstractive summary, a "decision link" from the DA to the decision sentence in the abstractive summary is added. For those extracted DAs that do not have any closely related sentence, the annotators are not obligated to specify a link. We then label the DAs that have one or more decision links as decision-related DAs.

In the 50 meetings we used for our experiments, annotators found on average four decisions per meeting and specified around two decision links for each decision sentence in the abstractive summary. Overall, 554 out of 37,400 DAs have been annotated as decision-related DAs, accounting for 1.4% of all DAs in the data set and 12.7% of the original extractive summaries (which consist of the extracted DAs). An earlier analysis established the intercoder reliability of the two-phase process at a kappa ranging from 0.5 to 0.8. In these experiments, for each meeting in the 50-meeting dataset we randomly choose the decision-related DA annotation of one annotator as the source of ground truth data.

2.2   Decision-Related Topic Segments

Decision-related topic segments are operationalized as the topic segments that contain one or more decision-related DAs. Overall, 198 out of 623 (31.78%) topic segments in the 50-meeting dataset are decision-related topic segments. As the meetings we use are driven by a scenario, we expect to find that interlocutors are more likely to reach decisions when certain topics listed in a predetermined agenda are brought up or when the discussions are related to the decisions made in previous meetings. For example, 80% of the segments labelled as Costing and 58% of those labelled Budget are decision-related topic segments, whereas only 7% of the Existing Product segments and none of the Trend-Watching segments are decision-related topic segments. (See Table 1 for a break-down of different types of decision-related segments.)

Table 1. Characteristics of topic segments that contain decision-related DAs

	ALL	TOP	SUB	FUNC
Percentage of Decision-related topicsegments per meeting (%)	33%	31%	35%	4%
Average number of decision-related dialogue acts per segment	3.7	4.5	2.76	3.83

3   AMI DecisionDetector

To locate decision-related information at the two levels of granularity, the AMI DecisionDetector consists of two components: (1) a decision-related DA detector which identifies important DAs pertaining to the decisions made, and (2) a decision-related topic segment detector which identifies the topic segments in which interlocutors have reached one or more decisions.

In the field of multiparty discourse understanding, previous research has commonly utilized a classification framework, in which variants of models are computed directly from data for classifying unseen instances. Models has been successfully trained for detecting the content topics [18], group activities [2,19,20], participant roles [21], addressees [22], and emotional effects (e.g., group level of interest [13], hot spots [16]). In this work, we have adopted a similar framework: the task of automatically detecting decision-related DAs is decomposed to a series of binary decisions [1]. A Maximum Entropy (MaxEnt) model is trained to automatically classify whether a DA is decision-related or not.

We evaluate the decision-related DA detector with a five-fold cross validation procedure using the set of 50 scenario-driven meetings. In each fold, a Maximum Entropy (MaxEnt) classifier is used to train models that can classify decision-related DAs on a subset of 40 meetings; next, the trained model is tested on the remaining 10 meetings that are unseen in the training phase. The decision-related topic segment detector leverages the set of outputs (i.e., binary decisions) from the decision-related DA detector to classify whether an unseen topic segment contains any decisions. The task of detecting decision-related topic segments thus can be viewed as that of recognizing decision-related DAs in a wider window.

3.1   Features Used

Previous work has shown that combining multiple knowledge sources (e.g., words, audio-video recordings, speaker intention) is important to automatically identifying different aspects of the argumentative process [18]. For example, paralinguistic features (e.g., prosody and the amount of disfluency) have been applied to detect deceptive speech [23]. Paralinguistic features have also been combined with features that indicate speaker intention (i.e., DA classes) to detect "hot spots"[2] [24,16]. Similarly, lexical features, such as occurrence counts of cue words, have been used to detect learning attitudes of students in a tutoring system [25] and to detect where speakers are agreeing with one another [12,14].

Here we are interested in examining the merits of multimodal feature combinations on the performance of AMI DecisionDetector. In particular, we examine the use of the following features:

Prosodic Features: Our previous work [1] has shown that there exist prominent acoustic characteristics of decision-related DAs. For example, when it comes to a decision point, interlocutors either speak very fast or very slowly; the pitch usually goes up first and then goes down in the midpoint of a dialogue act. In this work, we use the same set of prosodic features, i.e., duration, pause, speech rate, pitch contour, and energy level. For details of how to generate these features with Shriberg and Stolcke's direct modelling approach [26], please refer to [27]. An exploratory study has shown the benefits of including immediate prosodic contexts, and thus we also include prosodic features of the immediately preceding and following DA.

Lexical Features: Previous work has also shown the importance of the lexical characteristics of decision-related DAs. For example, interlocutors use "*We*" more than "*I*" and "*You*" when reaching a decision. Likewise, they also explicitly mention topical words, such as "*advanced chips*" and "*functional design*", and use fewer negative expressions, such as "*I don't think*" and "*I don't know*". Thus we also include lexical items in our feature sets. In each fold of cross validation, we compile a list of cue words, which have occurred more than once in the set of decision-related DAs in the "training" set of meetings. Each DA is then represented as a vector of unigrams in the list of cue words.

DA-related Features: These include DA classes and speaker roles (e.g., project manager, marketing expert). We also include DA classes of the immediately preceding and following DA. As mentioned in Section 2, we have grouped the 15 DA classes (15-Class) into five major groups (5-Class). We have also obtained the automatic 5-Class predictions for each DA [2]. The accuracy of the automatic DA

---

[2] Namely locations that exhibit a high level of affect in the voices of interlocutors.

class predictions is 59.1%. In the following experiment, we thus can evaluate the impact of the three different versions of DA class information: manual 15-Class, manual 5-Class, and automatic 5-Class.

Topical Features: As reported in Section 2.2, we find that interlocutors are more likely to reach decisions when certain topics are brought up. Also, we expect decision-making conversations to take place towards the end of a topic segment. Therefore, we include the following features: the label of the current topic segment, the position of the DA in a topic segment (measured in words, in seconds, and in %), the distance to the previous topic shift (both at the top-level and subtopic level)(measured in seconds), the duration of the current topic segment (both at the top-level and subtopic level)(measured in seconds).

4   Results

In Section 3.1, we described the four major types of features used in this study: prosodic (PROS), unigrams (LX1), DA-related (DA), and topical (TOPIC) features. As opposed to our previous work, which detects decision-related DAs on only the parts of meetings that have been identified as extract-worthy, we trained models to detect decision-related DAs directly from entire transcripts. We expect this task to be much more challenging as the imbalance between positive and negative cases is even more prominent. The proportion of positive cases has gone from 14% down to 2%. For comparison, we use the lexical models trained with the unigram lexical features (LX1) as our baseline.[3] The different combinations of features we used for training models can be divided into the three groups: (A) lexical features alone (BASELINE); (B) all available features except one of the four types of features (ALL-LX1, ALL-PROS, ALL-DA, ALL-TOPIC); and (C) all available features (ALL).

4.1   Experiment 1: Classifying Decision-Related Dialogue Acts and Topic Segments

Table 2 reports the performance on both the training (40 meetings) and the test set (10 meetings). Because previous work has shown that ambiguity exists in the assessment of the exact timing of decision-related DAs, the results in Table 2 are obtained using a lenient match measure, allowing a window of 20 seconds preceding and following a hypothesized decision-related DA for recognition. The task of detecting decision-related topic segments can be viewed as that of detecting decision-related DAs in a wider window. The right most three columns of the training set and test set results in Table 2 show the results of detecting decision-related topic segments.

The results demonstrate that, compared to the LX1 baseline, models trained with all features (ALL), including lexical, prosodic, DA-related and topical features, yield notably better precision on the task of decision-related topic segment

---

[3] Please note that the LX1 features used here are obtained on manual transcripts; so the lexical models can only be viewed as being trained semi-automatically.

Table 2. Effects of different combinations of features on detecting decision-related DAs and topic segments

	TRAIN SET						TEST SET					
Decision-Related	Dialogue Act			Topic Segment			Dialogue Act			Topic Segment		
Accuracy	P	R	F1	P	R	F1	P	R	F1	P	R	F1
BASELINE(LX1)	0.32	0.47	0.38	0.44	0.65	0.52	0.22	0.39	0.28	0.39	0.65	0.49
ALL-LX1	0.64	0.10	0.18	0.72	0.13	0.22	0.35	0.07	0.11	0.52	0.17	0.25
ALL-PROS	0.69	0.46	0.55	0.72	0.53	0.61	0.30	0.16	0.21	0.51	0.38	0.43
ALL-DA	0.70	0.48	0.57	0.72	0.56	0.63	0.32	0.24	0.26	0.49	0.44	0.46
ALL-TOPIC	0.64	0.36	0.46	0.70	0.48	0.57	0.24	0.18	0.20	0.49	0.41	0.44
ALL	0.72	0.38	0.49	0.74	0.45	0.55	0.35	0.19	0.24	0.56	0.38	0.44

prediction, 74% on the training set and 56% on the test set. However, in the test set, the overall accuracy (F1 score) of the combined models is relatively worse than the baseline, due to the substantially lower recall rate.

To study the relative effect of the different feature types, Rows 2-5 in the table report the performance of models in Group B, which are trained with all available features except LX1, PROS, DA and TOPIC, respectively. The amount of degradation in the overall accuracy (F1) of each of the models in relation to that of the ALL model indicates the contribution of the feature type that has been left out. For example, we find that the ALL model outperforms all except the model trained by leaving out DA-related features (ALL-DA). A closer investigation of the precision and recall of the ALL-DA model shows that including the DA-related features is detrimental to recall but beneficial for precision. This effect stems from the fact that decisions are more likely (1) to occur in certain types of dialogue acts, such as "Inform", "Suggest", "Elicit-Assessment", and "Elicit-Inform", and (2) to be preceded and followed by segmentation-type dialogue acts, such as "Stall" and "Fragment". Therefore, training models with DA-related features, such as the DA class of the current DA and its immediate context, helps eliminate incorrect predictions of decision-related DAs.

In sum, results suggest that (1) lexical features are the most predictive in terms of overall accuracy, despite low precision, (2) prosodic features have positive impacts on precision but not on recall, and (3) DA-related and topical features are both beneficial to precision but detrimental to recall.

4.2   Experiment 2: Exploring Automatically Generated DA Class Features

As our ultimate goal is to operate AMI DecisionDetector in an automatic fashion, we evaluate the impact of the automatically generated DA class features on the task of detecting decision-related DAs and topic segments. We have utilized the 5-class DA predictions (AUTO-5DA) generated in [2]. To understand whether the automatically generated features caused any degradation, we train models which combine all available lexical, prosodic and topical features with the AUTO-5DA features. We then evaluate the AUTO-5DA model against other models

Table 3. Effects of different versions of DA class features on detecting decision-related DAs and topic segments

	TRAIN SET						TEST SET					
Decision-Related	Dialogue Act			Topic Segment			Dialogue Act			Topic Segment		
Accuracy	P	R	F1	P	R	F1	P	R	F1	P	R	F1
EXTRACT (MANUAL-15DA)	0.73	0.61	0.66	0.74	0.66	0.69	0.37	0.38	0.36	0.50	0.53	0.50
EXTRACT (MANUAL-5DA)	0.70	0.77	0.70	0.68	0.73	0.70	0.36	0.44	0.39	0.49	0.62	0.54
EXTRACT (AUTO-5DA)	0.69	0.71	0.70	0.68	0.73	0.70	0.33	0.40	0.35	0.46	0.55	0.49

which combine the other features with the two types of manually annotated dialogue act class features: MANUAL-5DA and MANUAL-15DA. The results reported here are obtained by operating AMI DecisionDetector on the part of meetings that have been manually annotated as extract-worthy. This is because we want to focus on analyzing the impacts of the automatic DA features on the task of decision detection, rather than on that of extractive summarization.

Results in Table 3 show that our strategy that groups 15 DA classes into five major classes is beneficial to the models on the task of decision detection. It improves the recall of predicting decision-related topic segments by 18%. Replacing the manual 5-class DA features with the automatically generated version degrades the performance by 10%. However, the accuracy of prediction using the 5 automatically predicted DA classes (AUTO-5DA) compares favorably with the accuracy when using the 15 manually annotated DA classes (MANUAL-15DA).

5   Conclusions and Future Work

In this paper, we present AMI DecisionDetector, a system which performs automatic decision detection in meeting speech and provides visual aids for users who wish to review decisions. To avoid the costly requirement of operating on extractive summaries, we have examined how our computational models perform when detecting decisions directly from complete meeting transcripts. We have evaluated the models on the task of predicting decision-related discussions at two levels of granularity: dialogue acts and topic segments. To further overcome the problem of imbalanced class distribution (i.e., only 2% are positive cases), we have leveraged a variety of knowledge sources (e.g., words, prosody, DA-related contexts, topic annotations). Experimental results suggest that the model combining all the available knowledge sources performs substantially better, achieving 74% and 56% precision on the task of detecting decision-related topic segments in the training set and test set respectively. The framework we applied here can also be used to recover information for other aspects in the argumentation process, such as problems and action items.

We have also provided a quantitative account of the merits of different feature classes. Among features that are extracted from the widely ranging knowledge sources, lexical features are the most indispensable. Also, DA-related features can improve the precision of models but degrade the recall. These findings are consistent with the results of our previous experiment which operates AMI DecisionDetector on a selective set of dialogue acts in the transcripts.

However, there are also other findings that no longer hold true when our system is operated on complete transcripts. For example, [1] has shown topical features have a distinctive advantage for locating decision topic segments from extractive summaries. However, this is not the case when identifying decision points in entire transcripts. In addition, the model trained with lexical features alone outperforms the combined model in its recall rate. This is possibly because when attempting to detect decisions from the whole transcripts, the system needs to simultaneously disambiguate the extract-worthy and decision-related dialogue acts. Therefore, features that are good at disambiguating both will stand out, and features that fail in the extract-worthy DA detection task will be shown as weak features to the final performance of decision-related DA detection.

Another drawback of our previous approach is that many of the features used in this study require human intervention, such as manual transcriptions, annotated DA segmentations and labels, and other types of meeting-specific features (e.g., speaker role). However, these semi-automatic and manual features are not always available. Therefore, in this work we tested whether our system is robust to the noise introduced by the automatically generated versions of these features. An exploratory study has shown that the performance of our approach does not degrade considerably after replacing the reference words with the ASR words, despite word recognition errors. Our further investigation on the impacts of using an automatically generated version of the DA class features (as reported in [2]) shows that it is possible to include these automatic features in the model directly. It will not degrade the performance more than including the manually annotated 15-class DA features in the first place.

Also, our approach which automatically extracts decision-related DAs as summaries has some liabilities. First, the unconnected DAs in the extract result in semantic gaps that require contextualization to bridge. Second, anaphora and unexpected topic shifts between these extracted DAs also require context to resolve. Previously, we have attempted to provide such contexts by indicating the topic of the current discussion. However, a preliminary study has shown that the segment boundaries of decision-related discussions coincide with that of the topic segments less than 50% of the time. Last but not least, although it is our intuition that the decision-related DA extracts will assist users in finding and absorbing information in the meeting archives more efficiently and effectively, this assumption has yet to be tested with human subjects.

Therefore, we are now planning to conduct an extrinsic decision audit task-based evaluation on the utility of displaying decision-related DA information (as exemplified in Figure 2) to the users. We have also annotated decision-related discussion segmentation, which can be used to train computational models to find contexts that are needed for the interpretation of the identified decision points. Moreover, as we would like to disambiguate which sentence in the abstractive decision summary of a meeting is the most relevant to each of the identified decision points, the decision discussion segmentation annotations can also form a foundation for the development of the disambiguation model.

Acknowledgments

This work was supported by the European Union 6th FWP IST Integrated Project FP6-033812 AMIDA (Augmented Multiparty Interaction with Distance Access). We thank Alfred Dielmann for generating the 5-Class DA predictions for us, Jonathan Kilgour and Jean Carletta for their continuous support on the development of the AMI DecisionDetector system we demoed in MLMI 2007 at Brno, Czech Republic, and Theresa Wilson for her insightful feedbacks on the decision discussion segmentation annotation work. We also thank Steve Renals and the three anonymous reviewers for their enlightening comments on this paper.

References

1. Hsueh, P., Moore, J.: What decisions have you made: Automatic decision detection in conversational speech. In: Proceedings of NACCL/HLT 2007 (2007)
2. Dielmann, A., Renals, S.: DBN based joint dialogue act recognition of multiparty meetings. In: Proceedings of International Conference on Acoustics, Speech, and Signal Processing (2007)
3. Pallotta, V., Niekrasz, J., Purver, M.: Collaborative and argumentative models of meeting discussions. In: IJCAI 2005. Proceeding of CMNA-05 workshop on Computational Models of Natural Arguments (2005)
4. Rienks, R., Heylen, D., van der Weijden, E.: Argument diagramming of meeting conversations. In: Multimodal Multiparty Meeting Processing Workshop at the ICMI (2005)
5. Pallotta, V., Seretan, V., Ailomaa, M.: User requirements analysis for meeting information retrieval based on query elicitation. In: Proceedings of ACL 2007 (2007)
6. Banerjee, S., Rose, C., Rudnicky, A.I.: The necessity of a meeting recording and playback system, and the benefit of topic-level annotations to meeting browsing. In: Proc. of the International Conference on Human-Computer Interaction (2005)
7. Whittaker, S., Laban, R., Tucker, S.: Analysing meeting records: An ethnographic study and technological implications. In: Renals, S., Bengio, S. (eds.) MLMI 2005. LNCS, vol. 3869, Springer, Heidelberg (2006)
8. Post, W.M., Cremers, A.H., Henkemans, O.B.: A research environment for meeting behavior. In: Proceedings of the 3rd Workshop on Social Intelligence Design (2004)
9. Romano, N.C., Nunamaker, J.F.: Meeting analysis: Findings from research and practice. In: Proceedings of HICSS-34, IEEE Computer Society Press, Los Alamitos (2001)
10. Niekrasz, J., Purver, M., Dowding, J., Peters, S.: Ontology-based discourse understanding for a persistent meeting assistant. In: Proc. of the AAAI Spring Symposium (2005)
11. Marchand-Mailet, S.: Meeting record modeling for enhanced browsing. Technical report, Computer Vision and Multimedia Lab, Computer Centre, University of Geneva, Switzerland (2003)
12. Galley, M., McKeown, J., Hirschberg, J., Shriberg, E.: Identifying agreement and disagreement in conversational speech: Use of bayesian networks to model pragmatic dependencies. In: Proc. of the 42nd Annual Meeting of the ACL (2004)
13. Gatica-Perez, D., McCowan, I., Zhang, D., Bengio, S.: Detecting group interest level in meetings. In: ICASSP 2005. IEEE Int. Conf. on Acoustics, Speech, and Signal Processing, IEEE Computer Society Press, Los Alamitos (2005)

14. Hillard, D., Ostendorf, M., Shriberg, E.: Detection of agreement vs. disagreement in meetings: Training with unlabeled data. In: Proc. HLT-NAACL 2003 (2003)
15. Purver, M., Ehlen, P., Niekrasz, J.: Shallow discourse structure for action item detection. In: The Workshop of HLT-NAACL: Analyzing Conversations in Text and Speech, ACM Press, New York (2006)
16. Wrede, B., Shriberg, E.: Spotting hot spots in meetings: Human judgements and prosodic cues. In: Proceedings of EUROSPEECH 2003 (2003)
17. Carletta, J., et al.: The AMI meeting corpus: A pre-announcement. In: Proceedings of 2nd Joint Workshop on Multimodal Interaction and Related Machine Learning Algorithms (2005)
18. Hsueh, P., Moore, J.D.: Combining multiple knowledge sources for dialogue segmentation in multimedia archives. In: Proceedings of the 45th Annual Meeting of the ACL (2007)
19. McCowan, I., Gatica-Perez, D., Bengio, S., Lathoud, G., Barnard, M., Zhang, D.: Automatic analysis of multimodal group actions in meetings. IEEE Transactions on Pattern Analysis and Machine Intelligence (PAMI) 27(3), 305–317 (2005)
20. Zhang, D., Gatica-Perez, D., Bengio, S., McCowan., I.: Semi-supervised adapted hmms for unusual event detection. In: CVPR 2005. Proc. IEEE International Conference on Computer Vision and Pattern Recognition, IEEE Computer Society Press, Los Alamitos (2005)
21. Banerjee, S., Rudnicky, A.I.: You are what you say: Using meeting participants' speech to detect their roles and expertise. In: Proceedings of the Workshop of HLT-NAACL: Analyzing Conversations in Text and Speech, ACM Press, New York (2006)
22. Jovanovic, N., op den Akker, R., Nijholt, A.: Addressee identification in face-to-face meetings. In: The Proceedings of the 11th Conference of the European Chapter of the Association for Computational Linguistics (2006)
23. Graciarena, M., Shriberg, E., Stolcke, A., Enos, F., Hirschberg, J., Kajarekar, S.: Combining prosodic, lexical and cepstral systems for deceptive speech detection. In: Proc. IEEE ICASSP, IEEE Computer Society Press, Los Alamitos (2006)
24. Wrede, B., Shriberg, E.: The relationship between dialogue acts and hot spots in meetings. In: Proceedings of IEEE ASRU Workshop, IEEE Computer Society Press, Los Alamitos (2003)
25. Litman, D.J., Forbes-Riley, K.: Recognizing student emotions and attitudes on the basis of utterances in spoken tutoring dialogues with both human and computer tutors. Speech Communication (2006)
26. Shriberg, E., Stolcke, A.: Direct modeling of prosody: An overview of applications in automatic speech processing. In: International Conference on Speech Prosody 2004 (2001)
27. Murray, G., Renals, S., Taboada, M.: Prosodic correlates of rhetorical relations. In: Proceedings of HLT/NAACL ACTS Workshop (2006)

Czech Text-to-Sign Speech Synthesizer*

Zdeněk Krňoul, Jakub Kanis, Miloš Železný, and Luděk Müller

Univ. of West Bohemia, Faculty of Applied Sciences, Dept. of Cybernetics
Univerzitní 8, 306 14 Pilsen, Czech Republic
{zdkrnoul,jkanis,zelezny,muller}@kky.zcu.cz

Abstract. Recent research progress in developing of the Czech – Sign
Speech synthesizer is presented. The current goal is to improve the system for automatic synthesis to produce accurate synthesis of the Sign
Speech. The synthesis system converts written text to an animation of
an artificial human model (avatar). This includes translation of text to
sign phrases and their conversion to the animation of the avatar. The
animation is composed of movements and deformations of segments of
hands, a head and also a face. The system has been evaluated by two
initial perceptual tests. The perceptual tests indicate that the designed
synthesis system is capable to produce the intelligible Sign Speech.

Keywords: Sign speech, automatic synthesis, machine translation.

1 Introduction

In the scope of this paper, we use the term Sign Speech (SS) for both the
Czech Sign Language (CSE) and Signed Czech (SC). The CSE is a natural
and adequate communication form and a primary communication tool of the
hearing-impaired people in the Czech Republic. It is composed of the specific
visual-spatial resources, i.e. hand shapes (manual signals), movements, facial
expressions, head and upper part of the body positions (non-manual signals). It
is not derived from or based on any spoken language. CSE has basic language
attributes, i.e. system of signs, double articulation, peculiarity and historical
dimension, and has its own lexical and grammatical structure. On the other
hand, the SC was introduced as an artificial language system derived from the
spoken Czech language to facilitate communication between deaf and hearing
people. SC uses grammatical and lexical resources of the Czech language. The
Czech sentence is audibly or inaudibly articulated during the SC production and
the CSE signs of all individual words of the sentence are simultaneously signed
with the articulation.

The use of written language instead of spoken one is a wrong idea in the case
of the Deaf. Hence, the Deaf have problems with understanding the majority
language (the language used by hearing people) when they are reading a written
text. The majority language is the second language of the Deaf and its use by

---

* Support for this work was provided by the Grant Agency of Academy of Sciences of
the Czech Republic, project No. 1ET101470416 and MŠMT LC536.

A. Popescu-Belis, S. Renals, and H. Bourlard (Eds.): MLMI 2007, LNCS 4892, pp. 180–191, 2007.

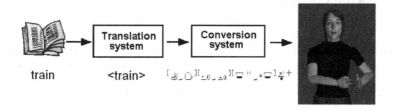

Fig. 1. The schema of the Sign Speech synthesis system

the deaf community is only particular. Thus, the majority language translation into the Sign Speech is highly important for better orientation of the Deaf in the majority language-speaking world. Currently, human interpreters provide this translation, but their service is expensive and not always available. A full dialog system (with ASR and Text-to-Sign-Speech (TTSS) systems on one side (from spoken to sign language) and Automatic-Sign-Speech-Recognition (ASSR) and TTS systems on the other side (from sign to spoken language)) represents a solution which does not intend to fully replace the interpreters, but its aim is to help in everyday communication in selected constrained domains such as the post office, health care, traveling, etc.

Our synthesis system consists of two parts: the translation and the conversion subsystem (see Figure 1). The translation system transfers Czech written text to its textual representation in the Sign Speech (textual sign representation). The conversion system then converts this textual sign representation to an animation of an artificial human model (avatar). The resulting animation then represents the corresponding utterance in the Sign Speech.

The translation system is an automatic phrase-based translation system. A Czech sentence is divided into phrases and these are then translated into corresponding Sign Speech phrases. The translated words are reordered and rescored using a language model at the end of the translation process. In our synthesizer we use own implementation of simple monotone phrase-based decoder - SiM-PaD. This decoder and its performance will be described in more details in next section.

The problem of translated phrase conversion is composed of the isolated sign animation and their concatenation. Each sign is expressed by the manual and non-manual component. The manual component represents necessary movements, orientations and shapes of hands. The non-manual component is composed of complemented movements of the upper half-body, face gestures or face articulation (lips and inner mouth organs). The symbolic notation Ham-NoSys[1] (Hamburg Notation System for Sign Languages) was applied for controlling of the manual component and the no-facial upper half-body movements. The synthesis of a mouth articulation is separately supplemented by a talking head system.

---

[1] Available at www.sign-lang.uni-hamburg.de/projects/HamNoSys.html.

2   Translation System

The machine translation model is based on the noisy channel model. When we apply the Bayes rule on the translation probability $p(\mathbf{t}|\mathbf{s})$ for translating a sentence $\mathbf{s}$ in a source language into a sentence $\mathbf{t}$ in a target language we obtain:

$$\text{argmax}_t p(\mathbf{t}|\mathbf{s}) = \text{argmax}_t p(\mathbf{s}|\mathbf{t}) p(\mathbf{t})$$

Thus the translation probability $p(\mathbf{t}|\mathbf{s})$ is decomposed into two separate models: a translation model $p(\mathbf{s}|\mathbf{t})$ and a language model $p(\mathbf{t})$ that can be modeled independently. In the case of phrase-based translation the source sentence $\mathbf{s}$ is segmented into a sequence of I phrases $\bar{\mathbf{s}}_1^I$ (all possible segmentations have the same probability). Each source phrase $\bar{s}_i, i = 1, 2, ..., I$ is translated into a target phrase $\bar{t}_i$ in the decoding process. This particular ith translation is modeled by a probability distribution $\phi(\bar{s}_i|\bar{t}_i)$. The target phrases can be reordered to get more precise translation. The reordering of the target phrases can be modeled by a relative distortion probability distribution $d(a_i - b_{i-1})$ as in [1], where a_i denotes the start position of the source phrase which was translated into the ith target phrase. b_{i-1} denotes the end position of the source phrase translated into the $(i-1)th$ target phrase. Also a simpler distortion model $d(a_i - b_{i-1}) = \alpha^{|a_i - b_{i-1}-1|}$ [1], where α is a predefined constant, can be employed. The best target output sentence $\mathbf{t_{best}}$ can then be for a given source sentence $\mathbf{s}$ acquired as:

$$\mathbf{t_{best}} = \text{argmax}_t p(\mathbf{t}|\mathbf{s}) = \prod_{i=1}^{I} [\phi(\bar{s}_i|\bar{t}_i) d(a_i - b_{i-1})] p_{LM}(\mathbf{t})$$

Where $p_{LM}(\mathbf{t})$ is a language model of the target language (usually a trigram model with some smoothing, built from a huge portion of target texts).

2.1   Comparison of Decoders

We will compare SiMPaDs performance with the performance of state-of-the-art phrase-based decoder Moses in case of Czech to Signed Czech translation. And we will introduce class-based language model and post-processing method which either improve results of translation.

Decoders. The first decoder is freely available state-of-the-art factored phrase-based beam-search decoder - **MOSES** [2]. Moses can work with factored representation of words (i.e. surface form, lemma, part-of-speech, etc.) and uses a beam-search algorithm, which solves a problem of the exponential number of possible translations (due to the exponential number of possible alignments between source and target translation) for efficient decoding. The training tools for extracting of phrases from the parallel corpus are also available, i.e. the whole translation system can be constructed given a parallel corpus only. For language modeling we use the SRILM[2] toolkit.

---

[2] Available at http://www.speech.sri.com/projects/srilm/download.html.

The second decoder is our simple monotone phrase-based decoder - **SiMPaD**. The monotonicity means using the monotone reordering model, i.e. no phrase reordering is permitted. In the decoding process we choose only one alignment with the longest phrase coverage (for example if there are three phrases: p_1, p_2, p_3 coverage three words: w_1, w_2, w_3, where $p_1 = w_1 + w_2$, $p_2 = w_3$, $p_3 = w_1 + w_2 + w_3$, we choose the alignment which contains phrase p_3 only). A standard Viterbi algorithm is used for the decoding. SiMPaD uses SRILM language models.

Data and Evaluation Criteria. The main resource for the statistical machine translation is a parallel corpus which contains parallel texts of the source and the target language. The acquisition of such a corpus in the case of SS is complicated by the absence of an official written form of both the CSE and the SC. Therefore we used a Czech to Signed Czech (CSC) parallel corpus [3] for training of decoders.

The CSC corpus contains 1130 dialogs from a telephone communication between customer and operator in a train timetable information center. The parallel corpus was created by semantic annotation of several hundred dialogs and by adding the SC translation of all the dialogs. A SC sentence is written as a sequence of CSE signs. The whole CSC corpus contains 16 066 parallel sentences, 110 033 running words and 109 572 running signs, 4082 unique words and 720 unique signs. Each sentence of the CSC corpus has assigned the written form of the SC translation, a type of the dialog act, and its semantic meaning in a form of semantic annotation. For example (we use English literal translation) for the Czech sentence: *good day I want to know how me it is going in Saturday morning to brno* we have the SC translation: *good_day I want know how _ _ go in Saturday morning to brno* and for the part: *good day* the dialog act: *conversational_domain= "frame" + speech_act= "opening"* and the semantic annotation: *semantics= "GREETING"* . The dialog act: *conversational_domain= "task" + speech_act= "request_info* and the semantic annotation: *semantics= "DEPARTURE(TIME, TO(STATION))"* is assigned to the rest of the sentence. The corpus contains also a handcrafted word alignment (added by annotators during the corpus creation) of every Czech – SC sentence pair. For more details about the CSC corpus see [3].

We use the following criteria for evaluation. The first criterion is **Sentence Error Rate (SER)**: It is a ratio of the number of incorrect sentence translations to the number of all translated sentences. The second criterion is **Word Error Rate (WER)**: This criterion is adopted from ASR area and is defined as the Levenshtein edit distance between the produced translation and the reference translation in percentage (a ratio of the number of all deleted, substituted and inserted produced words to the total number of reference words). The third criterion is **Position-independent Word Error Rate (PER)**: it is simply a ratio of the number of incorrect translated words to the total number of reference words (independent of the word order). The last criterion is **BLEU** score [4]: it counts a modified n-gram precision for the output translation with respect to the reference translation. A lower value of the first three criteria and a higher value of the last one indicate better i.e. more precise translation.

Experiment. Both decoders are trained on the CSC corpus. SiMPaD uses a phrase table of handcrafted phrases (acquired from the handcrafted alignment of the CSC corpus; a phrase translation probability is estimated by the relative frequency [1]) and phrase-based language model (a basic unit of the language model is a phrase instead of a word). Moses uses the phrase table of automatically acquired phrases and the standard language model. The phrases were acquired from Giza++ word alignment of the parallel corpus (the word alignment established by Giza++ toolkit[3]) by some heuristics (we used default heuristics). There are many parameters which can be specified in the training and decoding process of Moses. Unless otherwise stated, we used the default values of parameters (for more details see Moses' documentation in [2]).

To improve results of translation we used two enhancements - a class-based language model and post-processing method. As well as in the area of ASR, there are problems with out-of-vocabulary words (OOV) in the automatic translation area. We can translate only words which are in a translation vocabulary (we know their translation into the target language). By the analysis of the translation results we found that many OOV words are caused by missing a station or a personal name. Because the translation is limited to the domain of dialogs in the train timetable information center, we decided to solve the problem of OOV words similarly as in work [5], where the class-based language model was used for the real-time closed-captioning system of TV ice-hockey commentaries. The classes of player's names, nationalities and states were added into the standard language model in this work. Similarly, we added two classes into our language model - the class for all known station names: STATION and the class for all known personal names: PERSON. Because the semantic annotation of the corpus contains station and personal names, we can simply replace these names by relevant class in training and test data and collect a vocabulary of all station names for their translation (the personal names are always spelled).

The post-processing method includes two steps. Firstly, we can remove the words which are omitted in the translation process (they are translated into 'no translation' sign respectively) from the resulting translation. In any case, to keep these words in training data gives better results (more detailed translation and language models). Secondly, we can substitute OOV words by a finger-spelling sign, because the unknown words are finger spelled in the SC usually. The results of a decoder comparison are in Table 1. The results of SiMPaD and Moses decoder with the phrase-based and the standard trigram language model (suffix _LM(P)) are in the first and third column. The results of decoders with the class-based language model and the post-processing method (suffix _CLM(P)_PP) are in the second and forth column.

We compared the SiMPaD's results with the state-of-the-art phrase-based decoder Moses. We found that the SiMPaD's results are fully comparable with Moses's results while SiMPaD is almost 5 times faster than the Moses decoder. Hence, the SiMPaD is convenient for the real time translation, which is sufficient for the TTSL system. We introduced the class-based language model and the

---

[3] Available at http://www.isi.edu/~och/GIZA++.html.

Table 1. The results of SiMPaD and Moses decoder in Czech $\Longrightarrow$ Signed Czech translation

	SiMPaD_LMP	SiMPaD_CLMP_PP	Moses_LM	Moses_CLM_PP
SER[%]	44.84 ± 1.96	$\mathbf{40.59 \pm 2.06}$	45.30 ± 2.40	$\mathbf{41.97 \pm 2.20}$
BLEU	67.92 ± 1.93	$\mathbf{73.43 \pm 1.78}$	68.77 ± 1.72	$\mathbf{73.64 \pm 1.84}$
WER[%]	16.02 ± 1.08	$\mathbf{14.23 \pm 1.06}$	16.37 ± 1.02	$\mathbf{14.73 \pm 1.16}$
PER[%]	13.30 ± 0.91	$\mathbf{9.65 \pm 0.78}$	11.22 ± 0.82	$\mathbf{8.67 \pm 0.73}$

post-processing method which improved the translation results from about 8.1 % (BLEU) to about 27.4 % (PER) of a relative improvement in the case of SiMPaD decoder and from about 7.1 % (BLEU) to about 22.7 % (PER) of a relative improvement in the case of Moses decoder (the relative improvement is measured between the word-based model - _LM(P) and the class-based model with the post-processing - _CLM(P)_PP).

3 Conversion System

The conversion system is based on HamNoSys 3.0 notation. The notation is deterministic and suitable for the processing of the Sign Speech in a computer system. The methodology of the notation allows precise and also extensible expression of the sign. However an editor should be used for faster composition of correct selection of symbols to final string. Symbolic strings of notated signs are stored in a vocabulary.

3.1 Analysis of Symbols and Trajectories for Isolated Sign

The synthesis process is based on key frames and trajectories. Firstly, our synthesis system automatically carries out the analysis of symbolic string and generates a tree structure composed from key frames. Then, the feature and animation trajectories are created by the trajectory processing (see Figure 2).

The structure of the HamNoSys notation can be described by a context-free grammar. The grammar is defined by four symbol groups $G = (V_t; V_n; P; S)$. The inventory of terminal symbols V_t, non terminal symbols V_n, the set of parsing

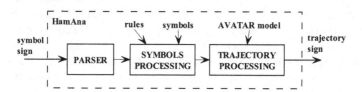

Fig. 2. The schema of our conversion system HamAna. The sign in symbolic meaning is transferred to parse tree. The symbol processing puts together the information from each symbol and the trajectory processing generates the animation trajectories.

the rule	the description
T (001) ∂	the symbol for hand shape ∂ is generalized on the symbol **T**
HT1 (026) T MT	the generalized hand shape **T** is modified by symbol **MT**
VHO (006) ⌈ HO2 ⌐ HO2 ⌉	the processing of orientation of left and right palm separately
HAMNOSYS (202) HH1	one of output rules, **HAMNOSYS** is the starting symbol

Fig. 3. The example of grammatical rules

rules P and starting symbol S were collected. The V_t symbols were directly derived from the HamNoSys symbols, the auxiliary symbols V_n were chosen according to the generalization of a HamNoSys notation structure and in the relation to the parsing rules P. The parsing rules were created to cover all HamNoSys notation variants. Example of 4 from all the 361 rules is in Figure 3. The form of the rule is following: one left non terminal symbol, a number of action and a right side of the rule (for example $HT1(026) \rightarrow T \quad MT$, where $HT1$ is left non terminal symbol, (026) is number of rule action and $T \quad MT$ is the right side). The right side is given by one or combination of V_t and V_n symbols. The number indicates the action joined with the rule. There are 28 rule actions for the symbol processing and 11 rule actions for the following trajectory processing. We implemented Earley's algorithm for the syntactic analysis. The structurally correct notations are then represented by a tree structure.

For accepted symbolic string, we have the parse tree and also the path to each leaf node. The path from a root of the tree to the leaf node of the particular symbol is given by the sequence of rule actions. The processing of nodes is carried out by several tree walks whilst the size of the tree is reduced. Each node is described by two key frames to separate dominant and non dominant hand. The structure of key frame is composed from specially designed items. These items are read for all leaf nodes from a symbol definition file. The list of all items and the example of definition file is depicted in Figure 4. Currently, the definition file covers 138 HamNoSys symbols. In the next processing, the key frames of the remaining nodes are joined and blended according to the rule actions.

The feature trajectory is created from key frames as the time sequence of feature frames in particular leaf nodes of the tree. The structure of a feature frame is derived from the key frame structure and contains only items for the static position of hand in space, the orientation of wrist and hand shape. The feature trajectories are created by next tree walk. Due to the notation of a superposition of notated movements, the relevant subtrees have to be marked by parallel flags. To overcome this, only the start and end feature frame of leaf nodes have to be precomputed. These frames are computed from the geometry of the animation model. In our approach, the location of hand is implicitly given by a position of wrist join in 3D space. In the case of the precise contact of dominant hand to other segments of animation model, the location of wrist for

the location	the pointer body segment the index of pointer segment the location segment name the array of location segment and indexes distance from location segment	***hamsym.dat***
		HAMSYM ˣ fingor 180.0 90.0 0.0
the motion	vector of relative translation (x,y,z) the type of motion the size of amplitude the amplitude gain the turn of motion amplitude the angles of circle sector	HAMSYM ☰ locsegname hanim_l5 idxloc 1 1 1 1 1 whichidxloc 2 distance 0.4
the orientation	the orientation of wrist (α,β,γ)	HAMSYM ~
the hand shape	the vector of hand shape (dim 21) the angles of three times finger flexion the mask of finger selection the shape of thumb	typemov zigzag turn 0.0 amplit 1.0

Fig. 4. On left is the list of all items. The relevant subsets are stored for each HamNoSys symbol. On right is the example of the stored items in the definition file: the orientation of hand, location on the body and the modifier of the direct movement.

the dominant hand is recomputed in the relation with the relevant pointer body segment. For this purpose, the algorithm computes firstly the location of non dominant arm thereafter the location of the dominant arm.

The following step of synthesis algorithm transfers the start and end frames to the relevant parent nodes. The transfer is controlled by the parallel flags and rule actions of the HamNoSys repetition modalities. In this state of the algorithm, the upper half-body movements or the some random motions of head have to be applied. The duration of trajectories is determined by number of frames. The frequency of frames is implicitly set to 25 frames per second. In order to get total duration of processed sign, the number of feature frames is inserted into leaf nodes and transferred into the root node. Here, the identical duration of trajectories in two parallel subtrees has to be taken into account.

The feature trajectories are computed for leaf nodes of the reduced tree according to type of a motion and other items in the relevant key frames. Next processing transfers these feature trajectories into the root node. The transferred trajectories are concatenated, merged, repeated or inversed according to rule actions of the trajectory processing. The trajectory for the dominant arm in root node is now complete and the trajectory for the non dominant arm is either empty or incomplete. If the symbols of a sign symmetry are notated then the trajectory for the non dominant arm is completed from the trajectory of the dominant arm or from the initial values. The final feature trajectories in the root node are transformed into the animation trajectories by an inverse kinematics technique to control the joints of animation model. The analysis of symbols allows the computation of trajectories only for hands and the upper half-body. Trajectories for the lip articulation and face gestures are produced by the "talking head" system separately.

3.2 Talking Head System

Trajectories for face gestures and also for articulation of lips, a tongue and jaws are created by visual synthesis carried out by the talking head subsystem. This visual synthesis is based on concatenation of phonetic units. Any word or phrase which represents an isolate sign in textual form, is here processed as a string of successive phones. The lip articulatory trajectories are concatenated by the visual unit selection method [6]. This synthesis method uses an inventory of phonetic units and the regression tree technique. It allows precise coverage of coarticulation effects. In the inventory of units, several realizations of phoneme are stored. Our synthesis method assumes that the lip and tongue shape is described by a linear model. The realization of a phoneme is described by 3 linear components for lip shape and 6 components for a tongue shape. The lip components represent a lip opening, a protrusion and an upper lip raise. The tongue components consist of a jaw height, a dorsum raise, a tongue body raise, a tip raise, a tip advance and a tongue width. The synthesis algorithm performs a selection of an appropriate phoneme candidate according to the phoneme context information. This information is built from a triphone context, the occurrence of a coarticulation resistant component (of the lip or the tongue) in adjacent phonemes and also from the time duration of neighboring speech segments. Final trajectories are computed by a cubic spline interpolation between the selected phoneme realizations.

These facial trajectories should be time-aligned with the timing of acoustic form of the relevant sign. This form is produced by an appropriate TTS system. The synthesis of face gesture trajectories is based on the concatenation and the combination of the neutral face expression and one of the 6 basic face gestures: happiness, anger, surprise, fear, sadness and disgust.

3.3 Synchrony of Facial Trajectories and Continuous Sign Speech Synthesis

The synchrony of the manual and non-manual components is crucial in the synthesis of continuous Sign Speech and ensures overall intelligibility. An asynchrony is caused by the different speech rate of the spoken speech and the Sign Speech. We designed an effective solution for Signed Czech. The synchrony method combines the basic concatenation technique with the time delay processing at the level of words. Firstly, for each isolated sign the animation trajectories from the analysis of symbols described in 3.1 and trajectories from the talking head system described in 3.2 are generated. The time delay processing determines the duration of both trajectories and selects the longest variant. The following step of processing evaluates an interpolation time for the concatenation of adjacent isolated signs in the synthesized utterance with regard to the longest variant. Thus added interpolation time ensures the fluent shift of a body pose. We select the simple linear interpolation of the frames on the boundaries of concatenated signs. The interpolation of a hand shape and its 3D position is determined by weight average, the finger direction and palm orientation is interpolated by the extension to the quaternion.

3.4   Animation Model

Our animation algorithm employs a 3D geometric animation model of the avatar in compliance with the H-Anim standard[4]. Our model is composed of 38 joints and body segments. These segments are represented by textured triangular surfaces. The setting of the correct shoulder and elbow rotations from a position of the wrist is solved by the inverse kinematics[5]. There are 7 degrees of a freedom for each limb. The settings of the remaining joints and local deformation of the triangular surfaces allows the animation of an arbitrary avatar pose. The local deformation of triangular surfaces is primarily used for the animation of the face and the tongue model. The surfaces are deformed according to animation schema based on the definition of several control points and splines functions [7]. The rendering of the animation model is implemented in C++ code and OpenGL. The animation model is shown in Figure 5.

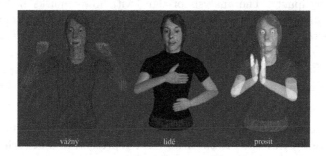

Fig. 5. The animation model for three signs: wire-frame topology, textured and blended rendering

4   Perceptual Evaluation

Two initial tests on an intelligibility of synthesized Sign Speech have been performed. The goal has been to evaluate a quality of our Sign Speech synthesizer. Two participants who are experts in the Sign Speech served as judges. We used the vocabulary of about 130 signs selected from the CSC corpus. We completed several video records of the avatar animation and also of the signing person. The video records of the signing person were taken from the electronic vocabulary [8]. The capturing of video records of our animation was prepared under two conditions.

4.1   Isolated Signs

The equivalence test was aimed at the comparison of animation movements of isolated signs with the movements of the signing person. Video records of 20

---

[4] Available at www.h-anim.org.
[5] Available at cg.cis.upenn.edu/hms/software/ikan/ikan.html.

pairs of randomly selected isolated signs were completed. The frontal view of the avatar model and the signing person was used in this test. The participants evaluated this equivalence by marks from 1 to 5. The meaning of the marks was:

- 1 totally perfect; the animation movements are equivalent to the signing person
- 2 the movements are good, the location of the hand, shapes or speed of the sign are a little different but the sign is intelligible
- 3 the sign is difficult to recognize; the animation includes mistakes
- 4 incorrectly animated movements
- 5 totally incorrect; it is a different sign

The results are shown in the left panel of Figure 6. The average mark of participant 1 is 2.25 and of participant 2 is 1.9. The average intelligibility is 70% (marks 1 and 2 indicate an intelligible sign). There was 65% mark agreement between participants. The analysis of signs with lower marks shows that the majority of mistakes are caused by the symbolic notation rather than inaccuracy in the conversion system. Thus, it is highly important to obtain as accurate symbolic notation of isolated signs as possible.

4.2 Continuous Speech

We created 20 video animation records of short utterances. The view of the avatar animation here was partially from the side. The participants judged the whole Sign Speech utterance. Subtitles (text representation of each sign) were added to the video records. Thus, the participants knew the meaning of the utterance and determined the overall intelligibility. The participants evaluate the intelligibility by marks from 1 to 5. The meaning of marks was:

- 1 the animation shows the signs from subtitles
- 2 the well intelligible utterance
- 3 the badly intelligible utterance
- 4 the almost unintelligible utterance
- 5 the totally unintelligible utterance

The results are shown in the right panel of Figure 6. All the utterances were evaluated by mark 1 or 2. On average, the animation of 70 % utterances shows

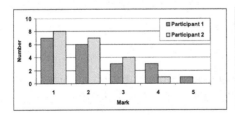 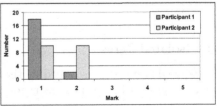

Fig. 6. Perceptual evaluation, left: isolated signs, right: continuous Sign Speech

the signs from subtitles. The results indicate that the synthesis of continuous speech is intelligible. The concatenation and synchrony method of isolated signs is sufficient.

5 Conclusion

The translation and conversion subsystems were introduced. The translation system is the phrase-based translation system which uses some heuristics (the monotone reordering and the longest phrase coverage) to speed up the translation process. The conversion system is based on the HamNoSys symbolic notation, which is capable to express the space configuration of each sign. The method of conversion the symbolic notation of sign to appropriate animation was presented.

The perceptual tests reveal that the synchrony on the level of word preserves the intelligibility for continuous Sign Speech. However the intelligibility of isolated signs highly depends on symbolic notation of particular signs in the vocabulary. Thus, it is necessary to concentrate on the acquisition of precise symbolic notation of isolated signs in future work.

References

1. Koehn, P., et al.: Statistical Phrase-Based Translation. In: HLT/NAACL (2003)
2. Koehn, P., et al.: Moses: Open Source Toolkit for Statistical Machine Translation. In: Annual Meeting of the Association for Computational Linguistics (ACL), demonstration session, Prague, Czech Republic (June 2007)
3. Kanis, J., et al.: Czech-Sign Speech Corpus for Semantic Based Machine Translation. In: Sojka, P., Kopeček, I., Pala, K. (eds.) TSD 2006. LNCS (LNAI), vol. 4188, pp. 613–620. Springer, Heidelberg (2006)
4. Papineni, K.A., et al.: Bleu: A method for automatic evaluation of machine translation, Technical Report RC22176 (W0109-022), IBM Research Division, Thomas J. Watson Research Center (2001)
5. Hoidekr, J., et al.: Benefit of a class-based language model for real-time closed-captioning of TV ice-hockey commentaries. In: Proceedings of LREC 2006. Paris: ELRA, pp. 2064–2067 (2006), ISBN 2-9517408-2-4
6. Krňoul, Z., Železný, M., Müller, L., Kanis, J.: Training of Coarticulation Models using Dominance Functions and Visual Unit Selection Methods for Audio-Visual Speech Synthesis. In: Proceedings of INTERSPEECH 2006 - ICSLP, Bonn (2006)
7. Krňoul, Z., Železný, M.: Realistic Face Animation for a Czech Talking Head. In: Proceedings of 7th International Conference on TEXT, SPEECH and DIALOGUE TSD 2004, Springer, Heidelberg (2004)
8. Langer, J., et al.: Znaková zásoba českého znakového jazyka. Palacký University Olomouc (2005)

Using Prosodic Features
in Language Models for Meetings

Songfang Huang and Steve Renals

The Centre for Speech Technology Research
University of Edinburgh
Edinburgh, EH8 9LW, UK
{s.f.huang,s.renals}@ed.ac.uk

Abstract. Prosody has been actively studied as an important knowledge source for speech recognition and understanding. In this paper, we are concerned with the question of exploiting prosody for language models to aid automatic speech recognition in the context of meetings. Using an automatic syllable detection algorithm, the syllable-based prosodic features are extracted to form the prosodic representation for each word. Two modeling approaches are then investigated. One is based on a factored language model, which directly uses the prosodic representation and treats it as a 'word'. Instead of direct association, the second approach provides a richer probabilistic structure within a hierarchical Bayesian framework by introducing an intermediate latent variable to represent similar prosodic patterns shared by groups of words. Four-fold cross-validation experiments on the ICSI Meeting Corpus show that exploiting prosody for language modeling can significantly reduce the perplexity, and also have marginal reductions in word error rate.

1 Introduction

Prosody has long been studied as a knowledge source for speech understanding, and has been successfully used for a variety of tasks, including topic segmentation [1], disfluency detection [2], speaker verification [3], and speech recognition [4,5,6].

Recently there has been an increasing research interest in multiparty conversations, such as group meetings. Speech in meetings is more natural and spontaneous than read or acted speech. The prosodic behaviours for speech in meetings are therefore much less regular. Can prosody aid the automatic processing of multiparty meetings? Shriberg et al. [2] gave the answer 'yes' to this question, from the evidence of successfully exploiting prosodic features for predicting punctuation, disfluencies, and overlappings in meetings. It has also been noted that prosodic features can serve as an efficient non-lexical feature stream for tasks such as dialogue acts (DA) segmentation and classification, speech summarization, and topic segmentation and classification in the meetings domain.

This paper is concerned with the question of exploiting prosody to aid automatic speech recognition (ASR) in the context of meetings. Three essential components in a state-of-the-art ASR system, namely the acoustic model, language

A. Popescu-Belis, S. Renals, and H. Bourlard (Eds.): MLMI 2007, LNCS 4892, pp. 192–203, 2007.
© Springer-Verlag Berlin Heidelberg 2007

model (LM), and lexicon, can all potentially serve to accommodate prosodic features. In this paper we are interested in exploiting prosodic features in language models for ASR in meetings.

The goal of a language model is to provide a predictive probability distribution for the next word conditioned on the strings seen so far, i.e., the immediately preceding $n-1$ words in a conventional n-gram model. In addition to the previous words, prosodic information associated with the audio stream, which is parallel to the word stream, can act as a complementary knowledge source for predicting words in LMs. This understanding is the initial motivation for this work.

Due to the large vocabulary size in LMs (typically greater than 10,000 words), incorporating prosodic information in language models is more difficult than in other situations such as DA classification which has a much smaller number of target classes (typically several tens). To exploit prosody for LMs, a central question is how the relationship between prosodic features F and the word types W, $P(W|F)$, may be modeled. In this paper, two models will be investigated, namely the factored language model (FLM) [7] and the hierarchical Bayesian model (HBM) [8]. In the FLM-based approach, conditional probabilities $P(W|F)$ are directly estimated from the co-occurrences of words and prosody features via maximum likelihood estimation (MLE). The HBM-based approach provides a richer probabilistic structure by introducing an intermediate latent variable—in place of a direct association between words and prosodic features—to represent similar prosodic patterns shared by groups of words. This work is characterised by an automatic and unsupervised modeling of prosodic features for LMs in two senses. First, the prosodic features, which are syllable-based, are automatically extracted from audio. Second, the association of words and prosodic features is learned in an unsupervised way.

The rest of this paper is organized as follows. The next section reviews some related work on exploiting prosody for ASR. The ICSI Meeting Corpus, used throughout this paper, is described in Sect.3. The extraction of prosodic features is discussed in Sect.4. Section 5 focuses on the modeling approaches, including FLM-based and HBM-based methods. Experiments and results are reported in Sect.6, followed by a discussion in the final section.

2   Related Work

It is well accepted that humans are able to understand prosodic structure without lexical cues. Sub-lexical prosodic analysis [9] attempts to mimic this human ability using syllable finding algorithms based on band pass energy. Prosodic features are then extracted at the syllable level. The extraction of syllable-based prosodic features is attractive, because the syllable is accepted as a means of structuring prosodic information. This approach was verified on DA and hotspot categorization [9], which encourages us to utilize syllable-based prosodic features in LMs for ASR.

A basic approach to incorporate prosodic features in acoustic models for ASR uses "early integration", in which the prosodic features are appended to the standard acoustic features [10]. Early work to utilize prosody in language models used prosodic features to evaluate possible parses for recognized words, which in turn would be the basis for reordering word hypotheses [11]. More recently, approaches that integrate prosodic features with LMs have emerged, in which LMs are conditioned on prosodic evidence by introducing intermediate categories. Taylor et al. [12] took the dialogue act types of utterances as the intermediate level, by first using prosodic cues to predict the DA type for an utterance and then using a DA-specific LM to constrain word recognition. Stolcke et al. [4] instead used prosodic cues to predict the hidden event types (filled pause, repetition, deletion, repair) at each word boundary with hidden event n-gram model, and then conditioned the word portion of the n-gram on those hidden events. Chan et al. [6] proposed to incorporate prosody into LMs using maximum entropy. However, the prosodic features they used were derived from manual ToBI transcriptions. An example of using prosody in the lexicon was provided by Chen et al. [5], where prosodic features, such as stress and phrase boundary, were included in the vocabulary. Each word had different variations corresponding to stress and whether or not it precedes a prosodic phrase boundary. This approach attempted to capture the effects of how prosodic features affect the spectral properties of the speech signal and the co-occurrence statistics of words.

Most research on using prosodic features for ASR has been applied to small and task-oriented databases. The goal of effectively using prosody for large-vocabulary speech recognition, such as recognition of meeting speech, still remains elusive. There has been little work in this direction in the meeting domain. One reason for this is due to the difficulty of modeling the relationship between symbolic words and normally non-symbolic prosodic features. Therefore, to find an approximate prosodic representation for each word in the vocabulary is one way to use prosodic features for ASR.

Realizing the difficulty of modeling prosody via intermediate representations, Shriberg et al. proposed direct modeling of prosodic features [13]. In this approach, prosodic features are extracted directly from the speech signal. Machine learning techniques (such as Gaussian Mixture Models, and decision trees) then determine a statistical model to use prosodic features in predicting the target classes of interest. No human annotation of prosodic events is required in this case. However, using prosodic features to predict very large number of target categories like words will again fail in capturing the prosodic discriminabilities.

3   Meeting Corpus

The experiments reported here were performed using the ICSI Meeting Corpus [14], which is a corpus of 75 naturally-occurring, unrestricted, and fairly unstructured research group meetings, each averaging about an hour in length. We performed our experiments using a four-fold cross-validation procedure in which we trained on 75% of the data and tested on the remaining 25%, rotating until all the data was tested on. The corpus was divided into four folds, first by

Table 1. The summary of the four-fold cross-validation setup on the ICSI Meeting Corpus used in this paper

Fold	Number of Sentences	Number of Tokens
0-fold	27,985	209,766
1-fold	27,981	208,554
2-fold	27,968	208,294
3-fold	27,975	205,944

ordering all the sentences in sequence, and then for each fold sequentially selecting every fourth sentence. After further removing the sentences that are too short in length to extract prosodic features, this procedure resulted in the data set summarised in Table 1.

4   Prosodic Feature

A notable aspect of the prosodic features used here is that they are syllable-based. It is reasonable to address prosodic structures at the syllable level, because prosodic features relating to the syllable reflect more clearly perceptions of accent, stress and prominence. The syllable segments were automatically detected based solely on the parallel acoustic signals using an automatic syllable detection algorithm. The framework for the extraction of syllable-based prosodic features is shown in Fig.1, which follows an approach to automatic syllable detection suggested by Howitt [15], which in turn was originated in work by Mermelstein [16].

1. **Front-end Processing.** The speech signal was first framed using a 16 ms Hamming window with a shift period of 10 ms. The raw energy before windowing and pre-emphasis was computed for each frame and saved in log magnitude representation for subsequent silence detection. A 256-point FFT was used to compute the power spectrum.
2. **Silence Detection.** The raw energy data was smoothed using a 6th-order low-pass 50 Hz filter. Each frame was classified into either speech or silence

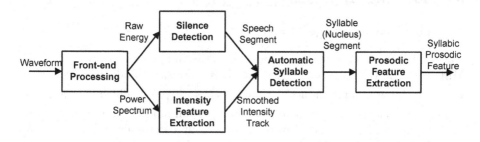

Fig. 1. The framework of the extraction of syllable-based prosodic features

based solely on whether or not the log frame energy was above a threshold. A running window consisting of 10 consecutive frames was used to detect the onsets of speech and silence. The detected speech segments, which were further extended by 5 frames at both sides, were fed into the following syllable detection.

3. **Intensity Feature Extraction.** A single measure of intensity was computed, following Howitt's adjusted features [15]. A 300–900 Hz band-pass filter was used to filter out energy not belonging to vowels. By a weighted summation (converted to magnitude squared forms) of the spectral bins within 300–900 Hz frequencies from the spectrogram, an intensity track (converted back to decibels) was computed for syllable detection, which again was smoothed by a low-pass 50 Hz filter to help reduce small peaks and noise.

4. **Automatic Syllable Detection.** The recursive convex hull algorithm [16], which is a straightforward and reliable syllable detection algorithm, was used to find the nuclei by detecting peaks and dips in the intensity track computed in the above step. The syllables were then obtained by extending the nuclei on both sides, until a silence or a boundary of adjacent nuclei is detected.

5. **Prosodic Feature Extraction.** Four prosodic features were extracted for each syllable consisting of the duration of syllable, the average energy, the average F0, and the slope of F0 contour. F0 information was obtained using the ESPS get_f0 program.

We ran vector quantization (VQ), with 16 codewords (labeled 's0' to 's15') over all the 892,911 observations of syllable-based prosodic features in the ICSI Meeting Corpus. Before running VQ, each feature was normalized to unit variance.

The syllables belonging to an individual word were obtained by aligning the word with the syllable stream according to a forced time alignment at the word level, and selecting those syllables whose centres were within the begin and end times of words. By concatenating relevant VQ indices for syllables, we obtained the symbolic representations of prosodic features at the word level, which can then serve as potential cues for language modeling. For example, the prosodic representation for word 'ACTUALLY' might be the symbol 's10s12s6', or 's10s15s6' in other contexts.

5   Modeling Approach

5.1   Factored Language Model

One straightforward method for modeling words and prosodic features is to use MLE based on the co-occurrences of words W and the prosodic representations F, i.e., training a unigram model $P(W|F) = \frac{\text{Count}(F,W)}{\text{Count}(F)}$. This unigram model can then be interpolated with conventional n-gram models. More generally, we can use the FLM [7] to model words and prosody deterministically. The FLM, initially developed to address the language modeling problems faced by morphologically rich or inflected languages, is a generalization of standard n-gram language models, in which each word w_t is decomposed into a bundle of K word-related features (called *factors*), $w_t \equiv f_t^{1:K} = \{f_t^1, f_t^1, \ldots, f_t^K\}$. Factors may include the word itself. Each word in an FLM is dependent not only on a single stream of its preceding words, but also on additional parallel streams of factors. Combining with interpolation or generalized parallel backoff (GPB) [7] strategies, multiple backoff paths may be used simultaneously. The FLM's factored representation can potentially accommodate the multimodal cues, in addition to words, for language modeling—in this case the prosodic representations. This configuration allows more efficient and robust probability estimation for those rarely observed word n-grams.

Supposing the word w_t itself is one of the factors $\{f_t^1, f_t^1, \ldots, f_t^K\}$, the joint probability distribution of a sequence of words $(w_1, w_2, \ldots, w_T)$ in FLMs can be represented as the formalism shown in (1), according to the chain rule of probability and the n-gram-like approximation.

$$P(w_1, w_2, \ldots, w_T) = P(f_1^{1:K}, f_2^{1:K}, \ldots, f_T^{1:K})$$
$$= \prod_{t=1}^{T} P(f_t^{1:K}|f_{t-1}^{1:K}, f_{t-2}^{1:K}, \ldots, f_1^{1:K})$$
$$\approx \prod_{t=1}^{T} P(w_t|f_{t-n+1:t-1}^{1:K}) \qquad (1)$$

There are two key steps to use FLMs. First an appropriate set of factor definitions must be chosen. We employed two factors: the word w_t itself and the syllable-based prosodic representation f_t, as shown in Fig.2(A). Second it is necessary to find the suitable FLM models (with appropriate model parameters and interpolation/GPB strategy) over those factors. Although this task can be described as an instance of the structure learning problem in graphical models, we heuristically designed the model structure for FLMs. It is convenient to regard this FLM-based model as an interpolation of two conventional n-gram models $P(w_t|w_{t-1}, w_{t-2})$ and $P(w_t|w_{t-1}, f_t)$:

$$P_{\text{FLM}}(w_t|w_{t-1}, w_{t-2}, f_t) = \lambda_{\text{FLM}} P(w_t|w_{t-1}, w_{t-2}) + (1-\lambda_{\text{FLM}}) P(w_t|w_{t-1}, f_t) \quad (2)$$

Figure 2(B) shows the parallel backoff graph used in the experiments for factors w_t and f_t. We perform the interpolation in a GPB framework, as depicted in Fig.2, manually forcing the backoff from $P(w_t|w_{t-1}, w_{t-2}, f_t)$ to two parallel paths by setting a very large value of $gtmin$ for $P(w_t|w_{t-1}, w_{t-2}, f_t)$.

5.2 Hierarchical Bayesian Model

We argue that it is essential but difficult to find intermediate symbolic representations to associate words and low-level prosodic features for language modeling. In this paper, we have categorized syllable-based prosodic features into 16 classes, and represented the prosodic features for each word as a concatenation of indices for syllables belonging to that word. The FLM-based approach uses

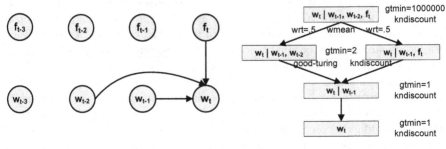

A. The Graphical Model Representation for FLM   **B. Generalized Parallel Backoff**

Fig. 2. (A) A directed graphical model representation for the factor configuration in a FLM over factors including words w_t, the prosodic representations f_t. (B) The generalized parallel backoff graph for w_t and f_t used in the experiments.

this prosodic information by directly associating word and prosodic representations. One limitation of this FLM-based approach is that there may be too many varieties of prosodic representations for individual words, due to the errors introduced by the automatic syllable detection and forced alignment. For example, the word 'ABSOLUTELY' in the ICSI Meeting Corpus has more than 100 different prosodic representations. Language models trained via MLE using such prosodic representations will be more likely to overfit to the training data. Rather than the direct association of words and prosodic representations, we introduce a latent variable between word and prosody and assume a generative model that generates words from prosodic representations through the latent variable. This probabilistic generative models is investigated within the framework of hierarchical Bayesian models [8].

Topic models have recently been proposed for document modeling to find the latent representation (*topic*) connecting documents and words. Latent Dirichlet allocation (LDA) [17] is one such topic model. LDA is a three-level hierarchical Bayesian model, in which each document is represented as a random mixture over latent topics, and each topic in turn is represented as a mixture over words. The topic mixture weights $\boldsymbol{\theta}$ are drawn from a prior Dirichlet distribution:

$$P(\boldsymbol{\theta}|\boldsymbol{\alpha}) = \frac{\Gamma(\sum_{i=1}^{K}\alpha_i)}{\prod_{i=1}^{K}\Gamma(\alpha_i)}\theta_1^{\alpha_1-1}\ldots\theta_K^{\alpha_K-1} \tag{3}$$

where $\boldsymbol{\alpha} = \{\alpha_1,\ldots,\alpha_K\}$ represents the prior observation count of the K latent topics with $\alpha_i > 0$. The LDA model is based on the "bag-of-words" assumption, that is, words in a document exchangeably co-occur with each other according to their coherent semantic meanings. In this sense, LDA can be considered as a probabilistic latent semantic analysis model. However what if we assume that words in a document exchangeably co-occur with each other according to their coherent prosodic patterns? This is the intuition of our use of LDA for the probabilistic association of words and prosody, which we call the prosody-topic model.

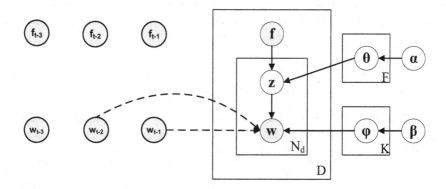

Fig. 3. The graphical model representation for the prosody-topic model (right) and its interaction with n-gram model (left), where shaded nodes denote observed random variables, while unshaded ones denote latent variables or parameters. The boxes are 'plates' representing the replications of a corresponding substructure.

In a prosody-topic model, a document in the corpus is composed by including all those words that have the same prosodic representation (i.e., 's10s12s6'). The prosodic representation is then served as the author of that document. If we apply LDA over this corpus, we can extract the latent 'topics' connecting words and prosodic representations. Each topic is expected to have coherent prosodic patterns. Considering our prosodic representations in this paper, for example, words in one individual topic are expected to have the same number of syllables whose pronunciations are similar. Unlike LDA, we need to explicitly retain the prosodic representations in the prosody-topic model. On the other hand, if we regard the prosodic representations as the 'authors' for corresponding documents, the prosody-topic model leads to the author-topic model [18], in which each document has only one unique author.

In short, the general idea of the prosody-topic model is that each prosodic representation is represented by a multinomial distribution over latent topics, and each topic is represented as a multinomial distribution over words. Prosody thus serves the same role as semantics, being the guideline to cluster co-occurring words in a document. The goal of a prosody-topic model is to learn the distribution of words for each topic, which therefore finds the latent representations association the word and prosodic representations. The graphical model for prosody-topic model is shown in Fig.3, and the generative process for each document d can be described as follows:

1. Select the unique prosodic representation (author) label f for document d.
2. Choose topic proportions $\theta|\{f, \theta_{1:F}\}$ for document d according to f, each $\theta_f \sim Dirichlet(\alpha)$
3. For each of the N_d words w_n in document d:
   (a) Choose a topic $z_n|\theta \sim Mult(\theta)$.
   (b) Choose a word $w_n|\{z_n, \phi_{1:K}\} \sim Mult(\phi_{z_n})$, $\phi_{z_n} \sim Dirichlet(\beta)$.

Since each document only has a single author, the probability of words w_t given prosodic representations f_t in a prosody-topic model can be easily obtained by integrating out latent topics, as shown in (4):

$$P_{\text{HBM}}(w_t|f_t) = \sum_{k=1}^{K} P(t_k|f_t)P(w_t|t_k) = \sum_{k=1}^{K} \theta_{f_t t_k} \phi_{t_k f_t} \qquad (4)$$

where t_k is one of the K topics, $\theta_{f_t t_k}$ and $\phi_{t_k f_t}$ can be learned by approximate inference methods, such as variational EM or Gibbs sampling. This unigram-like probability can be interpolated with conventional n-gram models:

$$P_{\text{HBM}}(w_t|w_{t-1}, w_{t-2}, f_t) = \lambda_{\text{HBM}} P(w_t|w_{t-1}, w_{t-2}) + (1 - \lambda_{\text{HBM}})P_{\text{HBM}}(w_t|f_t) \quad (5)$$

6 Experiment and Result

We evaluated the FLM- and HBM-based approaches on the 4-fold cross-validation ICSI Meeting Corpus as described in Sect.3, in terms of perplexity (PPL) and word error rate (WER) respectively.

The FLM models were trained using the SRILM [19] toolkit[1], which has an extension for FLMs. Some modifications were made to the FLM toolkit regarding the manner of dealing with some special symbols such as '<s>', '</s>', and 'NULL', e.g., we manually set $P(w_t|w_{t-1}, w_{t-2}, \text{NULL}) = P(w_t|w_{t-1}, w_{t-2})$, and scored the end-of-sentence '</s>' in perplexity calculations to account for the large number of short sentences in the meeting corpus. The FLM models share a common closed vocabulary of 50,000 word types with the AMI-ASR system [20]. The smoothing methods and parameters for FLM models are shown in Fig.2.

The prosody-topic models were trained using a publicly available Matlab topic modeling toolbox[2]. The algorithm for inference is Gibbs sampling [21], a Markov chain Monte Carlo algorithm to sample from the posterior distribution. We chose the number of topics $K = 100$, and ran the Gibbs sampling algorithm for 2500 iterations, which took around one hour to finish the inference on a 3-fold ICSI data. Instead of automatically estimating the hyperparameters α and β, we fixed these two parameters to be $50/K$ and 0.01 respectively, as in [18].

The PPL results were obtained by successively testing on the specific fold with the language model trained on the other three folds. The interpolation weights λ_{FLM} and λ_{HBM} were both set to 0.5. Table 2 shows the PPL results on the 4-fold cross-validation ICSI Meeting Corpus. Both FLM-based and HBM-based approaches produce some reduction in PPL, especially the HBM-based approach has over 10% relative reduction in PPL than the baseline trigram model. One interesting thing we found during analysing the PPL results sentence-by-sentence is that those having higher probabilities than baseline trigrams normally have reasonable prosodic representations for words, i.e., representing the right number of syllables in a word.

---

[1] http://www.speech.sri.com/projects/srilm/
[2] http://psiexp.ss.uci.edu/research/programs_data/toolbox.htm

Table 2. PPL results for 4-fold cross-validation experiments. **BASELINE-3G** denotes the baseline trigram results using the FLM toolkit. **FLM-3G-F** denotes the results for the FLM-based model, while **HBM-3G-F** for the HBM-based prosody-topic model.

TRAIN-TEST	BASELINE-3G	FLM-3G-F	HBM-3G-F
123 – 0	78.4	73.6	70.5
023 – 1	78.9	73.9	70.7
013 – 2	78.3	73.4	70.1
012 – 3	78.3	73.3	70.8
AVERAGE	78.5	73.5	70.5

Table 3. Word error rate results, which share the same notations as in Table 2, except that the **BASELINE-2G** column represents the baseline results from the first-pass AMI-ASR decoding using an interpolated bigram model.

TRAIN-TEST	BASELINE-2G	BASELINE-3G	FLM-3G-F	HBM-3G-F
123–0	29.8	29.5	29.2	29.1
023–1	29.6	29.3	29.1	29.0
013–2	29.5	29.2	29.0	28.9
012–3	29.4	29.2	29.1	29.0
AVERAGE	29.6	29.3	29.1	29.0

Table 3 shows the WER results of n-best rescoring on the ICSI Meeting Corpus. It should be noted that the BASELINE-2G WER results were obtained during the first-pass decoding of the AMI-ASR system using an interpolated bigram LM trained on seven text corpora including Hub4, Switchboard, ICSI Meeting, and a large volume (around 1GB in size) of web data. The lattices were generated using this interpolated bigram LM. By retaining the time information for candidate words, the lattices were then used to produce n-best lists with time stamps for subsequent rescoring experiments via the *lattice-tool* program in the SRILM toolkit. In our experiments, the 500-best lists were produced from the lattices, which were then aligned with the syllable streams to get prosodic representation for each word, and finally reordered according to scores of different interpolated LMs to search for the best hypothesis. Marginal reductions in WER were observed in our experiments.

7   Discussion and Future Work

In this paper we have investigated two unsupervised methods to exploit syllable-based prosodic features in language models for meetings. Experimental results on the ICSI Meeting Corpus showed our modeling approaches, both FLM-based and HBM-based, have significant reductions in PPL and marginal reductions in WER. The limited gains in WER may be partly caused by the following reasons. First, there are inevitably some errors in automatic syllable detection. It is hard for us to carry out evaluations on our syllable detection algorithm because of the lack of annotated data with syllable information. Second, additional errors are introduced by the forced alignment due to the overlapping cross-talk in meetings, which occasionally assigned an unreasonable number (i.e., more than 10) of syllables to a simple word. Third, the lattices were generated by an interpolated bigram model trained on a large corpus. This might prevent the recovery of more probable hypotheses from those n-best lists produced by a generalized LM, using specific LMs only trained on ICSI meeting data for rescoring.

Considering the two modeling approaches, we are more interested in the HBM-based method. Bayesian language models [22], which provide an internally coherent probabilistic models and fit well in the hierarchical Bayesian model framework, have been proved to have comparable performance to the conventional n-gram models. In future, we will consider more tighter incorporation rather than simple interpolation, i.e., investigating the prosody-topic model and (Bayesian) language models in one united generative model within the hierarchical Bayesian framework. Moreover, meeting-specific cues will be taken into consideration for the prosody-topic model. For example, prosody encodes some information for DAs. DA in meetings normally has well-defined types. It is interesting to extend the prosody-topic model by investigating the relationship between word, prosody, and DA in one generative model.

Acknowledgement

We thank the AMI-ASR team for providing the lattices for rescoring. This work is jointly supported by the Wolfson Microelectronics Scholarship and the European Union 6th FWP IST Integrated Project AMI (Augmented Multi-party Interaction, FP6-506811, publication AMI-243).

References

1. Shriberg, E., Stolcke, A., Hakkani-Tür, D.Z., Tür, G.: Prosody-based automatic segmentation of speech into sentences and topics. Speech Communication 32(1-2), 127–154 (2000), Special Issue on Accessing Information in Spoken Audio
2. Shriberg, E., Stolcke, A., Baron, D.: Can prosody aid the automatic processing of multi-party meetings? evidence from predicting punctuation, disfluencies, and overlapping speech. In: Proceedings of ISCA Tutorial and Research Workshop on Prosody in Speech Recognition and Understanding, Red Bank, NJ (2001)
3. Sönmez, K., Shriberg, E., Heck, L., Weintraub, M.: Modeling dynamic prosodic variation for speaker verification. In: Proceedings of 5th International Conference on Spoken Language Processing, Sydney, Australia, pp. 3189–3192 (1998)
4. Stolcke, A., Shriberg, E., Hakkani-Tür, D., Tür, G.: Modeling the prosody of hidden events for improved word recognition. In: Proceedings of 6th European Conference on Speech Communication and Technology, Budapest, Hungary, pp. 307–310 (1999)
5. Chen, K., Hasegawa-Johnson, M.: Improving the robustness of prosody dependent language modelling based on prosody syntax dependence. In: IEEE Workshop on Automatic Speech Recognition and Understanding, St. Thomas, U.S. Virgin Islands, pp. 435–440. IEEE Computer Society Press, Los Alamitos (2003)

6. Chan, O., Togneri, R.: Prosodic features for a maximum entropy language model. In: ICSLP 2006. Proceedings of Interspeech, Pittsburgh, US (2006)
7. Bilmes, J.A., Kirchhoff, K.: Factored language models and generalized parallel backoff. In: Proceedings of HLT/NACCL, pp. 4–6 (2003)
8. Gelman, A., Carlin, J.B., Stern, H.S., Rubin, D.B.: Bayesian Data Analysis. Chapman & Hall/CRC, London (1995)
9. Aylett, M.P.: Detecting high level dialog structure without lexical information. In: Proceedings of ICASSP 2006, Toulouse, France (2006)
10. Lei, X., Siu, M., Hwang, M.-Y., Ostendorf, M., Lee, T.: Improved tone modeling for mandarin broadcast news speech recognition. In: ICSLP 2006. Proceedings of Interspeech, Pittsburgh, US (2006)
11. Veilleux, N.M., Ostendorf, M.: Prosody/parse scoring and its applications in ATIS. In: Proceedings of ARPA HLT Workshop, Plainsboro, NJ, pp. 335–340 (1993)
12. Taylor, P., King, S., Isard, S., Wright, H.: Intonation and dialog context as constraints for speech recognition. Language and Speech 41(3-4), 489–508 (1998)
13. Shriberg, E., Stolcke, A.: Direct modeling of prosody: An overview of application in automatic speech processing. In: Proceedings of International Conference on Speech Prosody, Nara, Japan (2004)
14. Janin, A., Baron, D., Edwards, J., Ellis, D., Gelbart, D., Morgan, N., Peskin, B., Pfau, T., Shriberg, E., Stolcke, A., Wooters, C.: The ICSI meeting corpus. In: Proceedings of IEEE ICASSP, Hong Kong, China, IEEE Computer Society Press, Los Alamitos (2003)
15. Howitt, A.W.: Automatic Syllable Detection of Vowel Landmarks. PhD thesis, Massachusetts Institute of Technology (2000)
16. Mermelstein, P.: Automatic segmentation of speech into syllabic units. Journal Acoustical Society of America 58(4), 880–883 (1975)
17. Blei, D.M., Ng, A.Y., Jordan, M.I.: Latent Dirichlet allocation. Journal of Machine Learning Research 3 (2003)
18. Rosen-Zvi, M., Griffiths, T., Steyvers, M., Smyth, P.: The author-topic model for authors and documents. In: Proceedings of the 20th Conference on Uncertainty in Artificial Intelligence, Banff, Canada (2004)
19. Stolcke, A.: SRILM - an extensible language modeling toolkit. In: Proceedings of International Conference on Spoken Language Processing, Denver, Colorado (2002)
20. Hain, T., Dines, J., Gaurau, G., Karafiat, M., Moore, D., Wan, V., Ordelman, R.J.F., Renals, S.: The development of the AMI system for the transcription of speech in meetings. In: Proceedings of 2nd Joint Workshop on Multimodal Interaction and Related Machine Learning Algorithms, Edinburgh, UK (2005)
21. Griffiths, T.L., Steyvers, M.: Finding scientific topics. In: Proceedings of the National Academy of Sciences, 101 (suppl. 1), pp. 5228–5235 (2004)
22. Teh, Y.W.: A hierarchical Bayesian language model based on Pitman-Yor processes. In: Proceedings of the Annual Meeting of the ACL, vol. 44 (2006)

Posterior-Based Features and Distances in Template Matching for Speech Recognition

Guillermo Aradilla and Hervé Bourlard

IDIAP Research Institute,
Rue du Simplon 4, Martigny, Switzerland
{aradilla,bourlard}@idiap.ch
http://www.idiap.ch

Abstract. The use of large speech corpora in example-based approaches for speech recognition is mainly focused on increasing the number of examples. This strategy presents some difficulties because databases may not provide enough examples for some rare words. In this paper we present a different method to incorporate the information contained in such corpora in these example-based systems. A multilayer perceptron is trained on these databases to estimate speaker and task-independent phoneme posterior probabilities, which are used as speech features. By reducing the variability of features, fewer examples are needed to properly characterize a word. In this way, performance can be highly improved when limited number of examples is available. Moreover, we also study posterior-based local distances, these result more effective than traditional Euclidean distance. Experiments on Phonebook database support the idea that posterior features with a proper local distance can yield competitive results.

Keywords: Speech Recognition, Template Matching, Posterior Features, KL-divergence, Bhattacharyya, Multi-Layer Perceptron.

1 Introduction

Hidden Markov models (HMMs) constitute the dominant approach for automatic speech recognition (ASR) systems. Their success is mainly based on their efficient algorithms for training and testing. However, these algorithms rely on some assumptions about data that do not hold for speech signals, such as piecewise stationary or independence of the feature vectors given a state. Template matching (TM) is a different approach for ASR that relies on the fact that a class can be described by a set of examples (templates). Since templates are real utterances, they can better model the dynamics of the trajectories generated by the speech features compared with HMM states in currently used monophone or triphone models. Moreover, TM is preferred in those cases where simplicity and flexibility for training and testing must be considered.

As a non-parametric approach, TM requires more training data than parametric models, such as HMM-based systems, to obtain comparable performance. Given

A. Popescu-Belis, S. Renals, and H. Bourlard (Eds.): MLMI 2007, LNCS 4892, pp. 204–214, 2007.

the increase of large speech corpora and computational resources, TM has recently drawn new attention. Investigation on this approach has been focused on increasing the number of templates [1,2,3] and, hence, improving its generalization capabilities. Since no speech corpora can guarantee to provide many examples for each word, sub-word units are typically used to ensure that a large enough number of templates is available for each possible word. Pronunciation dictionaries are, in this case, needed for concatenating these sub-word units into words. However, pronunciation of the words is not always easy to obtain, e.g., proper names.

We propose a different method to use the information contained in large speech corpora. Traditional features used in TM are based on short-term spectrum. These features contain linguistic information but also information about the gender[1] and the environment, i.e., they are speaker and task-dependent. In this work, we investigate the use of posterior probabilities of subword units as speech features. These posteriors can be estimated from a multilayer perceptron (MLP) which has been trained on large speech corpora. In this way, the MLP can capture the information contained on large speech corpora to generate speaker and task-independent features. Given the discriminative training procedure of the MLP and the long acoustic context used as input, posterior features are known to be more stable and more robust to noise than spectral-based features [4]. Since these features only contain, in theory, linguistic information, fewer templates are required to represent a word. Hence, in those applications where the number of available templates is few, we can expect to improve the performance. Posteriors estimates from the MLP outputs have already been successfully applied as features for ASR using HMM/GMM as acoustic model, system known as Tandem [4,5].

TM-based approaches traditionally use Euclidean or Mahalanobis distance as local similarity measure between features. These distances implicitly assume that features follow a Gaussian distribution. This assumption does not hold when using posterior distributions as features. Since posterior features are probability distributions over the space of subword units, more appropriate distances can be considered. In this work, we investigate local distances between frames that take into account the discriminative properties of posterior features.

This work is an extension of a previous experiment where we already applied posterior features to a TM-based ASR system [6]. On that first experiment, posterior features were not task-independent because the data to train the MLP belonged to the same database as the test set. Kullback-Leibler (KL) divergence was applied as local distance for being a natural distance between distributions. In this work, the MLP is trained on a large speech corpus and used for a different recognition task. We also show that other types of local distances can be successfully applied to posterior features which obtain similar performance to KL-divergence but are faster to compute.

This paper is summarized as follows: Section 2 introduces the TM approach for speech recognition, Section 3 presents the posteriors features, Section 4 describes the local distances investigated in this work, Section 5 presents the experiments and results and finally, Section 6 draws some conclusions.

---

[1] For instance, speaker recognition systems use spectral-based features as inputs.

2 Template Matching

TM is a non-parametric classifier that relies on the idea that a class w can be identified by a set of N_w examples (templates) $\{\mathbf{Y}_n^w\}_{n=1}^{N_w}$ belonging to that class. Unlike parametric models, TM directly uses all training data at the decoding time and no explicit assumption is made about the data distribution. A test element $\mathbf{X}$ is associated to the same class as the closest sample based on a similarity function φ between samples defined as:

$$\text{class}(\mathbf{X}) = \arg\min_{\{w'\}} \min_{\mathbf{Y}' \in \{\mathbf{Y}_n^{w'}\}} \varphi(\mathbf{X}, \mathbf{Y}') \tag{1}$$

where $\{w'\}$ denotes the set of all possible classes. However, as any non-parametric technique, a large amount of training data is required to obtain a good classification performance. TM has recently received new attention in the ASR field because current computational resources and speech corpora allow to deal with large amount of training data in a practical computational time.

In the case of speech, templates are sequences of feature vectors that correspond to particular pronunciations of a word. When comparing with HMMs, templates can describe in more detail the dynamics of the trajectories defined by speech features because they represent real utterances, whereas HMMs are parametric representations that summarize the information contained on the speech trajectories. Furthermore, the explicit use of non-linguistic information such as gender or speech rate can be easily applied when using templates but this type of long-span information is more difficult to incorporate into a parametric model.

The similarity measure φ between sequences must deal with the fact that utterances usually have different lengths. This measure is based on dynamic time warping (DTW) [7] and it minimizes the global distortion between two temporal sequences. This global distortion is computed as the sum of local distances $d(\mathbf{x}, \mathbf{y})$ between the matched frames. This matching is performed by warping one of the two sequences. In speech, the template sequence is typically warped so that every template frame $\mathbf{y}_m$ matches a frame of the test sequence $\mathbf{x}_n$. Given a template sequence $\{\mathbf{y}_m\}_{m=1}^{M}$ and a test sequence $\{\mathbf{x}_n\}_{n=1}^{N}$, DTW-based distance can be expressed as

$$\varphi(\mathbf{X}, \mathbf{Y}) = \min_{\{\phi\}} \sum_{i=1}^{N} d(\mathbf{x}_i, \mathbf{y}_{\phi(i)}) \tag{2}$$

where $\{\phi\}$ denotes the set of all possible warping functions for the template sequence. The warping function must hold some constraints of continuity and boundaries to ensure that the resampled template sequence is realistic. Typical constrains in the ASR field are:

$$0 \le \phi(i) - \phi(i-1) \le 2$$
$$\phi(1) = 1 \tag{3}$$
$$\phi(M) = N$$

These conditions guarantee that no more than one frame from the template sequence will be skipped for each test frame and also, that every test frame will be related to only one template frame.

Although the computation of (2) implies searching among a large set of warping functions, it can be efficiently computed by dynamic programming.

The local distance $d(\mathbf{x}, \mathbf{y})$ is typically chosen as Euclidean or Mahalanobis distance since spectral-based features are normally used for representing the speech signal. However, other types of similarity measures between frames can also be applied depending on the properties of the features. In Section 4, a description of the local distances investigated in this work will be given.

As described before, recent investigation to improve the performance of TM-based ASR systems is to take advantage of the current large speech corpora and computational resources by increasing the number of templates. TM becomes then a search problem among all possible templates [1]. In order to increase the speed and efficiency of the search, non-linguistic information can be used at the decoding time [8]. As templates and HMMs convey different types of information since they are different types of models, investigation has also been carried out for combining both approaches [2,3] with successful results. However, this technique requires a large amount of samples per word (or linguistic unit). In this work, we will focus on the situation where a few samples are given for every word. In this case, the goal is to reduce as much as possible the variability within a class so that a few samples will be enough to represent a class word. This variability reduction will be performed at the feature level and will be explained in detail in the next section.

3   Posterior Features

The posterior probability $p(q_k|z_t)$ of a phoneme q_k given a spectral-based acoustic feature z_t at time t can be estimated from a MLP. A posterior vector x_t can then be obtained where each dimension corresponds to a phoneme posterior $x_t = \{p(q_k|z_t)\}_{k=1}^{K}$. K corresponds to the total number of phonemes and is also the number of MLP outputs[2].

If posterior estimates were correct, these features could be considered as optimal speech features by assuming that words are formed by phonemes since, in theory, they only carry linguistic information and also, they can be seen as optimal phone detectors as it is demonstrated in [9]. This reduction of the undesirable information makes posterior features more stable as it is illustrated in Figure 1.

Traditional features, like MFCC [10] or PLP [11], contain information about the spectrum and hence, about the speaker and its environment. However, posterior features can be considered speaker and task-independent since they only contain information about the phoneme that has been pronounced. Rigorously

---

[2] We are using this notation for the sake of simplicity, but in fact an acoustic context (typically 4 frames) is used as input of the MLP, hence, rigorous notation should be $p(q_k|z_{t-\Delta}^{t+\Delta})$.

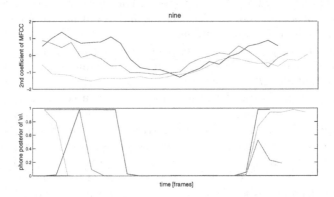

Fig. 1. The value of the second component of the feature vector in the case of MFCC features and phone posterior corresponding to the phoneme /n/ are plotted for three different templates of the word "nine". It can be seen that values from spectral-based feature vectors are more variable within a phone than posterior features, which follow a more stationary trajectory.

speaking, posterior features are not task-independent since the MLP is implicitly learning the prior probability of each phoneme, which will be dependent of the database. However, when using large vocabulary corpora, these probabilities converge to phoneme priors of the language of the database. In this way, posterior features are language-dependent.

4   Local Distance

From (2), it can be observed that DTW-based distance requires a distance $d(\mathbf{x}, \mathbf{y})$ between reference and test samples of the observation space. Since any local distance assumes a particular geometry of the observation space, the choice of the local distance plays a crucial role on the performance of the system. Traditionally, these distances are based on Euclidean and Mahalanobis distances. In the TM-based approach, investigation has been recently carried out to estimate the parameters of the weighting matrix of the Mahalanobis distance to improve the performance. A maximum-likelihood estimation was described in [12] and a discriminative procedure was presented in [13]. However, these methods require a large amount of data to properly estimate the weights.

Since posterior vectors can be seen as distributions over the space of subword units (e.g., phonemes), measures from the information theory field can be applied. These measures can capture higher order statistics from the data than Euclidean-based distances. Furthermore, they can explicitly consider the particular properties of posterior vectors (i.e., values must be non-negative and sum must be equal to one).

In the following, we will consider that y represents a frame from the template and x denotes a frame from the test sequence. As explained before, x and y can be considered discrete distribution on the $\mathbb{R}^K$ space (i.e. there are K different phonemes).

In addition, local distance directly affects the decoding time since computing the local distance is the most frequent operation on the DTW algorithm. Hence, the choice of the local distance should also take into account its computational time.

4.1   Squared Euclidean Distance

This is the traditional distance used as local distance between frames. However, it is related with the Gaussian distribution. Indeed, when taking the logarithm of a Gaussian distribution with unity covariance matrix, it becomes the squared Euclidean distance plus a constant factor.

$$D_{Eucl}(x, y) = \sum_{k=1}^{K} (x(k) - y(k))^2 \tag{4}$$

However, when measuring the similarity between posterior features, Euclidean distance is not very appropriate since posterior space holds some special properties which are not taken into account by this distance.

4.2   Kullback-Leibler Divergence

KL divergence (or relative entropy) comes from the information theory field and can be interpreted as the amount of extra bits that are needed to code a message generated by the a reference distribution y, when the code is optimal for a given test distribution x [14].

$$D_{KL}(x \,\|\, y) = \sum_{k=1}^{K} y(k) \log \frac{y(k)}{x(k)} \tag{5}$$

KL-divergence is a natural measure between distributions. The fact that it is not symmetric must not affect its application to DTW algorithm. In this case, the reference distribution y is considered to be the template frame whereas x corresponds to the test frame.

4.3   Bhattacharyya Distance

This distance was initially motivated by geometrical considerations since it computes the cosine between two distributions [15]. It is also a particular case of the Chernoff bound (an upper bound for the Bayes error) [16].

$$D_{Bhatt}(x, y) = -\log \sum_{k=1}^{K} \sqrt{x(k)y(k)} \tag{6}$$

Bhattacharyya distance is symmetric and also it is faster to compute than KL divergence because less logarithms must be computed. This distance has been used already in speech processing for phone clustering [17].

4.4 Distance Based on Bayes Risk

Bhattacharyya distance is originated from an upper bound of the Bayes risk. However, the exact probability of error can be easily computed for discrete distributions [18]:

$$\text{Bayes Error} = \sum_{k=1}^{K} \min\{x(k), y(k)\} \tag{7}$$

A distance can be derived similar to Bhattacharyya distance by taking the negative logarithm:

$$D_{Bayes}(x, y) = -\log \sum_{k=1}^{K} \min\{x(k), y(k)\} \tag{8}$$

This distance is even simpler to compute than (6) because it avoids the square root function.

5   Experiments

5.1 Description

In this work, Phonebook database has been used to carry out word recognition experiments using the TM-based approach. This database consists of 47455 utterances of isolated words. There are 3992 different words pronounced by around 12 different speaker in average. Experiments with different lexicon sizes have been carried out: 5, 10, 20, 50 and 100 different words were selected randomly from the global lexicon. For each experiment and each word, one or two utterances have been selected as templates and the rest of utterances containing the selected words have been used for test. Since lexicon has been selected at random, experiments have been repeated ten times using a different lexicon at each time. Results have been consistent, i.e., similar results have been obtained at each time and average results are shown.

Two types of features have been considered: PLP and phoneme posterior probabilities. PLP features also contain delta features. Posterior features have been obtained from a MLP trained on 30 hours of the CTS database following the MRASTA procedure [19]. The MLP contains 2000 hidden units and 46 phonemes (including silence) have been considered.

Constraints for DTW are the same as described in Formula 3. Euclidean, KL-divergence, Bhattacharyya and Bayes-based distance are considered as local distances. PLP features only use Euclidean distance (the rest of local distance can only be applied to discrete posterior vectors).

Experiments on decoding time have been carried out on a workstation with a Athlon64 4000+ processor.

5.2 Results

Results on Table 1 show the effectiveness in using posterior features for TM. PLP features contain information about the speaker and since the task is speaker-independent, results when using these spectral-based features are far from being competitive. This explains why TM is mainly focused on speaker-dependent tasks with small vocabulary. On the other hand, posterior features have been estimated by taking into account the information captured by the MLP from the large speech corpus used for training. This, jointly with the discriminative training of the MLP make posterior features robust to speaker and environment conditions.

Table 1. System accuracy when using one or two templates per word. The size of the lexicon has been varied to observe the effect of increasing the lexicon. The last column shows the average number of test utterances.

lexicon size	one template					two templates					# test utts
	PLP Eucl	Posteriors Eucl	KL	Bhatt	Bayes	PLP Eucl	Posteriors Eucl	KL	Bhatt	Bayes	
5	79.3	93.2	98.2	98.7	98.0	90.8	96.6	98.9	98.9	98.5	55
10	74.7	91.9	97.8	98.3	97.5	85.4	95.7	98.9	98.9	98.4	104
20	69.8	89.5	95.6	96.5	95.7	81.9	94.2	98.4	97.9	97.5	212
50	59.7	83.1	92.9	94.1	92.9	74.2	90.2	96.6	96.8	96.1	545
100	53.2	78.5	89.7	91.4	89.7	68.0	87.5	94.9	95.1	94.2	1079

Moreover, posterior-based distances such as KL divergence, Bhattacharyya and Bayes-based distance yield better results than traditional Euclidean distance since they explicitly deal with the space topology of the posterior features.

Figure 2 plots the system accuracy with two templates per word and also shows the effect of increasing the size of the lexicon. When using 100 different words, the performance of the system is still around 95%, which is reasonable result given the complexity of the task and the limited amount of samples per word[3].

Experiments have been carried out to investigate the effect of the local distance on the decoding time. Results are shown in Figure 3. It can be observed that KL-divergence takes a long time for decoding because of the logarithm function. Bhattacharyya distance replaces the logarithm function by a square root function, which takes less time than the logarithm. Bayes-based distance is faster than the previous since selecting the minimum value is a very simple operation. Finally, Euclidean distance is faster than the rest but its accuracy is significantly worse than the other distances.

---

[3] Experiments comparing templates and hybrid HMM/MLP [20] have been carried out using the test set described in [21]. There are 8 different test sets consisting each one of 75 different words. In this case, we obtained similar results in both systems, i.e. around 95% accuracy.

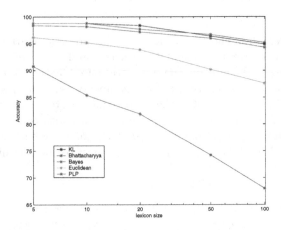

Fig. 2. Accuracy of the system using 2 templates per word

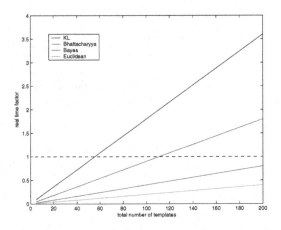

Fig. 3. This figure shows the real time factor depending on the total number of templates. Real time factors is defined as the ratio between the decoding time and the duration of the test sequence. A dashed line indicates when the decoding time is equal to the duration of the sequence.

6   Conclusion

In this work we have tested the effectiveness of posterior features on a TM-based approach. Since these features have been trained using a large vocabulary database, they can be considered speaker and task-independent. These properties make these features very suitable for those conditions where a word must be represented by a few examples. Moreover, the choice of the local distance has been investigated since it both assumes a topology on the feature space and also directly affects the decoding time. Though KL-divergence is a very appropriate local distance when using posterior features, it takes a long time to be computed because it requires a logarithm function for each dimension of the posterior vector. Other types of distances based on the probability of error have also been investigated which are simpler to compute and yield similar performance.

Future work should be focused on investigating other ways to incorporate information of large speech corpora on TM-based approach. A possible way would be to combine the posterior features from different MLPs. Initial experiments have already been carried out with successful results.

Acknowledgments. This work was supported by the EU 6th FWP IST integrated project AMI (FP6-506811). The authors want to thank the Swiss National Science Foundation for supporting this work through the National Centre of Competence in Research (NCCR) on "Interactive Multimodal Information Management (IM2)".

References

1. Wachter, M.D., Demuynck, K., Compernolle, D.V., Wambacq, P.: Data Driven Example Based Continuous Speech Recognition. In: Proceedings of Eurospeech, pp. 1133–1136 (2003)
2. Aradilla, G., Vepa, J., Bourlard, H.: Improving Speech Recognition Using a Data-Driven Approach. In: Proceedings of Interspeech, pp. 3333–3336 (2005)
3. Axelrod, S., Maison, B.: Combination of Hidden Markov Models with Dynamic Time Warping for Speech Recognition. In: ICASSP 2004. Proceedings of International Conference on Acoustics, Speech and Signal Processing, vol. 1, pp. 173–176 (2004)
4. Q. Zhu, B.C., Morgan, N., Stolcke, A.: On Using MLP features in LVCSR. Proceedings of International Conference on Spoken Language Processing (ICSLP) (2004)
5. Hermansky, H., Ellis, D., Sharma, S.: Tandem Connectionist Feature Extraction for Conventional HMM Systems. In: ICASSP 2000. Proceedings of International Conference on Acoustics, Speech and Signal Processing (2000)
6. Aradilla, G., Vepa, J., Bourlard, H.: Using Posterior-Based Features in Template Matching for Speech Recognition. In: ICSLP 2006. Proceedings of International Conference on Spoken Language Processing (2006)
7. Rabiner, L., Juang, B.H.: Fundamentals of Speech Recognition. Prentice-Hall, Englewood Cliffs (1993)
8. Aradilla, G., Vepa, J., Bourlard, H.: Using Pitch as Prior Knowledge in Template-Based Speech Recognition. In: ICASSP 2006. Proceedings of International Conference on Acoustics, Speech, and Signal Processing (2006)
9. Niyogi, P., Sondhi, M.M.: Detecting Stop Consonants in Continuous Speech. The Journal of the Acoustic Society of America 111(2), 1063–1076 (2002)
10. Davis, S.B., Mermelstein, P.: Comparison of parametric representations for monosyllabic word recognition in continuously spoken sentences. IEEE Transactions on Audio, Speech and Signal Processing 28, 357–366 (1980)
11. Hermansky, H.: Perceptual Linear Predictive (PLP) Analysis of Speech. The Journal of the Acoustic Society of America 87 (1990)

12. Wachter, M.D., Demuynck, K., Wambacq, P., Compernolle, D.V.: A Locally Weighted Distance Measure For Example Based Speech Recognition. In: ICASSP 2004. Proceedings of International Conference on Acoustics, Speech and Signal Processing, pp. 181–184 (2004)

13. Matton, M., Wachter, M.D., Compernolle, D.V., Cools, R.: A Discriminative Locally Weighted Distance Measure for Speaker Independent Template Based Speech Recognition. In: ICSLP 2004. Proceedings of International Conference on Spoken Language Processing (2004)

14. Cover, T.M., Thomas, J.A.: Information Theory. John Wiley, Chichester (1991)

15. Bhattacharyya, A.: On a Measure of Divergence between Two Statistical Populations Defined by their probability distributions. Bull. Calcutta Math. Soc. 35, 99–109 (1943)

16. Fukunaga, K.: Introduction to Statistical Pattern Recogntion. Academic Press, London (1990)

17. Mak, B., Barnard, E.: Phone Clustering Using the Bhattacharyya Distance. In: ICSLP 1996. Proceedings of International Conference on Spoken Language Processing, pp. 2005–2008 (1996)

18. Duda, R.O., Hart, P.E., Stork, D.G.: Pattern Classification. Wiley, Chichester (2001)

19. Hermansky, H., Fousek, P.: Multi-Resolution RASTA Filtering for TANDEM-based ASR. In: Proceedings of Interspeech (2005)

20. Bourlard, H., Morgan, N.: Connectionist Speech Recognition: A Hybrid Approach, vol. 247. Kluwer Academic Publishers, Boston (1993)

21. Dupont, S., Bourlard, H., Deroo, O., Fontaine, V., Boite, J.M.: Hybrid HMM/ANN Systems for Training Independent Tasks: Experiments on Phonebook and Related Improvements. In: ICASSP 1997. Proceedings of International Conference on Acoustics, Speech and Signal Processing (1997)

A Study of Phoneme and Grapheme Based Context-Dependent ASR Systems

John Dines[1] and Mathew Magimai Doss[1,2]

[1] IDIAP Research Institute
P.O. Box 592, Martigny, CH-1920, Switzerland
[2] International Computer Science Institute, 1947 Center St, Berkeley, CA 94704
{dines,mathew}@idiap.ch

Abstract. In this paper we present a study of automatic speech recognition systems using context-dependent phonemes and graphemes as subword units based on the conventional HMM/GMM system as well as tandem system. Experimental studies conducted on three different continuous speech recognition tasks show that systems using only context-dependent graphemes can yield competitive performance on small to medium vocabulary tasks when compared to a context-dependent phoneme-based automatic speech recognition system. In particular, we demonstrate the utility of tandem features that use an MLP trained to estimate phoneme posterior probabilities in improving grapheme based recognition system performance by implicitly incorporating phonemic knowledge into the system without having to define a phonetically transcribed lexicon.

1 Introduction

State-of-the art automatic speech recognition (ASR) systems represent words as a sequence of sub-word units, typically phonemes which have a strong correlation with the acoustic observations. In recent studies, attention has been drawn toward speech recognition systems using grapheme as sub-word units [1,2,3,4]. The main advantages of using grapheme as sub-word units are (1) the definition of lexicon is easy (orthographic transcription), and (2) the pronunciation models are relatively noise free. The main drawback of using graphemes as sub-word units is that a single grapheme can map onto many different phonemes, i.e. there is often a weak correspondence between graphemes and acoustic observations, particularly in the English language.

Schukat-Talamazzaini et al. were one of the first to present results in speech recognition based on graphemes [4]. They used "polygraph" sub-word units for word modelling, which is essentially letters-in-context similar to polyphones (phonemic units allowing preceding and following context of arbitrary length). Experimental studies conducted on continuous speech recognition task and isolated word recognition showed that good results (better than context-independent phone) can be obtained using "polygraph" as sub-word units.

A. Popescu-Belis, S. Renals, and H. Bourlard (Eds.): MLMI 2007, LNCS 4892, pp. 215–226, 2007.

In a recent study, the approach of mapping orthographic transcription to a phonetic one has been investigated in the context of speech recognition [1]. In this approach, the orthographic transcription of the words are used to map them onto acoustic hidden Markov model (HMM) state models using phonetically motivated decision tree questions. For instance, a grapheme is assigned to a phonetic question if the grapheme maps to the phoneme. Recognition studies performed on Dutch, German and English yielded performances comparable to phoneme-based ASR system for the Dutch and German languages, and fairly poor performance for the English language.

Killer et al. have investigated context dependent grapheme based speech recognition, where the context is modelled through a decision tree based clustering procedure [2]. Experimental studies conducted on English, German and Spanish languages yielded competitive results compared to phoneme-based system for German and Spanish languages, but once again fairly poor performance for the English language.

In [5,3], we proposed a phoneme-grapheme based system that jointly models both the phoneme and grapheme sub-word units during training. During decoding, recognition is done either using one or both sets of sub-word units. This system was investigated in the framework of a hybrid hidden Markov model/artificial neural network (HMM/ANN) system. Improvements were obtained over a context-independent phoneme based system using both sub-word units in recognition on two different tasks: isolated word recognition [3] and recognition of numbers [5].

In this paper, we present a study of context-dependent phonemes and graphemes as sub-word for English ASR systems. We analyse the use of grapheme as sub-word units for English ASR by comparing it with the standard phoneme based system using two different features (standard PLP cepstral feature [6] and tandem feature [7]) on three tasks of increasingly complexity: OGI Numbers95 (NU95) [8], DARPA resource management (RM) [9] and continuous telephone speech (CTS) [10]. Our studies show that on tasks of smaller complexity such as NU95 the grapheme based ASR system can perform as good as the phoneme based ASR system. At the same time, on tasks of increased complexity such as RM and CTS the performance difference between the two systems, phoneme based system and grapheme based system, becomes more pronounced with the phoneme based system being the better one. Our studies also show that on these tasks of increased complexity the difference between the two systems is greatly reduced when using tandem features.

2   Background

Lexical representations play a critical role in ASR. In all but the most constrained tasks, it is necessary to represent words by a sequence of sub-word units (the so called 'beads-on-a-string' paradigm), in order to give a compact representation of the lexicon that still provides good correspondence between the words and acoustic observations. Most commonly, sub-word units take the form of phonemes, as they are limited in number (of the order of 45 for English) and show good correspondence with the acoustic observations. One disadvantage of the use of phonemes is that mapping from words to phonemes is generally a knowledge driven process that is difficult to automate with a high level of fidelity, thus making it an expensive process in terms of development time and effort. Automatic means for deriving pronunciations in text-to-speech synthesis exist in order to enable such systems to handle out-of-vocabulary text, but generally such mechanisms are not employed in ASR. Interestingly enough, if we examine two of the most commonly used techniques in letter-to-sound mapping and ASR lexical representations, we see that context-dependency plays a critical role. In this section, we describe the use of context dependent sub-word units in letter-to-sound mapping and acoustic modelling for ASR, drawing attention to the similarities between the two. We also briefly describe the tandem acoustic features, which feature significantly in our studies.

2.1 Letter-to-Sound Mapping Using Decision Trees

In text-to-speech synthesis it is often necessary to produce pronunciations for words that lie outside the pronunciation dictionary of the system. Such systems employ letter-to-sound (LTS) mapping techniques to automatically generate pronunciations. A common approach to this problem is to use decision trees [11]. The decision tree approach is carried out by first aligning grapheme and phoneme symbols from a pronunciation dictionary that is to be used for training[1]. For each grapheme occurrence the graphemes surrounding it (a context window of N to the left and right) are recorded as well as the phoneme which has been aligned to the grapheme. The decision tree is trained from this data by pooling all of the instances of a particular grapheme together then successively splitting the data according to the grapheme context that gives rise to the largest decrease in leaf node impurity (entropy of the leaf's phoneme distribution times number of sample points). By building a decision tree in this manner a set of rules is derived that use a grapheme's context to determine its pronunciation.

2.2 Context Dependent Modelling of Sub-word Units

Word pronunciations can differ greatly from their lexical form, particularly due to the effects of coarticulation, making it common practice to explicitly model each sub-word unit according to the context in which it occurs. Due to the limitations of data coverage and decoding complexity, a single phone context to the left and right (the 'triphone') is generally used. Even then, a large quantity of data is required in order to independently learn the statistics of each context dependent unit, hence, a parameter sharing scheme is needed. The most commonly employed parameter sharing scheme is the decision tree-tying approach [12], which pools all of the data for a particular sub-word unit into a single

---

[1] Extra measures need to be taken to deal with words which have fewer/more phonemes than graphemes.

root node and performs tree growth by selecting the question at each split that maximises the increase in likelihood of the acoustic models over the training data. The decision tree approach not only achieves more robust modelling of seen contexts, but also enables the synthesis of unseen contexts.

The questions used to split the data may be singleton (each question relates to only a single sub-word unit), knowledge based (eg. phonemic: "is the left phone context a VOWEL") or data driven [13]. In general, the knowledge based approach is used as it gives both good data utilisation and generalisation, in particular for unseen contexts, but clearly, for grapheme based systems, only singleton and data driven can be used. Killer et. al. [2] explored different approaches to question set derivation for context-dependent grapheme based speech recognition and demonstrated that, in fact, the singleton questions sets gave best results, though with the disadvantage of inefficient data utilisation compared to data-driven approaches. It should also be noted that context dependent modelling of grapheme-based sub-word units displays strong similarities with the letter-to-sound mapping described in the previous section, since we are learning a mapping from the graphemic representation to the acoustic feature space, which is much more strongly correlated with the phonemic representation.

2.3   Tandem Acoustic Features

Tandem systems have been shown to yield state-of-the-art performance in ASR [7]. A tandem system combines the discriminative training of an ANN with Gaussian mixture modelling by using the processed posterior probabilities generated by an MLP as the input feature for the HMM/GMM-based system. It has been demonstrated that tandem features exhibit greater robustness to unwanted variabilities [14,15]. This is due to the ability of the ANN to project multiple frames of acoustic features onto dimensions carrying information most pertinent to the speech recognition task.

A tandem based system can also be viewed as a cascade of classifiers, thus, permitting the integration of decisions made in an earlier classification stage into later stages. Tandem acoustic features are of interest in this study as they present a means of introducing phonetic knowledge into a grapheme based system through the use of an MLP trained on phonemic targets, without the need for explicit specification of a phonemic pronunciation dictionary (though phonemic targets are still required for the training of the MLP, this can be performed on a corpus where phonetic transcriptions are available).

3   Empirical Studies

3.1   Experimental Setup

Our studies were conducted on three well known speech corpora that comprise tasks of varying complexity with regard to training data, lexicon and language model. The major features of each corpora are listed in Table 1, highlighting their respective differences.

Table 1. Summary of the three corpora used in our studies. CI: context independent, CD: content dependent.

Name	Component	Description	Statistic
OGI Numbers95	Audio data	Quantity of data	
		Train:	90 mins
		Test:	30 mins
	Lexicon	Closed[2]	
		Words:	31
		Phoneme (CI/CD):	24/81
		Graphemes (CI/CD):	19/85
	Acoustic model	Word internal, context dependent	
	Language model	Wordloop	
DARPA RM	Audio data	Quantity of data	
		Train:	3.8 hrs
		Test:	1.1 hrs
	Lexicon	Closed	
		Words:	991
		Phonemes (CI/CD):	42/2269
		Graphemes (CI/CD):	29/1912
	Acoustic model	Word internal, context dependent	
	Language model	Wordpair	
CTS[3]	Audio data	Quantity of data	
		Train:	32 hrs
		Test:	1.3 hrs
	Lexicon	Open	
		Words:	1000
		Phonemes (CI/CD):	47/20k
		Graphemes (CI/CD):	36/9k
	Acoustic model	Cross-word, context dependent	
	Language model	Bigram	

Acoustic models were trained for the three corpora using the hidden Markov model toolkit (HTK) from both PLP and tandem-features [16] . In each case, the acoustic models were trained through: 8 iterations of re-estimation on context-independent models, 2 iterations of re-estimation on context-dependent models followed by model tying, 7 iterations of re-estimation on tied context-dependent models and finally increment of mixtures from 1 to 8 in multiples of two with 3 iterations of re-estimation at each increment step. In these studies we investigated singleton, knowledge-based and data driven question sets for state tying. We used a fixed log-likelihood threshold to control decision tree growth, thus models were allowed to achieve differing levels of complexity based on the sub-word units, features, and question sets used.

---

[2] Meaning that the same words appear in train and test data.

[3] We use the CTS task as defined in [10], which has been designed to have reduced complexity for training and evaluating ASR systems on CTS data.

In making comparisons between systems on the same task and on different tasks we make the following caveats. First of all, we are primarily interested on differentiating between systems based on the sub-word units and how this is affected by features and question sets. We note that due to the approach taken with respect to decision tree tying, our systems will have a different number of parameters, but also point out that allowing clustering to proceed with a lower log-likelihood threshold would not likely contribute to ASR performance (though may have yielded models with similar number of states, hence, parameters). Conversely, by focusing on the result of the clustering procedure, rather than being concerned with model complexity, we are able to make some observations on the role of features and questions sets on the clustering process itself.

PLP feature extraction comprised 13th order PLP cepstral coefficients and their deltas and delta-deltas. The features were computed every 10ms over a window of 30 ms. For the tandem-features, an MLP was trained on the PLP features with output units corresponding to context-independent phonemes. The phoneme targets for MLP training were derived from a forced alignment of the training data using the PLP based acoustic models. We extracted the tandem-features using the MLP's phoneme log-posterior estimates followed by Karhunen-Loeve transformation. In the grapheme dictionary, the numbers and abbreviated words were replaced by their graphemic representation eg. 45 $\Rightarrow$ FORTY FIVE.

3.2   OGI Numbers95

The OGI numbers95 (NU95) database comprises a limited vocabulary task that employs a word-loop language model. In our experiments we used the definition of the training set, validation set, and test set based on that defined in [17]. For the purposes of investigating different lexical representations, this is a very simple task. In comparing the ASR systems produced from context dependent phoneme and grapheme models shown in Table 2 we can see that the complexity of the acoustic models is quite similar with the grapheme system having slightly more models/states than its phoneme based counterpart. This is reflected in the overall performance of the grapheme system, which has slightly lower error rates than the phoneme system. The tandem based systems had the same performance on this task, this being significantly better than that obtained from PLP features. While these results suggest that phoneme and grapheme system can achieve equivalent performance, it is clear that this is because both the context dependent grapheme and phoneme acoustic models have an almost one-is-to-one mapping to their corresponding lexical entry.

3.3   DARPA Resource Management

We next performed ASR evaluations on the DARPA resource management (RM) corpus. This corpus is also of relatively low complexity compared to state-of-the-art evaluation tasks, but is still quite a step up from the OGI numbers task. In particular, the lexicon is greatly increased from 31 to almost 1000, thus context dependent models may no longer have a unique mapping to a single word. The

Table 2. ASR results on OGI Numbers95 task

Unit	Feature	Quest	Log. Models	Phy. Models	Log. States	Phys. States	WER (in %)
Phoneme	PLP	Phonemic	81	74	241	191	6.3
	Tandem	Phonemic	81	74	241	193	4.4
Grapheme	PLP	Singleton	85	79	256	198	5.9
	Tandem	Singleton	85	78	256	196	4.4

Table 3. ASR results on DARPA resource management task

Unit	Feature	Quest	Log. Models	Phy. Models	Log. States	Phys. States	WER (in %)
Phoneme	PLP	Singleton	2269	1501	6729	1477	5.7
	Tandem	Singleton	2269	1628	6729	2013	5.7
Grapheme	PLP	Singleton	1912	1298	5727	1369	7.3
	Tandem	Singleton	1912	1360	5727	1985	6.3
Merged	PLP	Singleton	4181	2799	12456	2846	5.5
	Tandem	Singleton	4181	2988	12456	3998	5.1

lexicon is still closed, thus it is not necessary for the acoustic models to generalise to words not seen in training, nor is it necessary to synthesise unseen contexts.

The results from the experiments on the RM corpus are shown in Table 3. We originally extended our analysis of the RM corpus in order to better compare the systems by building systems using both singleton and data driven questions sets (according to [13]). We only report the results for singleton questions sets here as the data driven approach was not found to provide significantly different results on this task.

A number of observations can be made from these results. In particular we can note that for both PLP and tandem features the number of physical states in the grapheme and phoneme systems is roughly equivalent, despite there being fewer actual (logical) states for the grapheme system. This demonstrates that the decision tree growth for grapheme based models needs to be deeper (more questions) in order to model the more complex relationship between graphemes and the feature space. In particular, the grapheme based context dependent modelling must account for the many-to-one mapping associated with LTS, in addition to the challenges associated with conventional phoneme based modelling such as coarticulation.

In comparing the PLP and tandem feature based systems we see that tandem features provide a significant improvement for the grapheme based system, although, for this task, it still remains behind that of the phoneme based system. We also observe that tandem based features lead to a greater number of states, mostly likely due to there being less unwanted variability in the tandem features, which leads to better separation of context-dependent state distributions and thus more effective clustering. This is particularly important for the grapheme system where co-articulatory effects further complicate the task of learning the relationship between the feature space and the context dependent grapheme models.

In a last test we also merged phoneme and grapheme acoustic models and lexica (without retraining), thus enabling a mixture of grapheme and phoneme based models to be used during recognition. We see that this gives a slight improvement over both phoneme and grapheme systems, suggesting that the grapheme models, while giving overall inferior performance to the phoneme system, still manage to achieve some degree of complementarity. I.e. grapheme modelling is not just an inferior alternative to phoneme modelling. Further analysis performed using the merged models and dictionaries on the development set of DARPA RM task showed that grapheme models were more preferred for function words which short in terms of length (number of graphemes). We also measured the mutual information between context independent and dependent phoneme and grapheme labels at the frame level, but the analysis of results did not provide any additional insights.

3.4 Conversational Telephone Speech

The final evaluation carried out as part of this study was with the conversational telephone speech (CTS) corpus. This corpus is significantly more complex than those previously described in that it is an open lexicon (meaning that words may appear in testing that do not appear during training), although for the task definition that we chose the lexicon is of similar size to that used in RM [10]. Furthermore, the acoustic conditions are significantly more challenging as the audio is taken from a telephone channel. In training the context-dependent models on the CTS corpus, we made one change to the training procedure, which was to allow for cross-word context dependency. Due to the increased complexity of the task we have only conducted limited investigations on the CTS task, namely the male part of the corpus. The results are detailed in Table 4.

One of the first points that stands out from these results is the discrepancy between the number of logical models and physical states in the phoneme and grapheme systems. The phoneme system has twice the number of logical models (by virtue of the fact that there are more phonemes than graphemes), but conversely half as many physical states. This is partly due to the fact

Table 4. Preliminary ASR results on male part of the CTS task

Unit	Feature	Quest	Log. Models	Phy. Models	Log. States	Phys. States	WER (in %)
Phoneme	PLP	Phonemic	20810	5601	62430	1325	45.7
	Tandem	Phonemic	20640	7370	61920	1786	45.3
Grapheme	PLP	Singleton	9309	4435	27927	2602	53.0
	Tandem	Singleton	9278	4125	27834	2885	50.3

that the singleton question set will naturally lead to deeper decision trees, but can also be attributed to the greater complexity required in modelling context-dependent graphemes. This is consistent with the findings in Black et. al. [11], who demonstrated that using an early stopping criterion to prevent over-fitting of decision tree learning of letter-to-sound mappings was actually detrimental to performance.

Further observations from these results may also be noted. First of all, once again the tandem features appear to provide some improvement in both phoneme and grapheme systems, particularly in the grapheme case. Unfortunately though, the grapheme tandem system still lags significantly behind the phoneme system. In addition to the inherent difficulties in using grapheme based sub-word units, we can also attribute additional factors to this loss in performance. The use of cross-word context dependent models made the grapheme based system significantly disadvantaged in that cross-word contexts are likely to be counter-productive for letter-to-sound mapping. In addition, the open nature of the vocabulary demands that the grapheme based system be able to generalise to unseen words and contexts, which is considerably more challenging than for the phoneme system. While these issues could be addressed to some extent by (for example) the use of special symbols to disambiguate word internal and cross-word contexts the problem of generalisation may not be easily solved (and at the least may require significantly more training data than for the phoneme based system).

4   Conclusions

In this paper we have studied the use of context-dependent phonemes and graphemes as sub-word units for automatic speech recognition. ASR studies conducted on different tasks show that by using context-dependent graphemes as sub-word units, performance similar to the state-of-the-art context-dependent phoneme based ASR system can be achieved on constrained tasks. Analysis demonstrates that the contextual modelling of grapheme units gives behaviour similar to phonemes and is achieved in a similar fashion to that observed in letter-to-sound mapping techniques.

In OGI Numbers95 studies we obtained better performance using graphemes when the acoustic models were trained with PLP features and similar performance when trained with tandem features. In the DARPA RM task studies we observed a marked difference between ASR systems using phoneme and grapheme when trained with PLP features. However, this difference is reduced when using tandem features. An explanation for this can be that the tandem system is able to implicitly incorporate phonetic knowledge while still having no requirement for the specification of a phonetic lexicon. In the much more complex CTS task we also observed improvements thanks to tandem features, though not to the same extent to that observed on the simpler tasks. These observations are summarised in Table 5

Table 5. Summary of findings from our studies. ≈ means comparable, ↑/↓ means somewhat greater/reduced, ⇑/⇓ means significantly greater/reduced

Lexicon	Cross-word Modelling	System Complexity	Phoneme-Grapheme Correspondence	Performance (grapheme)	Tandem vs PLP
small (closed)	no	≈	≈	≈	↑
medium (closed)	no	↑	↓	↓	↑
medium (open)	yes	⇑	⇓	⇓	↑

In both OGI Numbers95 task and DARPA RM task the words that are present in the dictionary are present in both training data and test data. In other words, there were no unseen contexts unlike in the CTS task. It is likely that this played a large role in the significantly reduced performance of the grapheme based CTS system compared with the phoneme based system. Further research will need to look at how to overcome this, either through improved parameter sharing approaches or by drawing upon non-acoustical data such as existing pronunciation lexica (which may not provide full coverage of the acoustic training data). It may also be interesting to look at a wider sub-word unit context in the framework of either WFST based decoding or lattice rescoring.

We also carried out an experiment on the RM corpus in which we merged grapheme and phoneme models and lexica and showed improved performance over either system alone. This suggests that the grapheme based models are complimentary to the phoneme models. In order further validate this hypothesis on a more challenging task such as CTS, it is clear that there are a number of hurdles that would first need to be overcome. Firstly, the use of cross-word models would require that we merge models in a less naive fashion as the current approach does not support cross-word contexts between phoneme and grapheme systems.

While it may be difficult to justify the use of grapheme based sub-word modelling for the English language, which is rich in linguistic resources and which exhibits poor grapheme-phoneme correspondence, however, we expect that the findings of our research to be of value for resource poor languages. In particular, the adoption of a tandem based scheme with grapheme modelling provides the possibility of incorporating phonetic knowledge from a resource rich language such as English into a resource language, while avoiding the need to develop a pronunciation dictionary, as supported by previous studies of tandem features [18].

Acknowledgements

This work was supported by the European Union 6th FWP IST Integrated Project AMIDA (Augmented Multi-party Interaction with Distance Access, FP6-033812) and the Swiss National Center of Competence in Research (NCCR) on Interactive Multi-modal Information Management (IM2).

References

1. Kanthak, S., Ney, H.: Context-dependent acoustic modeling using graphemes for large vocabulary speech recognition. In: Proceedings of Int. Conf. Acoustics, Speech and Signal Processing (ICASSP), pp. 845–848 (2002)
2. Killer, M., Stüker, S., Schultz, T.: Grapheme based speech recognition. In: Proceedings of Eurospeech, pp. 3141–3144 (2003)
3. Magimai.-Doss, M., Stephenson, T.A., Bourlard, H., Bengio, S.: Phoneme-Grapheme based automatic speech recognition system. In: Proceedings of Workshop on Automatic Speech Recognition and Understanding (ASRU), pp. 94–98 (2003)
4. Schukat-Talamazzini, E.G., Niemann, H., Eckert, W., Kuhn, T., Rieck, S.: Automatic speech recognition without phonemes. In: Eurospeech, pp. 129–132 (1993)
5. Magimai.-Doss, M., Bengio, S., Bourlard, H.: Joint decoding for phoneme-grapheme continuous speech recognition. In: ICASSP. Proceedings of Int. Conf. Acoustics, Speech and Signal Processing, pp. I–177–I–180 (2004)
6. Hermansky, H.: Perceptual Linear Predictive (PLP) analysis of speech. Journal of Acoustical Society of America 87(4), 1738–1752 (1990)
7. Hermansky, H., Ellis, D., Sharma, S.: Tandem connectionist feature stream extraction for conventional HMM systems. In: ICASSP. Proceedings of Int. Conf. Acoustics, Speech and Signal Processing, pp. III–1635–1638 (2000)
8. Cole, R.A., Fanty, M., Noel, M., Lander, T.: Telephone speech corpus development at CSLU. In: ICSLP 1994. Proceedings of Int. Conf. Spoken Language Processing (1994)
9. Price, P.J., Fisher, W., Bernstein, J.: A database for continuous speech recognition in a 1000 word domain. In: ICASSP 1988. Proceedings of Int. Conf. Acoustics, Speech and Signal Processing, vol. 1, pp. 651–654 (1988)
10. Chen, B., Çetin, Ö., Doddington, G., Morgan, N., Ostendorf, M., Shinozaki, T., Zhu, Q.: A CTS task for meaningful fast-turnaround experiments. In: Proceedings of Rich Transcription Fall Workshop, Palisades, NY (2004)
11. Black, A.W., Lenzo, K., Pagel, V.: Issues in building general letter to sound rules. In: Proceedings of 3rd ESCA Workshop on Speech Synthesis, Jenolan Caves, Australia, pp. 77–80 (1998)
12. Odell, J.J.: The use of context in large vocabulary continuous speech recognition. PhD thesis, Queens College, University of Cambridge (1995)
13. Ciprian, C., Morton, R.: Mutual information phone clustering for decision tree induction. In: ICSLP 2002. Proceedings of Int. Conf. Spoken Language Processing, Denver, Collorado (2002)
14. Zhu, Q., Chen, B., Morgan, N., Stolcke, A.: On using MLP features in lvcsr. In: ICSLP 2004. Proceedings of Int. Conf. Spoken Language Processing, Korea (2004)
15. Ikbal, S., Misra, H., Sivadas, S., Hermansky, H., Bourlard, H.: Entropy based combination of tandem representations for robust speech recognition. In: ICSLP 2004. Proceedings of Int. Conf. Spoken Language Processing, Korea (2004)
16. Young, S., Odell, J., Ollason, D., Valtchev, V., Woodland, P.: Hidden Markov model toolkit V3.2.1 reference manual. Technical report, Speech group, Engineering Department, Cambridge University, UK (2002)

17. Mirghafori, N., Morgan, N.: Combining connectionist multi-band and full-band probability streams for speech recognition of natural numbers. In: Proceedings of Int. Conf. Spoken Language Processing, pp. 743–746 (1998)
18. Stolcke, A., Grézl, F., Hwang, M.Y., Lei, X., Morgan, N., Vergyri, D.: Cross-domain and cross-language portability of acoustic features estimated by multilayer perceptrons. In: ICASSP 2006. Proceedings of Int. Conf. on Acoustics, Speech and Signal Processing, Toulouse, France (2006)

Transfer Learning for Tandem ASR Feature Extraction

Joe Frankel[1,2], Özgür Çetin[2], and Nelson Morgan[2]

[1] University of Edinburgh
[2] International Computer Science Institute
joe@cstr.ed.ac.uk

Abstract. Tandem automatic speech recognition (ASR), in which one or an ensemble of multi-layer perceptrons (MLPs) is used to provide a non-linear transform of the acoustic parameters, has become a standard technique in a number of state-of-the-art systems. In this paper, we examine the question of how to transfer learning from out-of-domain data to new tasks.

Our primary focus is to develop tandem features for recognition of speech from the meetings domain. We show that adapting MLPs originally trained on conversational telephone speech leads to lower word error rates than training MLPs solely on the target data. Multi-task learning, in which a single MLP is trained to perform a secondary task (in this case a speech enhancement mapping from farfield to nearfield signals) is also shown to be advantageous.

We also present recognition experiments on broadcast news data which suggest that structure learned from English speech can be adapted to Mandarin Chinese. The performance of tandem MLPs trained on 440 hours of Mandarin speech with a random initialization was achieved by adapted MLPs using about 97 hours of data in the target language.

1 Introduction

This work is concerned with the use of multi-layer perceptrons (MLPs) to provide non-linear transformations of acoustic features for use in automatic speech recognition (ASR). This approach is known as tandem ASR [1], and has become a common addition to modern ASR systems. For example, several of the meeting ASR systems presented at the NIST Rich Transcription 2006 spring evaluation (RT06s) included the use of non-linear feature transforms using MLPs.

The process of producing tandem features is sketched in Figure 1. Multiple frames of acoustic parameters are fed into one or an ensemble of MLPs. Rather than interpreting the outputs as phone class posteriors as in hybrid artificial neural network (ANN)/HMM ASR [2], they are subjected to a logarithm transformation and dimensionality reduction, then treated as observations. Once computed, they are appended to standard acoustic parameters in a hidden Markov model (HMM) system with Gaussian mixture model (GMM) observation distributions. The power of tandem ASR is two-fold. Firstly, multiple frames of acoustic features

A. Popescu-Belis, S. Renals, and H. Bourlard (Eds.): MLMI 2007, LNCS 4892, pp. 227–236, 2007.

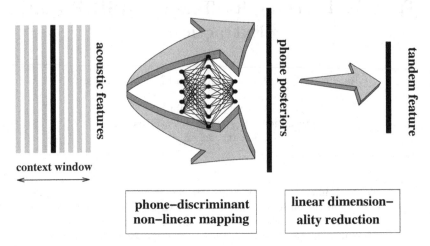

Fig. 1. Schematic of tandem feature mappings

are used as input to the MLP, which introduces longer-span contextual information. Secondly, the non-linear mapping is trained against phone targets, which has the effect that separation between phone classes is maximized in the output space. This separation leads to improved discrimination by the GMM which describes the output space associated with each HMM state. In addition, tandem features have been shown to exhibit some cross-task and cross-language portability [3]. Recent work [4] has also shown that the tandem feature extraction is complimentary to discriminative feature transformations such as function minimum phone error (fMPE).

Given the ever-increasing amounts of data on which systems are trained, and the considerable computational expense of training large MLPs, it is of interest to determine if learning can be transferred across domains and tasks. In this paper, we examine this question using experimental work within two scenarios. The first is recognition of farfield speech signals within the meetings domain. It was shown in [3] that adapting MLPs trained on out-of-domain data lead to better performance compared with unadapted MLPs, and that a slight improvement was given by adaptation of the Gaussians associated with HMM states onto the new features. In this paper we consider whether it is advantageous to adapt a previously-trained MLP or to retrain from scratch on the target data. Additionally, we present experiments on training MLPs using multi-task learning (MTL) [5], in which a single MLP is trained to perform a number of tasks. The second scenario we consider is the building a set of MLP-based features for a Mandarin broadcast news system.

2   Tandem ASR

Using MLP-based features in the tandem paradigm has been the subject of endeavour by a number of groups, including AMI [6], IDIAP [7] and ICSI [8].

The approach adopted at ICSI, which is also taken in this work, is to use a combination of the posteriors from two sets of MLPs.

The first of these is PLP-MLP, in which a 9 frame window of perceptual linear prediction (PLP) cepstra is used as input to a 3-layer MLP. The hidden units have sigmoid activation functions, and there is a softmax over the outputs.

The second is known as hidden activation temporal patterns (HATS) [9]. Under this configuration, separate small (e.g., 60 hidden unit) 3-layer MLPs are trained for phone classification with log critical band energies as input (one MLP per critical band with 50 frame windows). The hidden activations from each of these individual MLPs are then fed into a larger merger MLP, which is again trained on the task of phone classification.

The performance gains from using tandem and HATS feature independently are similar, though the additional information appears to be complementary, as the best results have been found using posterior combinations of the two [10].

Prior to the Fall 2004 DARPA speech-to-text (STT) evaluation, which focused on English continuous telephone speech (CTS), significant quantities of data (close to 2000 hours) became available in the form of the Fisher corpus. Training of MLPs on this data is described in [10], a process which took approximately six weeks despite the use of code optimized to run on a multiple-core machine, and introducing techniques specifically designed to reduce the training time. Training the MLPs on an order of magnitude more data than had been done previously led to improved performance. The focus of this paper is to explore methods by which the information encoded by these MLPs may be transferred to other domains.

3   Meetings Domain

The meetings domain offers a particular set of challenges due to the nature of spontaneous multiparty speech in which speakers frequently overlap. When the participants' speech is recorded using individual headset microphones (IHM), the high signal-to-noise ratio means that recognition has a word error rate (WER) of around 20%. However, it is not always possible or practical to have participants wearing individual microphones. In that case, tabletop recording is required, which creates a new set of problems due to reduced signal-to-noise ratio and presence of effects such as reverberation. These lead to significantly higher WERs, in the region 30-35%. The NIST RT06s meeting ASR evaluation specified three different farfield conditions:

- single distant microphone (SDM) - a single tabletop microphone source.
- multiple distant microphone (MDM) - tabletop microphone array with between 4 and 8 nodes.
- all distant microphones (ADM) - all channels used, which may include multiple microphone arrays.

In this work we use data from the MDM condition. The input waveforms were subject to delay-and-sum as described in [11].

All systems are gender-dependent, and employ many decoding stages including speaker adaptation, lattice generation, consensus decoding, n-best list rescoring, and cross-adaptation. For a full description, see [11].

3.1 Adaptation Procedure

Since the targets against which the MLPs are trained are the English phone set as used to train the CTS MLPs, the adaptation procedure we adopt is to carry out a few epochs of further training from the CTS-trained nets.

CTS MLPs were trained on 8kHz data, and the original CTS front-end configurations were preserved when generating input features for the meeting data. Both tandem (3-layer 9-frame PLP input) and HATS (15 critical band MLPs with 51 frame input followed by merger on hidden activations) were adapted. For the HATS, only the merger MLP was adapted.

The time-aligned phone segmentation which is used to provide training targets was produced by segmenting against the nearfield signals, which have a higher signal-to-noise ratio, and therefore were assumed to produce a more accurate and consistent segmentation. Any regions of overlapped speech were removed, and targets were generated for the farfield signals by matching each frame against the nearfield frame closest in time.

Since the targets were generated using nearfield alignments, the nearfield cuttings can be considered as clean versions of the farfield data. Previous experiments showed that adapting MLPs to nearfield data improved farfield WER, so a single epoch of adaptation to nearfield data was carried out first. This was followed by 3 epochs of adaptation to the farfield or combined farfield and nearfield signals. The MLPs from the epoch which gave the highest cross-validation (CV) accuracy during training were used for experimentation.

For the farfield MLPs, a single channel was selected at random to provide the data for each segment, though input normalizations were calculated over all segments for any given speaker/channel combination. The starting learn rates were equal to those in the last epoch of training of the CTS MLPs.

3.2 Multi-task Learning

Our goal is to produce features for use in a farfield system, though we consider methods by which the matched nearfield data might be used during the training phase, assuming it will not be available at test time. As discussed above, one possibility is simply to use nearfield parameters as input to the MLP during training.

Alternatively, we can use the nearfield data as an *output* of the MLP. This is a particular application of transfer learning known as multi-task learning (MTL) [5], in which a single MLP is trained to perform multiple related tasks. In our case, the MLP will be learning a speech enhancement mapping from farfield to nearfield speech parameters in addition to the usual phone posterior estimation.

The rationale is that by using a shared representation, related tasks can act as a prompts for each other. Additionally, given that local minima of the error

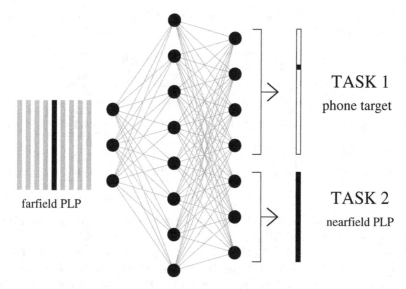

Fig. 2. Multi-task MLP for tandem ASR. The MLP learns a speech enhancement mapping from farfield to nearfield PLP features in addition to the usual phone posterior estimation.

function are unlikely to fall at the same location for multiple tasks, the risk of over-training is reduced.

Figure 2 shows the multitask PLP-MLP configuration. Farfield PLPs provide the inputs to a fully-connected hidden layer. The output layer is divided into two regions. The first is a group of 46 units with a softmax activation function, trained to perform phone posterior estimation. The second is a set of 39 sigmoid units which map to the frame of nearfield PLPs matching that from the centre frame of the input window. The target PLPs are scaled to be in the range [0.0, 1.0] in order to match the output range of sigmoid units. In the case of HATS, it is the merger MLP which is trained to be multi-task.

3.3 Experiments

Once the MLPs had been adapted to the target domain, features were generated to match the training data, and the Gaussians associated with HMM states were adapted to the new features. Unlike for MLP training, the data from all channels for each farfield segment was used as adaptation material. Once adapted, features were generated to correspond to the test data, and a multi-pass decoding pass was initiated, as described in [11]. The MLP features are used in the first stage of decoding, and are appended to Mel frequency cepstral coefficients (MFCCs).

The results presented in Table 1 are the word error rates (WERs) for the NIST RT05s evaluation data. The WER is shown for each MLP type after the first stage of decoding (one-pass), and at completion of the final multi-pass sweep. A baseline system in which no MLP features were used gives WERs of 50.2%

Table 1. Results (WER) on the NIST RT05s MDM evaluation data for a number of different adapted MLPs

	MLP train data	WER (%)	
		one-pass	multi-pass
no MLP	N/A	50.2	36.7
random	farfield	44.0	35.2
initialization	farfield MTL	44.1	34.3
adapted from CTS	nearfield	52.7	41.2
	farfield	43.3	33.2
	nearfield+farfield	42.7	33.0

and 36.7% for one-pass and multi-pass decoding respectively. Using the features from randomly-initialized MLPs, trained only on the farfield data, there is a substantial reduction in the WER after the first pass, from 50.2% to 44.0%, and at completion of all passes, the gain is reduced, though still evident with the WER reducing from 36.7% to 35.2%. For the MLPs which were trained in a multi-task setting, there is no benefit evident for the one-pass decoding, though at completion, the WER is reduced to 34.3%. These results suggest that guiding learning with the addition of a speech enhancement task is of benefit.

The next set of results correspond to MLPs which have been adapted from the CTS MLPs which were originally trained on 2000 hours of speech. Using the MLPs which were adapted on nearfield signals leads to increases in WER compared with a system with no MLP features. This can be attributed to mismatch of the nearfield and farfield signals. The MLPs adapted on only farfield data led to WERs of 43.3% and 33.2% after one and multiple decoding passes respectively. These results show substantial improvement over the baseline non-MLP system, and further reductions over the MTL-trained MLPs. Finally, adapting MLPs on pooled nearfield and farfield data gives the best results overall, with WERs of 42.7% and 33.0% for one-pass and multi-pass decoding.

One possible explanation for the superior performance of adapted MLPs is their size. Each of the CTS MLPs has 8 million parameters, which translates to 20,800 hidden units for each of the male and female tandem nets. By contrast, the male and female tandem nets which were trained from scratch had 7125 and 1825 hidden units respectively, and the MTL-trained MLPs slightly fewer in order to maintain an equal number of free parameters. The numbers of hidden units were determined by setting the total number of free parameters to equal 15% of the number of training frames.

4   Mandarin Broadcast News

In this section, we consider the problem of deriving tandem MLPs for use in a Mandarin Chinese system. Approximately 440 hours of Chinese broadcast news (BN) data was available, 97 of which had careful transcripts. For the remainder,

Table 2. Cross validation accuracies on Mandarin broadcast news data for tandem and HATS MLPs both with random initialization, and adapting from English CTS. Accuracies are shown for randomly-initialized MLPs trained on the 97-hour subset, with the number of free parameters set to be 5% and 10% of the total training frames.

initialization	Cross-validation accuracy	
	tandem	HATS
random 5% free params	74.1%	75.5%
random 10% free params	74.8%	76.2%
adapted, 3-layer	76.8%	77.2%
adapted, 4-layer	75.5%	76.5%

transcripts derived from a forced alignment of closed captions were available. Framewise labels were in terms of a set of 71 tonemes. In addition to training tandem and HATS MLPs from random initialization for the Mandarin data, the English CTS MLPs were used as a starting point for cross-lingual adaptation. Two strategies were explored: in the first (3-layer), the English CTS input-hidden layer was used in conjunction with a randomly initialized hidden-output layer. In the second (4-layer), an extra layer was added to the CTS MLPs. In both cases, training then proceeded with all weights being updated at each epoch. For HATS, it was the merger MLP which was adapted, and the critical-band MLPs used unchanged.

Whilst the English CTS system and tandem MLPS were gender dependent, the Mandarin system is gender independent. The female CTS nets were chosen as a basis for adaptation as they were originally trained on more data.

Table 2 shows cross-validation (CV) accuracies for the Mandarin MLPs both with random initialization, and for the 3-layer and 4-layer MLPs which are initialized from English CTS MLPs. In all cases the training data is the 97-hour subset. For the randomly-initialized MLPs, accuracies are shown for the number of free parameters set to be 5% and 10% of the total training frames. We find that for both tandem and HATS MLPs, it is the adapted versions which give higher CV accuracy. Additionally, it was found that these MLPs converged much more rapidly, achieving close to highest CV accuracy within the first epoch of training.

Despite the higher CV accuracy of the 10% version, it is the 5% MLPs which lead to the lowest WERs, and are used for the recognition results presented below in Table 3. The results are from a first-pass decoding using a three-gram LM with a lexicon of about 49,000 words. See [12] for more details about the Mandarin system.

Table 3 gives recognition WERs for the 2004 development and evaluation sets for the GALE Mandarin Broadcast News recognition task. The development set was used to tune decoding parameters, including language model scale factor and a scaling of Gaussian likelihoods.

Performance for the system without MLP-based features is 9.5% and 19.5% WER for the dev and eval sets respectively. The results using MLPs trained

Table 3. Results (WER) on the dev-04 and eval-04 data sets. Adaptation uses the 3-layer MLPs.

	WER (%)	
	dev-04	eval-04
no MLP	9.5	19.5
Chinese (97 hours)	8.2	18.2
Chinese (440 hours)	8.0	17.9
adapted (97 hours)	7.7	18.0

on 97 and 440 hours of data from a random initialization are given, with those trained on 440 hours giving a slightly lower error rate of 8.0% and 17.9% on the dev and eval sets respectively. The adapted MLPs give the lowest error of any on the development set, and all results are very similar on the evaluation data.

5   Discussion

Mismatch of training and testing data frequently has a significant impact on the performance of ASR systems. However, when porting a system from one domain to another, it is usual to take advantage of previously trained models and adapt the Gaussian mixture models associated with each state to the new domain. In this work we have shown that similarly, it is advantageous to utilize a previously-trained MLP and adapt to the new domain. For example, at completion of all decoding passes, the WER using MLPs trained from a random initialization was 35.2%, compared to 33.2% when adapting from CTS. In addition, the MTL MLPs gave a lower WER, from 35.2% to 34.3%. This suggests that strategies which combine adaption with multi-task learning may prove useful. For example, extra output and (possibly hidden) units could be added to the CTS MLPs to provide a speech enhancement mapping prior to training on the target data.

The results were less conclusive on the cross-lingual adaptation to Mandarin Chinese, though there is some evidence that given a relatively small amount of target data (e.g., 97 instead of 440 hours), it is preferable to adapt from English rather than train MLPs with random initialization. This suggests that there may be potential for sharing hidden representations between languages in order to increase available data.

Additionally, for the MLPs trained on the Mandarin data from a random initialization, it was found that smaller MLPs (free parameters set to be 5% rather than 10% of the total training frames) gave the best performance. This results in an MLP with a hidden layer of 4232 units, compared with 20800 in the adapted version. In this type of adaptation, we are considering the input-hidden layer as a general speech pattern classifier, and the hidden-output as a mapping to the particular set of outputs. Initializing with a ready-trained input-hidden layer makes it possible to train many more free parameters.

The results presented in this paper show that whilst mismatch in training and testing domains leads to performance degradations (e.g. nearfield MLPs on farfield data), there are sufficient commonalities to find a benefit from transfer learning. This may be due to the larger MLP structures which can be supported by pooling data from various sources.

Acknowledgements

Many thanks to Andreas Stolcke and others at ICSI for their input, and to SRI for the use of the Decipher ASR system.

This work was partly supported by the European Union 6th FWP IST Integrated Project AMI (Augmented Multi-party Interaction, FP6-506811), and by the Swiss National Science Foundation through NCCR's IM2 project.

This material is also partly based upon work supported by the Defence Advanced Research Projects Agency (DARPA) under Contract No. HR0011-06-C-0023. Any opinions, findings, and conclusions or recommendations expressed in this material are those of the authors and do not necessarily reflect the views of DARPA.

References

1. Hermansky, H., Ellis, D., Sharma, S.: Tandem connectionist feature stream extraction for conventional hmm systems. In: Proc ICASSP, Istanbul, Turkey, vol. III, pp. 1635–1638 (2000)
2. Trentin, E., Gori, M.: A survey of hybrid ANN/HMM models for automatic speech recognition. Neurocomputing 37(1), 91–126 (2001)
3. Stolcke, A., Grezl, F., Hwang, M.Y., Lei, X., Morgan, N., Vergyri, D.: Cross-domain and cross-language portability of acoustic features estimated by multilayer perceptrons. In: Proc. ICASSP, Toulouse, France (2006)
4. Zheng, J., Çetin, O., Hwang, M.Y., Lei, X., Stolcke, A., Morgan, N.: Combining discriminative feature, transform, and model training for large vocabulary speech recognition. In: Proc. ICASSP, Honolulu (2007)
5. Caruana, R.: Multitask learning. Machine Learning 28(1), 41–75 (1997)
6. Hain, T., Burget, L., Dines, J., Garau, G., Karafiat, M., Lincoln, M., Vepa, J., Wan, V.: The AMI meeting transcription system: Progress and performance. In: NIST RT 2006 Workshop (2006)
7. Hermansky, H.: TRAP-TANDEM: Data-driven extraction of temporal features from speech. In: IDIAP-RR 50, IDIAP, Martigny, Switzerland (2003)
8. Morgan, N., Zhu, Q., Stolcke, A., Sonmez, K., Sivadas, S., Shinozaki, T., Ostendorf, M., Jain, J., Hermansky, H., Ellis, D., Doddington, G., Chen, B., Çetin, O., Bourlard, H., Athineos, M.: Pushing the Envelope - Aside. IEEE Signal Processing Magazine 22(5), 81–88 (2005)
9. Chen, B., Zhu, Q., Morgan, N.: Learning long term temporal feature in LVCSR using neural networks. In: Proc. ICSLP, pp. 612–615 (2004)

10. Zhu, Q., Stolcke, A., Chen, B.Y., Morgan, N.: Using MLP features in SRI's conversational speech recognition system. In: Proc. Eurospeech, Portugal (2005)
11. Janin, A., Stolcke, A., Anguera, X., Boakye, K., Çetin, O., Frankel, J., Zheng, J.: The ICSI-SRI spring 2006 meeting recognition system. In: Proc. MLMI, Washington DC, USA (2006)
12. Hwang, M.Y., Wang, W., Lei, X., Zheng, J., Çetin, O., Peng, G.: Advances in Mandarin broadcast speech recognition. In: Proc.Interspeech, Antwerp (2007)

Spoken Term Detection System Based on Combination of LVCSR and Phonetic Search

Igor Szöke, Michal Fapšo, Martin Karafiát, Lukáš Burget, František Grézl,
Petr Schwarz, Ondřej Glembek, Pavel Matějka, Jiří Kopecký,
and Jan "Honza" Černocký

Speech@FIT, Faculty of Information Technology, Brno University of Technology
`speech@fit.vutbr.cz`

Abstract. The paper presents the Brno University of Technology (BUT)
system for indexing and search of speech, combining LVCSR and phonetic
approach. It brings a complete description of individual building blocks
of the system from signal processing, through the recognizers, indexing
and search until the normalization of detection scores. It also describes the
data used in the first edition of NIST Spoken term detection (STD) evaluation. The results are presented on three US-English conditions - meetings,
broadcast news and conversational telephone speech, in terms of detection
error trade-off (DET) curves and term-weighted values (TWV) metrics
defined by NIST.

1 Introduction

Search in speech is an important subfield of speech processing with numerous
applications in multi-modal storage and accessing of meetings, eLearning and
defense and security. This paper describes the Brno University of Technology
(BUT) system for indexing and search of speech, combining two techniques:

- Large vocabulary continuous speech recognition, where the recognition and
  indexing unit is a word.
- Phoneme recognition, where phonemes are recognized, and, for faster access,
  tri-phoneme sequences are indexed.

The theoretical basis of the search were described in [1] and we do not deal with
them in detail in this paper. Here, we concentrate on the BUT submission for
the NIST Spoken Term Detection (STD) Evaluations organized for the first time
in 2006. The paper contains a complete step-by-step description of BUT system
and discusses the results on NIST STD data. Moreover, new evaluation measures
defined by NIST are described and we deal in detail with the normalization of
scores needed in the case, where a single detection threshold is used for scoring
the results for all the searched terms.

The paper is organized as follows: section 2 describes the data and evaluation
metrics introduced for the 2006 NIST STD evaluation. Section 3 details the
BUT system from the recognizers to the normalization of final scores. Section 4
present and discusses the results and 5 concludes the paper.

A. Popescu-Belis, S. Renals, and H. Bourlard (Eds.): MLMI 2007, LNCS 4892, pp. 237–247, 2007.

2   NIST STD Evaluations 2006

The first edition of Spoken term detection evaluation was organized to facilitate research and development of technology for finding short word sequences rapidly and accurately in large heterogeneous audio archives [2]. In this paper, we will deal with STD for US English[1].

2.1   Data

There were three kinds of data with the following amounts available for both the development and evaluation:

- broadcast news (BCN) – 2.2 hours,
- conversational telephone speech (CTS) – 3 hours
- meeting speech (MTG) recorded over multiple distant microphones (MDM) – 2 hours.

For all sets, NIST has defined 1100 search-terms[2] having 1, 2, 3 and 4 words:

- 42 of them do not appear in any of BCN, CTS and MTG data
- 898 of 1100 appear in BCN with ≈4900 occurrences
- 411 of 1100 appear in CTS with ≈5900 occurrences
- 241 of 1100 appear in MTG with ≈3700 occurrences
- 160 of 1100 appear in all three BCN, CTS and MTG.

Examples of terms are:
"dr. carol lippa", "bush's father george bush", "thousand kurdish", "senator charles", "nato chief", "every evening", "kostunica", "audio", "okay".

2.2   Evaluation Metrics

The main mean for comparison of different systems were detection error trade-off (DET) curves, displaying, for various detection thresholds θ, the false alarm probability $P_{FA}(\theta)$ on x-axis and miss probability $P_{MISS}(\theta)$ on the y-axis:

$$P_{MISS}(\theta) = \underset{term}{avg}\left\{1 - N_{correct}(term, \theta)/N_{true}(term)\right\} \qquad (1)$$

$$P_{FA}(\theta) = \underset{term}{avg}\left\{N_{spurious}(term, \theta)/N_{NT}(term)\right\} \qquad (2)$$

where $N_{correct}(term, \theta)$ is the number of correct detections of $term$ with a score greater or equal to θ, $N_{spurious}(term, \theta)$ is the number of spurious (incorrect) detections of $term$ with a score greater or equal to θ, $N_{true}(term)$ is the number of occurrences of $term$ in corpus and $N_{NT}(term)$ is the number of opportunities

---

[1] Arabic and Mandarin were the two other languages analyzed in this evaluation.

[2] "quoted" queries where "quoted" refers to Google and similar search engines and means that no other word(s) can appear inside the query.

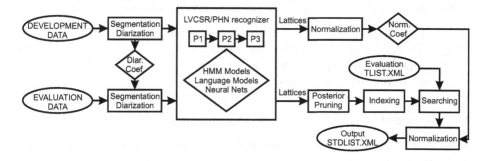

Fig. 1. Scheme of BUT system for NIST STD 2006 evaluations

for incorrect detection of *term* which is equal to length of the corpus in seconds minus $N_{true}(term)$.

NIST defined so called Term-Weighted Value $TWV(\theta)$ metric to "score" a system by one number. Term weighted value is evaluated by first computing the miss and false alarm probabilities for each term separately, then using these and a pre-determined prior probability to compute term-specific values, and finally averaging these term-specific values over all terms to produce an overall system value:

$$TWV(\theta) = 1 - \underset{term}{avg}\{P_{MISS}(term, \theta) + 99.9\, P_{FA}(term, \theta)\}$$

The threshold θ_M is found on development data by maximizing $TWV(\theta)$. $TWV(\theta_M)$ is then computed on evaluation data with θ_M threshold and denoted as $ATWV$ (see evaluation plan [2] for further details).

3   BUT System

The overall scheme of BUT system is on Figure 1. The following subsections detail the individual components of the system as well as the training data used.

3.1   Signal Processing

First, all NIST speech files were converted to raw format using sox. Segmenting speech into speech and silence was done by our neural net based phoneme recognizer [6]. All phoneme classes were linked to 'speech' class. CTS data were segmented according to energy in channels and speech/non-speech segmentation. The diarization for BCN and MTG data was done by David van Leeuwen. He used a Bayesian Information Criterion (BIC) based speaker segmentation and clustering system developed for the AMI RT06s speaker diarization evaluation [3]. 12 Perceptual Liner Prediction (PLP) features plus log energy were used as features, and he modeled clusters using a single Gaussian with full covariance matrix.

The data was split into shorter segments using the following heuristics:

1. in silences longer than 0.5s (output of speech/non-speech detector),
2. when speaker changed (in BCN and MTG),
3. if a segment was longer than 1 minute, it was split into 2 parts in silence closest to the center of segment.

3.2 Recognition

Segmented data was than processed by word (LVCSR) and phoneme (PHN) recognizers.

LVCSR – the general scheme. The STD 2006 LVCSR system is a simplified version of AMI LVCSR system[3] used for NIST RT 2006 evaluations [4]. It has has same structure for all tasks: CTS, BCN and MTG; the differences lie in acoustic and language models only. The scheme of LVCSR is on Fig. 2. The system operates in 3 passes of feature extraction and recognition:

In the first pass (P1), the front-end converts the segmented recordings into feature streams, with vectors comprised of 12 Mel-frequency Perceptual Liner Prediction (MF-PLP) features and raw log energy, first and second order derivatives are added. After, a cepstral mean and variance normalization (CMN/CVN) is performed on a per-channel basis with given segmentation. The first decoding pass yields initial transcripts that are subsequently used for estimation of vocal tract length normalization (VTLN) warping factors. The feature vectors and CMN and CVN are re-computed.

The second pass (P2) processes the new features and its output is used to adapt models with maximum likelihood linear regression (MLLR). Bigram lattices are produced and re-scored by trigram and four-gram language model.

In the third pass (P3), posterior features [6,5] are generated. The output from the second pass is used to adapt models with Constrained MLLR (CMLLR) and MLLR. The bigram lattices with posterior features are produced and finally re-scored with trigram and four-gram language model.

Feature extraction and acoustic modeling. All systems use standard crossword tied states HMM using Mel-PLP's generated in classical way with:

- 23 filter-bank channels for BCN and MTG system
- 15 filter-bank channels for CTS,

The resulting number of cepstral coefficients is always 13. The following techniques are used in HMM training:

- CMN/CVN is applied per speaker
- VTLN warping factors are computed using Brent search method and features are recomputed

---

[3] The LVCSR was developed in cooperation with AMI-project partners, see http://www.amiproject.org.

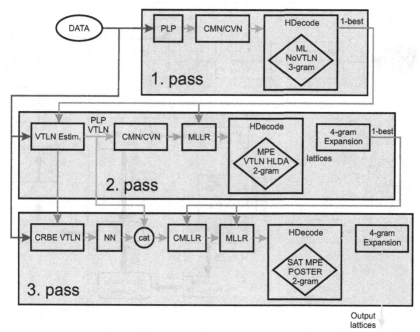

Fig. 2. Three passes of the recognition

- Deltas, double-and triple-deltas are added into the basic PLP feature stream, so that the feature vector has 52 dimensions. Heteroscedastic linear discriminant analysis (HLDA) is estimated with Gaussian components as classes. HLDA is estimated to reduce the dimensionality to 39.
- Posterior features - two kinds of posterior features are used:

LC-RC Posterior features. The LC-RC system [6] splits 310 ms temporal context in each filter-bank output into two halves and each half is processed by one neural net (NN) producing phoneme-state posteriors. These are merged by the third neural net. The resulting vector size is 135 (45 phonemes each with 3 states). After log and dimensionality-reduction by Karhunen-Loeve transform (KLT) to 70 dimensions (this step was necessary to fit the following HLDA statistics into memory), HLDA is estimated with Gaussian components as classes. HLDA was estimated to reduce the dimensionality to 25.

The resulting features are concatenated with PLP feature stream (25+39=64) and mean and variance normalized. The procedure is outlined in Fig. 3.

Bottle-neck LC-RC features. Bottleneck LC-RC differ from basic LC-RC in the last NN: the merger. It is a 5-Layer NN with middle layer containing 35 neurons only [5]. Non-linearly compressed information here is used as output. The HLDA is estimated to de-correlate and to reduce the dimensionality from 35 to 25.

Again, the resulting features are concatenated with PLP features (25+39=64) and mean and variance normalized.

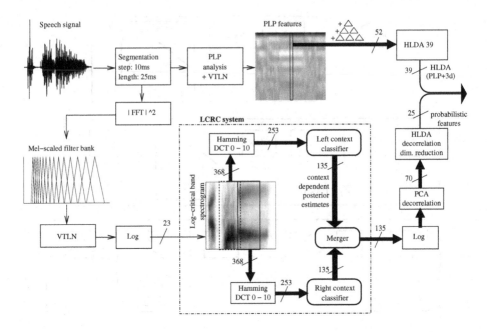

Fig. 3. Features used in the recognition system

Training of posterior features. At first, the neural network training with CMN/ CVN at the input was done on 30h of VTLN normalized data used for training of LVCSR acoustic models. Using these nets, full features were generated for all the data. The output was concatenated with PLP VTLN HLDA feature stream. The CMN/CVN were recomputed again and the models were trained by single-pass re-training. Further, the models were re-clustered and trained by the mixing-up procedure from 1 to N Gaussians. The optimal numbers of Gaussians were tuned for each task independently, the resulting numbers of Gaussians are 18 for MTG and BCN, 26 for CTS.

Speaker-adaptive training (SAT). One single CMLLR transform was trained per each meeting channel. Features were mapped to unique SAT space by CM-LLR and 8 iterations of ML-training (standard Baum-Welch) were run. After, new CMLLR transforms were trained, features transformed and 8 ML-iterations followed. And once more, so that the number of CMLLR+re-training macro-iterations was 3.

Discriminative training. The models were re-trained in 15 iterations of Minimum Phone-Error (MPE) training [8]. The alternative hypotheses for MPE were generated by much simpler system including just ML-trained models on PLP+HLDA without any adaptation. In case of SAT-MPE-training, we did not re-train the CMLLR transforms.

Table 1 outlines the acoustic models used in P1–P3 for different tasks.

Table 1. The acoustic models used in different steps for each task

task	P1	P2	P3
BCN	Basic PLP HMM	VTLN HLDA MPE	VTLN LC-RC SAT MPE
MTG	Basic PLP HMM	VTLN HLDA MPE	VTLN Bottleneck-LC-RC SAT MPE
CTS	HLDA	VTLN HLDA MPE	VTLN Bottleneck-LC-RC SAT MPE

Language models. The training of 4-gram language models was done at University of Sheffield by Vincent Wan. See below for the training data used. All language models were trained using the same data. The perplexity was maximized for each task independently.

Phoneme models. Phoneme recognition was based on the same features and models as LVCSR. Only the recognition network was changed to context dependent phoneme (triphone) loop (with context independent output ie. the output is phonemes) with phoneme bigram language model.

Decoding and posterior pruning. The decoding was performed using the standard LVCSR decoder `HDecode` from University of Cambridge. Generated lattices took significant space, so the posterior pruning was used for lattice size reduction. LVCSR and PHN lattices were pruned using different pruning factors.

3.3 Indexing

Word lattices are converted to forward index: each word-hypothesis (the word, its confidence, time and nodeID in the lattice file) is stored in a hit list. Forward index is then converted to inverted index which is sorted by words and by confidences of hypothesis. To save space and gain in speed of access, lattices are converted to binary format [1]. Phoneme lattices are also converted to forward index, the indexing units are phoneme trigrams (tri-phonemes). Forward index is also sorted to inverted index and lattice are converted to binary format.

3.4 Search

The term is first split to words (tokens). These are checked against the LVCSR dictionary and divided into in-vocabulary (IV) and out-of-vocabulary (OOV).

IV tokens are searched in inverted word-index to estimate their position in latices and then they are verified in the lattice (using token passing).

OOV tokens are converted to phonemes. Automatic grapheme-to-phoneme (G2P) tool based on rules is used for the conversion. Then the phoneme string is split to a train of overlapped tri-phonemes. Then they are also searched in inverted index (phoneme) and verified in lattice (phoneme). OOVs shorter than 3 phonemes (in total) are not searched and are dropped.

If all tokens are successfully verified, the time and score is produced. Score is computed as the sum of IV (LVCSR) part and OOV (PHN) part. IV scores are computed (by Viterbi approximation) using likelihood ratio in word lattice and then normalized. OOV scores are computed (by Viterbi approximation) using likelihood ratio in phoneme lattice and then normalized.

3.5 Summary of Training Data

The training data for acoustic models was the following:

- for BCN, the ihmtrain05 training set from NIST RT'06 evaluations [4] was used - it is a mixture of four meeting corpora, the NIST, ISL, ICSI and a preliminary release of the AMI corpus. In total, there are 112h of data. No BCN data were used.
- for MTG, the mdmtrain05 training set from NIST RT'06 evaluations [4] was used. The crosstalk parts were removed and beam-forming to one super-channel was done. In total, there are 63h of speech.
- for CTS, ctstrain04 - a subset of h5train03 set defined at Cambridge was used, in total 277h.

For language model training, done by Vincent Wan at the University of Sheffield, several resources were used (the numbers give the size of the corpus in megawords): Swbd/CHE 3.5, Fisher 10.5, Web (Swbd) 163, Web (Fisher) 484, Web (Fisher topics) 156, BBC - THISL 33, HUB4-LM96 152, SDR99-Newswire 39, Enron email 152, ICSI/ISL/NIST/AMI 1.5, Web (ICSI) 128, Web (AMI) 100, Web (CHIL) 70.

Grapheme to phoneme transcription rules were trained on AMI and BEEP pronunciation dictionaries.

The phoneme recognizer for segmentation was trained on Hungarian Speech-hDat-E [7] for BCN, ihmtrain05 for BCN and mdmtrain05 for MTG. LC-RC and Bottle-neck nets for generation of posterior features used the same training data as acoustic models.

3.6 Normalization

The normalization serves to make scores of different queries comparable (note that NIST scores STD systems with *one single threshold*). Our normalization is based on contributions of phonemes to normalization factors:

$$s_N(KW) = s(KW) - G - F\,len(KW) - P_1|p_1| + ... + P_K|p_K|,$$

where $s(KW)$ is raw score of the keyword, $s_N(KW)$ is the normalized score, $len(KW)$ is length of the keyword and $|p_1|...|p_K|$ are counts of individual phonemes in the keyword. G (a constant), F (length-dependent factor) and $P_1...P_N$ (phoneme-dependent factors) need to be trained: First, for large set of keywords, we derive scores for hits and false alarms (FA) on the development set. The scores corresponding to each keyword are used to construct pairs of (HIT, FA). For each pair, an equation is generated:

$$\frac{s(HIT) + s(FA)}{2} = G + F\,len(HIT) + P_1|p_1| + ... + P_K|p_K|,$$

where the left side represents an optimal threshold for given (HIT, FA) pair. We solve the over-defined set of equations in minimum square error sense and use the resulting factors to normalize scores.

The normalization coefficients were trained on the respective (BCN, CTS, MTG) part of NIST STD 2006 development data.

4   Results

The results of LVCSR systems for different tasks in terms of word error rate (WER) evaluated on the development sets, are the following: BCN 21.03%, CTS 22.83% and MTG 46.65%. The oracle results obtained by scoring the path in lattice that matches the best the reference, are respectively: BCN 9.06%, CTS 8.32% and MTG 21.79%. It is obvious that while BCN and CTS results are good and comparable to the state-of-the-art, the recognition on meetings is worse. This is due to the MDM condition, for which all the systems in NIST RT'06 evaluation performed quite poorly.

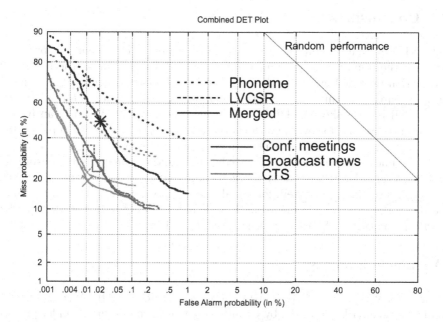

Fig. 4. DET curves for development US English data: LVCSR, phoneme-based and merged systems

Table 2. Minimum (M) TWV and actual (A) TWV values for individual and merged systems

task	EVAL ATWV Merged	EVAL MTWV Merged	EVAL MTWV LVCSR	DEVEL MTWV Merged
BCN	0.654	0.655	0.630	0.702
CTS	0.523	0.534	0.530	0.558
MTG	0.054	0.073	0.069	0.295

The STD results on all three conditions in terms of DET curves on development data can be seen in Fig. 4 and the results in terms of TWV are summarized in Table 2. First, we can see that the results on meetings are even worse than for the development data suggesting a problem with the data. Unfortunately, we are not able to analyze this in detail, as NIST does not intend to provide word transcriptions for the evaluation data.

In the other tasks, the results were satisfactory and we have seen the actual TWV not differing substantially from minimum TWV – a sign of good estimation of the optimal threshold.

Except for BCN, we see minimum effect of merging phonetic search with LVCSR, this is however caused by the term-lists provided – in CTS data, we have counted only 6 OOVs out of all 1100 requested terms.

5   Conclusions

The STD evaluation confirmed the usability of our STD system and provided us with the opportunity to compare it to other labs working in the field. The evaluation provided us also with several technical lessons, such as that using 4-gram expansion is only slightly better than 3-gram expansion, posterior pruning of LVCSR lattices shortens DET but does not decreases TWV significantly, etc.

In future, we need to work on the normalization - the scheme we implemented is a basic one, we can experiment with NN, calibration methods, etc.

CPU time and memory footprint needed are also the primary issue – despite its good accuracy, our system was far too slow compared to the other in the evaluation.

When designing the system for a real oriented user, we also need to take into account other user requirements, such as signal pre-processing, entering queries and combination with other speech search modalities.

Acknowledgments

This work was partly supported by European projects AMIDA (IST-033812) and Caretaker (FP6-027231), by Grant Agency of Czech Republic under project No. 102/05/0278 and by Czech Ministry of Education under projects Nos. MSM 0021630528 and LC06008. The hardware used in this work was partially provided by CESNET under projects Nos. 119/2004, 162/2005 and 201/2006. Lukáš

Burget was supported by Grant Agency of Czech Republic under project No. GP102/06/383.

We are grateful to Thomas Hain (University of Sheffield, UK) for leading the AMI and AMIDA LVCSR efforts, David van Leeuwen (TNO, the Netherlands) for the diarization of NIST STD data, Vinny Wan (University of Sheffield, UK) for language model training and to IDIAP (Switzerland) for beam-forming.

References

1. Burget, L., Černocký, J., Fapšo, M., Karafiát, M., Matějka, P., Schwarz, P., Smrž, P., Szöke, I.: Indexing and search methods for spoken documents. In: Sojka, P., Kopeček, I., Pala, K. (eds.) TSD 2006. LNCS (LNAI), vol. 4188, pp. 351–358. Springer, Heidelberg (2006)
2. NIST Spoken Term Detection Evaluation, http://www.nist.gov/speech/tests/std/
3. van Leeuwen, D.A., Huijbregts, M.: The AMI Speaker Diarization System for NIST RT06s Meeting Data. In: Renals, S., Bengio, S., Fiscus, J.G. (eds.) MLMI 2006. LNCS, vol. 4299, pp. 371–384. Springer, Heidelberg (2006)
4. Hain, T., et al.: The AMI Meeting Transcription System. In: Proc. NIST Rich Transcription 2006 Spring Meeting Recognition Evaluation Worskhop, p. 12. Washington D.C., USA (2006)
5. Grézl, F., Karafiát, M., Kontár, S., Černocký, J.: Probabilistic and bottle-neck features for LVCSR of meetings. In: Proc. ICASSP 2007, Hawaii (2007)
6. Schwarz, P., Matějka, P., Černocký, J.: Towards Lower Error Rates in Phoneme Recognition. In: Sojka, P., Kopeček, I., Pala, K. (eds.) TSD 2004. LNCS (LNAI), vol. 3206, Springer, Heidelberg (2004)
7. Eastern European Speech Databases for Creation of Voice Driven Teleservices, http://www.fee.vutbr.cz/SPEECHDAT-E/
8. Povey, D.: Discriminative Training for Large Vocabulary Speech, Recognition, PhD. Thesis, Cambridge University (July 2004)

Frequency Domain Linear Prediction for QMF Sub-bands and Applications to Audio Coding

Petr Motlicek[1], Sriram Ganapathy[1,2], Hynek Hermansky[1,2],
and Harinath Garudadri[3]

[1] IDIAP Research Institute,
Av. des Prés-Beudin 20, CH-1920, Martigny, Switzerland
{motlicek,ganapathy,hynek}@idiap.ch
[2] École Polytechnique Fédérale de Lausanne (EPFL), Switzerland
[3] Qualcomm Inc., San Diego, California, USA
hgarudad@qualcomm.com

Abstract. This paper proposes an analysis technique for wide-band audio applications based on the predictability of the temporal evolution of Quadrature Mirror Filter (QMF) sub-band signals. The input audio signal is first decomposed into 64 sub-band signals using QMF decomposition. The temporal envelopes in critically sampled QMF sub-bands are approximated using frequency domain linear prediction applied over relatively long time segments (e.g. 1000 ms). Line Spectral Frequency parameters related to autoregressive models are computed and quantized in each frequency sub-band. The sub-band residuals are quantized in the frequency domain using a combination of split Vector Quantization (VQ) (for magnitudes) and uniform scalar quantization (for phases). In the decoder, the sub-band signal is reconstructed using the quantized residual and the corresponding quantized envelope. Finally, application of inverse QMF reconstructs the audio signal. Even with simple quantization techniques and without any sophisticated modules, the proposed audio coder provides encouraging results in objective quality tests. Also, the proposed coder is easily scalable across a wide range of bit-rates.

Index Terms: Audio coding, Frequency Domain Linear Prediction (FDLP), Perceptual Evaluation of Audio Quality (PEAQ).

1 Introduction

Digital audio representation brings many advantages including unprecedented high fidelity, dynamic range and robustness in mobile and media coding applications. Due to the success provided by first generation digital audio applications, such as CD and DAT (digital audio tape), end-users expect CD-quality audio reproduction from any digital system.

Furthermore, emerging digital audio applications for network, wireless, and multimedia computing systems face a series of constraints such as reduced and variable channel bandwidth, limited storage capacity and low cost.

A. Popescu-Belis, S. Renals, and H. Bourlard (Eds.): MLMI 2007, LNCS 4892, pp. 248–258, 2007.
© Springer-Verlag Berlin Heidelberg 2007

IP networks, as a modern service platform, introduce new possibilities for the customer. As a consequence, new non-traditional services such as live audio and video streaming applications become popular (e.g. radio and TV broadcast over IP, multicast of a lecture, etc.). For example, unlike the present situation, audio and video consumed only 2% of Internet traffic in 2000. There are many reasons for this strong increase of audio and video traffic, such as faster Internet access, increased popularity of peer-to-peer applications (70% of the current Internet traffic), digital radio broadcasting on the web, technological advancements in soundcards and speakers, and development of high-quality compression techniques.

Interactive applications such as videophone or interactive games have a real-time constraint. This imposes a maximum acceptable end-to-end latency of the transmitted information, where end-to-end is defined as: capture, encode, transmit, receive, decode and display. The maximum acceptable latency depends on the application, but often is of the order of 150 ms. However, non-interactive applications have looser latency constraints, for example even a few seconds. The critical constraints are reduced errors in transmission, lesser breaks in continuity, and the overall signal quality.

This paper mainly focuses on audio coding for non-interactive applications. A novel speech coding system, proposed recently [1], exploits the predictability of the temporal evolution of spectral envelopes of a speech signal using Frequency-Domain Linear Prediction (FDLP) [2,3]. Unlike [2], the approach proposed in [1] applies FDLP to approximate relatively long (up to 1000 ms) segments of the Hilbert envelopes in individual frequency sub-bands. This approach was extended for wide-band applications (from 8 kHz up to 48 kHz) by including higher frequency sub-bands [4]. However, many difficulties arise specifically due to the need to pre-process and encode 1000 ms of a full-sampled input signal in each frequency sub-band. Efficient transmission of sub-band residual signals is also a challenge. This paper attempts to address most of these issues.

In contrast to the previous approaches [1,4], this paper proposes the use of FDLP on the sub-band signal instead of the full-band signal. First, the input

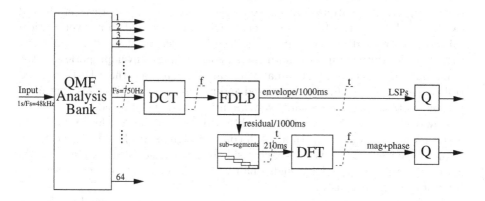

Fig. 1. QMF-FDLP encoder structure (f - frequency domain, t - time domain)

signal is decomposed into frequency sub-bands using the maximally decimated QMF bank. Then, FDLP technique is applied in each critically sampled frequency sub-band independently. The Line Spectral Frequencies (LSFs) as well as the spectral components of the residual sub-band signals are quantized. The interesting properties of this novel coding scheme are shown and the first version of a variable bit-rate audio encoder based on QMF - FDLP techniques is proposed.

The rest of the paper is organized as follows: Section 2 provides a brief overview of the FDLP principle. Section 3 describes the QMF - FDLP codec. Simulation results and audio quality evaluations are given in Section 4, followed by conclusions and discussions in Section 5.

2   FDLP

The novelty of the proposed audio coding approach is the employment of FDLP method to parameterize the Hilbert envelope (squared magnitude of an analytic signal) of the input signal [2,3]. FDLP can be seen as a method analogous to Temporal Domain Linear Prediction (TDLP). In the case of TDLP, the AR model approximates the power spectrum of the input signal. The FDLP fits an AR model to the squared Hilbert envelope of the input signal. Using FDLP, we can adaptively capture fine temporal nuances with high temporal resolution while at the same time summarizing the signal's gross temporal evolution at time scales of hundreds of milliseconds. In our system, we employ the FDLP technique to approximate the temporal envelope of sub-band signals in QMF sub-bands.

3   Structure of the Codec

The first version of the coder based on FDLP for very low bit-rate narrowband applications was proposed in [1]. The input speech signal was split into non-overlapping segments (hundreds of ms long). Then, each segment was DCT transformed and partitioned into unequal segments to obtain critical band-sized sub-bands. Finally, the DCT components which correspond to a given critical sub-band were used for calculating the FDLP model for that band. Since the FDLP model did not approximate the squared Hilbert envelope perfectly, the remaining sub-band residual signal (the carrier signal for the FDLP-encoded Hilbert envelope) was further processed and its frequency components (obtained by Fourier transform) were selectively quantized and transmitted.

However, the sub-band residuals, obtained by the the employed sub-band decomposition described in [1], are still complex signals. High coding efficiencies cannot be achieved (e.g. the sub-band residuals are oversampled) and henceforth, large bit-rate is required for high quality coding. In addition, this system is computationally expensive.

The following experiments, performed with audio signals sampled at 48 kHz, were motivated by the MPEG-1 architecture [5,6]. In the MPEG-1 encoder,

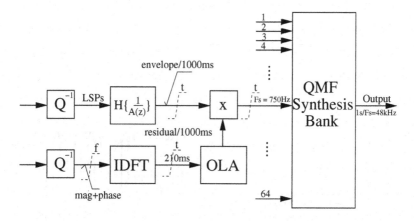

Fig. 2. QMF-FDLP decoder structure (f - frequency domain, t - time domain)

the input signal is first decomposed into critically sub-sampled frequency sub-bands using a QMF bank (a polyphase realization), whose channels are uniformly spaced. In our system, the same operation is performed on the input signal. The band-pass outputs are decimated by a factor M (the number of sub-bands), yielding the sub-band sequences which form a critically sampled and maximally decimated signal representation (i.e. the number of sub-band samples is equal to the number of input samples). The QMF filters have a flat pass-band response, which is advantageous for exploiting the predictability of frequency components of sub-band signals.

In the proposed coder, we employ 64 band decomposition compared to MPEG-1 standard having 32 frequency sub-bands. The parameters representing the FDLP model in each sub-band are not expensive from the final bit-rate point of view. The use of a higher number of sub-bands provides the following advantages:

1. Sub-band residuals are more frequency limited, and are easier to quantize using split VQ.
2. It is more advantageous when a psychoacoustic model (in the future) is employed to attenuate perceptually irrelevant frequency sub-bands.
3. Slightly better objective results.

The structure of the encoder and the decoder is depicted in figure 1 and 2, respectively.

3.1   Time-Frequency Analysis

The input signal is split into 1000 ms long frames. Each frame is decomposed into 64 sub-bands by QMF. A 99-th order prototype filter with the direct-form FIR polyphase structure is used for the frequency decomposition. The prototype filter was designed for high sidelobe attenuation in the stop-band of each analysis channel (around 78 dB), which ensures that intraband aliasing due to quantization noise remains negligible. The magnitude and phase frequency response of

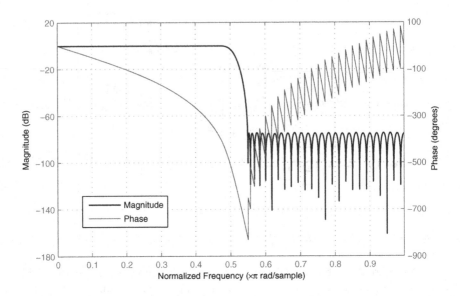

Fig. 3. Magnitude and phase frequency response of the QMF prototype filter

the FIR filter is depicted in figure 3. In order to perform decomposition into 64 sub-bands, a cascade implementation of 2 band QMF decomposition provided by the prototype filter is utilized. The algorithmic delay of the implementation is around 130 ms.

3.2   Critically Sampled Sub-band Processing

Each sub-band is DCT transformed which yields the input to the FDLP module. The magnitude frequency response of AR model, computed through the autocorrelation LPC analysis on the DCT transformed sub-band signal, approximates the squared Hilbert envelope of the 1000 ms sub-band signal. Spectral Transform Linear Prediction (STLP) technique [7] is used to control the fit of AR model to the Hilbert envelope of the input. The associated FDLP-LSF parameters are quantized and then transmitted.

The 1000 ms residual signal in each critically sampled sub-band, which represents the Hilbert carrier signal for the FDLP-encoded Hilbert envelope, is split into five overlapping sub-segments 210 ms long with 10 ms overlap. This is to take into account the non-stationarity of the Hilbert carrier over the 1000 ms frame. An overlap length of 10 ms ensures smooth transitions when the sub-segments of the residual signal are concatenated in the decoder. Finally, each sub-segment is DFT transformed which results in 79 complex spectral components distributed over 0 Hz to $F_s/2$ (= 375 Hz) with a frequency resolution of 4.75 Hz. The magnitude and phase components of the complex spectral representations are then quantized and transmitted to the decoder.

3.3 Quantization of Parameters

Quantization of LSFs: The LSFs corresponding to the AR model in a given frequency sub-band over the 1000 ms input signal are vector quantized. In the experiments, a 20th order all-pole model is used. The total contribution of the FDLP models to the bit-rate for all the sub-bands is around 5 kbps.

Quantization of the magnitude components of the DFT transformed sub-segment residual: The magnitude spectral components are vector quantized. Since a full-search VQ in this high dimensional space would be computationally infeasible, split VQ approach is employed. Although the split VQ approach is suboptimal, it reduces computational complexity and memory requirements to manageable limits without severely affecting the VQ performance. We divide the input vector of spectral magnitudes into separate partitions of a lower dimension. The VQ codebooks are trained (on a large audio database) for each partition using the LBG algorithm. Quantization of the magnitude components using the split VQ takes around 30 kbps for all the sub-bands.

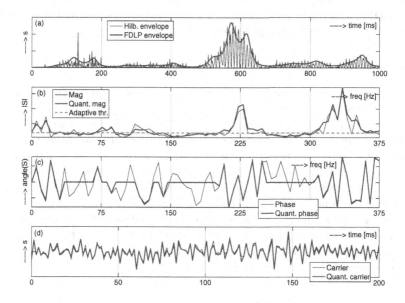

Fig. 4. Time-frequency characteristics obtained from a randomly selected audio example: (a) 1000 ms segment of the squared Hilbert envelope (estimated from the squared magnitude of an analytic signal) computed for the 3rd QMF sub-band, and its FDLP approximation. (b) Magnitude Fourier spectral components of the 200 ms residual sub-segment, and its reconstructed version, the adaptive threshold is also shown. (c) Phase Fourier spectral components of the 200 ms residual sub-segment, and its reconstructed version (the components which are not selected for quantization by the adaptive threshold are set to zero). (d) Original 200 ms sub-segment of the residual signal (carrier) of the FDLP envelope, and its reconstructed version.

Quantization of the phase components of the DFT transformed sub-segment residual: It was found that the phase components are uncorrelated across time. The phase components have a distribution close to uniform, and therefore, have a high entropy. Hence, we apply a 4 bit uniform scalar quantization for the phase components. To prevent excessive consumption of bits to represent phase coefficients, those corresponding to relatively low magnitude spectral components are not transmitted, i.e., the codebook vector selected from the magnitude codebook is processed by adaptive thresholding in the encoder as well as in the decoder. Only the spectral phase components whose magnitudes are above the threshold are transmitted. The threshold is adapted dynamically to meet a required number of spectral phase components (bit-rate). The options for reconstructing the signal at the decoder are:

1. Fill the incomplete phase components with uniformly distributed white noise.
2. Fill the incomplete phase components with zeros.
3. Set the magnitude components corresponding to incomplete phase components to zero.

In objective quality tests, it was found that the third option performed the best. Figure 4 shows time-frequency characteristics of the original signal and the reconstructed signal for the proposed codec.

3.4   Decoding

In order to reconstruct the input signal, the residual in each sub-band needs to be reproduced and then modulated by temporal envelope given by FDLP model.

The transmitted VQ codebook indices are used to select appropriate codebook vectors for the magnitude spectral components. Then, the adaptive threshold is applied on the reconstructed magnitudes and the transmitted scalar quantized phase spectral components are assigned to the corresponding magnitudes lying above the adaptive threshold. 210 ms sub-segment of the sub-band residual is created in the time domain from its spectral magnitude and phase information. The Overlap-add (OLA) technique is applied to obtain 1000 ms residual signal, which is then modulated by the FDLP envelope to obtain the reconstructed sub-band signal. Finally, a QMF synthesis bank is applied on the reconstructed sub-band signals to produce the output signal.

4   Experiments and Results

The qualitative performance of the proposed algorithm was evaluated using Perceptual Evaluation of Audio Quality (PEAQ) distortion measure [8]. In general, the perceptual degradation of the test signal with respect to the reference signal is measured, based on the ITU-R BS.1387 (PEAQ) standard. The output combines a number of model output variables (MOV's) into a single measure, the Objective Difference Grade (ODG) score, which is an impairment scale with meanings shown in table 1.

Table 1. PEAQ scores and their meanings

ODG Scores	Quality
0	imperceptible
−1	perceptible but not annoying
−2	slightly annoying
−3	annoying
−4	very annoying

Table 2. PEAQ results on audio samples in terms of ODG scores for speech, speech over music, music

bit-rate	-	-	-	124	100	64	48
File	G-27h	Q-32h	Q-64h	Q-64	Q-64	MP3-LAME	AAC+ v1
SPEECH							
es01_s	-3.579	-1.380	-1.075	-0.811	-1.334	-2.054	-1.054
es02_s	-3.345	-1.826	-1.360	-0.820	-1.396	-1.432	-1.175
louis_raquin_1	-3.710	-2.785	-2.626	-1.670	-2.484	-1.931	-1.793
te19	-3.093	-1.829	-1.514	-1.016	-1.518	-1.140	-1.152
SPEECH OVER MUSIC							
Arirang_ms	-2.709	-2.056	-2.149	-1.741	-2.703	-1.750	-1.215
Green_sm	-2.656	-2.588	-2.332	-2.096	-2.765	-1.703	-1.147
noodleking	-2.312	-0.777	-0.677	-0.485	-0.705	-1.388	-0.912
te16_fe49	-2.316	-1.028	-1.106	-1.099	-1.678	-1.346	-1.558
te1_mg54	-2.668	-1.343	-1.340	-0.949	-1.487	-1.341	-1.319
twinkle_ff51	-2.174	-2.070	-1.705	-1.557	-2.162	-1.519	-0.822
MUSIC							
brahms	-2.635	-1.157	-1.038	-1.495	-1.788	-1.204	-1.777
dongwoo	-1.684	-0.675	-0.583	-0.471	-0.658	-1.369	-0.753
phi2	-2.194	-0.973	-0.696	-0.445	-0.834	-1.735	-0.748
phi3	-2.598	-1.263	-1.058	-0.774	-1.215	-1.432	-0.937
phi7	-3.762	-1.668	-1.635	-3.356	-3.624	-2.914	-1.551
te09	-2.997	-1.353	-1.239	-0.841	-1.490	-1.734	-1.579
te15	-2.006	-0.670	-0.644	-0.545	-0.845	-1.726	-0.995
trilogy	-2.002	-0.439	-0.461	-0.509	-0.694	-2.064	-0.951
AVERAGE	-2.691	-1.438	-1.291	**-1.149**	**-1.632**	**-1.655**	**-1.191**

The test was performed on 18 audio signals sampled at 48 kHz. These audio samples are part of the framework for exploration of speech and audio coding defined in [9]. They are comprised of speech, music and speech over music recordings. The ODG scores are presented in table 2.

First, the narrow-band speech coder [1] was extended for audio coding on 48 kHz sampled signal [4], where Gaussian sub-band decomposition (into 27 bands uniformly spaced in the Bark scale) is employed. The FDLP model was applied on these sub-band signals and corresponding sub-band residuals were processed by adaptive threshold (half of the spectral components were preserved). This adaptive thresholding is performed to simulate the quantization process. The ODG scores for this technique are denoted as G-27h.

The results for QMF decomposition into 32 and 64 sub-bands are denoted as Q-32h and Q-64h, respectively. As in the previous case, the adaptive threshold is fixed to select half of the spectral components of the sub-band residuals. Without quantization, Q-32h performs better than G-27h with approximately the same number of parameters to be encoded. In a similar manner, Q-64h slightly outperforms Q-32h.

ODG scores for Q-64 with quantization of spectral components of the sub-band residuals are presented for bit-rates ~ 124 and ~ 100 kbps. In both these experiments, magnitude spectral components are quantized using split VQ (10 codebooks each of dimension 8 with size of 12 bits). Phase spectral components are scalarly quantized using 4 bits. To reduce the final bit-rates, the number of phase components to be transmitted is reduced using adaptive threshold (90% and 60% of phase spectral components resulting in 124 and 100 kbps respectively).

Finally, the ODG scores for standard audio coders: MPEG-1 Layer-3 LAME [6,10] and MPEG-4 HE AACplus-v1 [11,12], at bit-rates 64 and 48 kbps respectively, are also presented. The AACplus-v1 coder is the combination of Spectral Band Replication (SBR) [13] and Advanced Audio Coding (AAC) [14] and was standardized as High-Efficiency AAC (HE-AAC) in Extension 1 of MPEG-4 Audio [15].

5   Conclusions and Discussions

With reference to the ODG scores in table 2, the proposed codec needs to operate at 100 kbps in order to achieve the similar average quality as the MPEG-1 Layer-3 LAME standard at 64 kbps. In a similar manner, the proposed codec requires 124 kbps to perform as good as AACplus-v1 codec at 48kbps.

Even without any sophisticated modules like psychacoustic models, spectral band replication module and entropy coding, the proposed method (at the expense of higher bit-rate) is able to give objective scores comparable to the state-of-the-art codecs for the bit-rates presented. Furthermore, by modifying the adaptive threshold, the proposed technique allows to scale the bit rates, while for example in MPEG-1 layer-3 such operation is computationally intensive.

The presented codec and achieved objective performances will serve as baseline for future work, where we will concentrate on compression efficiency. Due to this reason, we compared the proposed codec with state-of-the-art systems which provide similar objective qualities. The succeeding version of the codec, having perceptual bit allocation algorithms (for the carrier spectral components), is expected to reduce the bit rate considerably yet maintaining the same quality.

Eventhough, the coder does not employ block switching scheme (thus avoiding structural complications), the FDLP technique is able to address temporal masking problems (e.g. pre-echo effect) in an efficient way [4]. Another important advantage of the proposed coder is its resiliency to packet loss. This results from reduced sensitivity of the human auditory system to drop-outs in a frequency band as compared to loss of a short-time frame.

Acknowledgements

This work was partially supported by grants from ICSI Berkeley, USA; the Swiss National Center of Competence in Research (NCCR) on "Interactive Multimodal Information Management (IM)2"; managed by the IDIAP Research Institute on behalf of the Swiss Federal Authorities, and by the European Commission 6th Framework DIRAC Integrated Project.

References

1. Motlicek, P., Hermansky, H., Garudadri, H., Srinivasamurthy, N.: Speech Coding Based on Spectral Dynamics. In: Sojka, P., Kopeček, I., Pala, K. (eds.) TSD 2006. LNCS (LNAI), vol. 4188, Springer, Heidelberg (2006)
2. Herre, J., Johnston, J.H.: Enhancing the performance of perceptual audio coders by using temporal noise shaping (TNS). In: 101st Conv. Aud. Eng. Soc. (1996)
3. Athineos, M., Hermansky, H., Ellis, D.P.W.: LP-TRAP: Linear predictive temporal patterns. In: Proc. of ICSLP, pp. 1154-1157, Jeju, S. Korea (October 2004)
4. Motlicek, P., Ullal, V., Hermansky, H.: Wide-Band Perceptual Audio Coding based on Frequency-domain Linear Prediction. In: Proc. of ICASSP, Honolulu, USA (April 2007)
5. Pan, D.: A Tutorial on MPEG/Audio Compression. IEEE Multimedia Journal , 60–74 (1995)
6. Brandenburg, K.: ISO-MPEG-1 Audio: A Generic Standard for Coding of High-Quality Digital Audio. J. Audio Eng. Soc. 42, 780–792 (1994)
7. Hermansky, H., Fujisaki, H., Sato, Y.: Analysis and Synthesis of Speech based on Spectral Transform Linear Predictive Method. In: Proc. of ICASSP, Boston, USA, vol. 8, pp. 777–780 (April 1983)
8. Thiede, T., Treurniet, W.C., Bitto, R., Schmidmer, C., Sporer, T., Beerends, J.G., Colomes, C., Keyhl, M., Stoll, G., Brandenburg, K., Feiten, B.: PEAQ – The ITU Standard for Objective Measurement of Perceived Audio Quality. J. Audio Eng. Soc. 48, 3–29 (2000)
9. ISO/IEC JTC1/SC29/WG11, Framework for Exploration of Speech and Audio Coding, MPEG, /N9254, July 2007, Lausanne, CH (2007)
10. LAME MP3 codec, http://lame.sourceforge.net
11. 3GPP TS 26.401: Enhanced aacPlus general audio codec; General Description
12. Brandenburg, K., Kunz, O., Sugiyama, A.: MPEG-4 Natural Audio Coding. Signal Processing: Image Communication 15(4), 423–444 (2000)
13. Dietz M., Liljeryd L., Kjorling K., Kunz O., Spectral Band Replication, a novel approach in audio coding. In: AES 112th Convention, Munich, DE, May 2002, Preprint 5553.

14. Bosi, M., Brandenburg, K., Quackenbush, S., Fielder, L., Akagiri, K., Fuchs, H., Dietz, M., Herre, J., Davidson, G., Oikawa, Y.: ISO/IEC MPEG-2 Advanced Audio Coding. J. Audio Eng. Soc. 45(10), 789–814 (1994)
15. ISO/IEC, Coding of audio-visual objects Part 3: Audio, AMENDMENT 1: Bandwidth Extension, ISO/IEC Int. Std. 14496-3:2001/Amd.1:2003 (2003)

Modeling Vocal Interaction
for Segmentation in Meeting Recognition

Kornel Laskowski and Tanja Schultz

interACT, Carnegie Mellon University, Pittsburgh PA, USA
{kornel,tanja}@cs.cmu.edu

Abstract. Automatic segmentation is an important technology for both automatic speech recognition and automatic speech understanding. In meetings, participants typically vocalize for only a fraction of the recorded time, but standard vocal activity detection algorithms for close-talk microphones in meetings continue to treat participants independently. In this work we present a multispeaker segmentation system which models a particular aspect of human-human communication, that of vocal interaction or the interdependence between participants' on-off speech patterns. We describe our vocal interaction model, its training, and its use during vocal activity decoding. Our experiments show that this approach almost completely eliminates the problem of crosstalk, and word error rates on our development set are lower than those obtained with human-generatated reference segmentation. We also observe significant performance improvements on unseen data.

1 Introduction

Vocal activity detection (VAD) is an important technology for any application with an automatic speech recognition (ASR) front end. In meetings, participants typically vocalize for only a fraction of the recorded time. Their temporally contiguous contributions should be identified prior to speech recognition in order to associate recognized output with specific speakers (who said what) and to leverage speaker adaptation schemes. Segmentation into such contributions is primarily informed by vocal activity detection on a frame-by-frame basis.

This work focuses on VAD for meetings in which each participant is instrumented with a close-talk microphone, a task which remains challenging primarily due to crosstalk from other participants (regardless of whether the latter have their own microphones). State-of-the-art meeting VAD systems which attempt to account for crosstalk rely on Viterbi decoding in a binary speech/non-speech space [12], assuming independence among participants. They employ traditional Mel-ceptral features as used by ASR, with Gaussian mixture models [1] or multi-layer perceptrons [6]. Increasingly, such systems are integrating new features, designed specifically for discriminating between nearfield and farfield speech, or speaker overlap and no-overlap situations [14]. Research in this field is being fueled in large part by the Rich Transcription (RT) Meeting Recognition

A. Popescu-Belis, S. Renals, and H. Bourlard (Eds.): MLMI 2007, LNCS 4892, pp. 259–270, 2007.
© Springer-Verlag Berlin Heidelberg 2007

evaluations organized by NIST[1]. Generally reported ASR word error rates (WERs) on NIST RT corpora are still at least 2-3% absolute higher with automatically generated segments than with manual segmentation [1], a difference which is significant in the context of overall transcription system performance.

This paper describes an automatic segmentation system which is an extension to the segmentation component in our NIST RT-06s Speech-to-Text submission system in the individual head-mounted microphone (IMH) condition for conference meetings [8]. Both segmentation systems implement a fundamentally different approach from those used in other state-of-the-art transcription systems, in three main ways. First, we have chosen to address the crosstalk problem by explicitly modeling the correlation between all channels. This results in a feature vector whose length is a function of the number of meeting participants, which may vary from test meeting to test meeting. Because a variable feature vector length precludes the direct use of exclusively supervised acoustic models, we have proposed an unsupervised joint-participant acoustic modeling approach [10]. Second, we employ a model of multi-participant vocal interaction, which allows us to explicitly model the fact that starting to speak while other participants are speaking is dispreferred to starting in silence. Finally, as a consequence of our fully-connected, ergodic hidden Markov model architecture, state duration cannot be modeled directly. Our analysis window size, an order of magnitude larger than that in other state-of-the-art systems, is a trade-off between the desired endpoint granularity and minimum expected talkspurt duration.

Following a description of the new system in Sections 2, 3 and particularly 4, we compare the system to our NIST RT-06s segmentation system. We show that our final segmentation system outperforms manual segmentation on our development set, effectively treats uninstrumented participants, and leads to WERs only 2.2% absolute higher on unseen data than with manual segmentation.

2   Computational Framework

The VAD system we use as our baseline was introduced in [10]. Rather than detecting the 2-state speech ($\mathcal{V}$) vs. non-speech ($\mathcal{N}$) activity of each partipant independently, the baseline implements a Viterbi search for the best path through a 2^K-state vocal interaction space, where K is the number of participants. Our state vector, $\mathbf{q}_t$, formed by concatenating the concurrent binary vocal activity states $\mathbf{q}_t[k]$, $1 \geq k \geq K$, of all participants, is allowed to evolve freely over the vocal interaction space hypercube, under stochastic transition constraints imposed by a fully-connected, ergodic hidden Markov model (eHMM). Once the best vocal interaction state path $\mathbf{q}^*$ is found, we index out the corresponding best vocal activity state path $\mathbf{q}^*[k]$ for each participant k. The underlying motivation for this approach is that it allows us to model the constraints that participants exert on one another; it is generally accepted that participants are more likely to begin vocalizing in silence than when someone else is already vocalizing [4].

---

[1] http://www.nist.gov/speech/tests/rt/

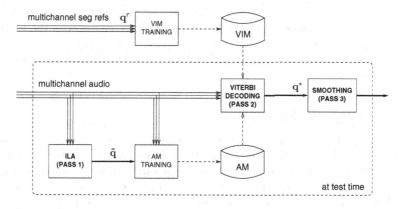

Fig. 1. Segmentation system architecture

The architecture of the proposed segmentation system is depicted in Figure 1. Tasks associated with its operation, shown as rectangles in the figure, include:

1. VIM TRAINING: training of a conversation-independent vocal interaction model (Section 4);
2. **PASS 1**: initial label assignment (ILA) for the test audio (Section 3.1);
3. AM TRAINING: training of conversation-specific acoustic models (Section 3.2) using the *test* audio and the labels from (2);
4. **PASS 2**: simultaneous Viterbi decoding of all participant channels, using the vocal interaction model from (1) and the acoustic models from (3); and
5. **PASS 3**: smoothing VAD output to produce a segmentation suitable for ASR.

Space constraints prohibit a comprehensive description of each task or component. We only briefly describe the multiparticipant IHM acoustic model in the following section. In Section 4, we detail the structure of the proposed vocal interaction model, and outline its training and use during decoding.

3   Unsupervised Multispeaker IHM Acoustic Modeling

3.1   Initial Label Assignment

We perform an unsupervised initial assignment of state labels to multichannel frames of audio using the heuristic

$$\tilde{\mathbf{q}}[k] = \begin{cases} \mathcal{V}, \text{ if } \sum_{j \neq k} \log \left(\dfrac{\max_{\tau} \phi_{jk}(\tau)}{\phi_{jj}(0)} \right) > 0 \\ \mathcal{N}, \text{ otherwise} \end{cases}, \tag{1}$$

where $\phi_{jk}(\tau)$ is the crosscorrelation between IHM channels j and k at lag τ, and $\tilde{\mathbf{q}}[k]$ is the initial label assigned to the frame in question. We have shown, in [11], that under certain assumptions the criterion in Equation 1 is equivalent to declaring a participant as vocalizing when the distance between the location of the dominant sound source and that participant's microphone is smaller than the geometric mean of the distances from the source to each of the remaining microphones.

3.2 Acoustic Model Training

The initial label assignment described in Equation 1 produces a partitioning of the multichannel test audio. The labeled frames are used to train a single, full-covariance Gaussian for each of the 2^K states in our search space, over a feature space of $2K$ features: a log-energy and a normalized zero-crossing rate for each IHM channel. Features are computed using 110 ms non-overlapping windows, following signal preemphasis $(1 - z^{-1})$.

For certain participants, and especially for frames in which more than one participant vocalizes, the ILA may identify too few frames in the test meeting for standard acoustic model training. To address this problem, we have proposed and evaluated two methods: feature space rotation, and sample-level overlap synthesis. Due to space constraints, we refer the reader to [10] for a description. We only mention here that the methods are controlled by three parameters, $\{\lambda_G, \lambda_R, \lambda_S\}$, whose magnitudes empirically appear to depend on the number of features per channel and on the overall test meeting duration.

4 Vocal Interaction Modeling

The role of the vocal interaction model during decoding is to provide estimates of $P(\mathbf{q}_{t+1} = \mathbf{S}_j \mid \mathbf{q}_t = \mathbf{S}_i)$, the probability of transitioning to a state $\mathbf{S}_j$ at time $t+1$ from a state $\mathbf{S}_i$ at time t. The complete description of the conversation, when modeled as a first-order Markov process, is an $N \times N$ matrix, where $N \equiv 2^K$. When participants are assumed to behave independently of one another, this probability reduces to $\prod_{k=1}^{K} P(\mathbf{q}_{t+1}[k] = \mathbf{S}_j[k] \mid \mathbf{q}_t[k] = \mathbf{S}_i[k])$. As a result, a participant-independent description consists of a 2×2 matrix.

In this work, we have chosen to not assume that participants behave independently. Descriptive studies of conversation [13] and of meetings [4], as well as computational models in various fields [2][5], have unequivocally demonstrated that an assumption of independence is patently false. To our knowledge, however, suitable models of multiparty vocal interaction have not been designed for or applied to the task of detecting vocal activity for automatic speech recognition in meetings. A main difficulty is the need to collapse the $2^K \times 2^K$ transition probability matrix in a conversation-independent and participant-independent manner, such that model parameters learned in one conversation will generalize to unseen conversations, even when the participants are different, and/or when the number of participants in the train meetings does not match the number of participants in the test meeting.

4.1   Model Structure

To address this issue, we have proposed the following model of vocal interaction:

$$P\left(\mathbf{q}_{t+1} = \mathbf{S}_j \,|\, \mathbf{q}_t = \mathbf{S}_i\right) = \tag{2}$$
$$P\left(\|\mathbf{q}_{t+1}\| = n_j \,,\, \|\mathbf{q}_{t+1} \cdot \mathbf{q}_t\| = o_{ij} \,|\, \|\mathbf{q}_t\| = n_i\right) \times$$
$$P\left(\mathbf{q}_{t+1} = \mathbf{S}_j \,|\, \|\mathbf{q}_{t+1}\| = n_j \,,\, \|\mathbf{q}_{t+1} \cdot \mathbf{q}_t\| = o_{ij} \,,\, \|\mathbf{q}_t\| = n_i\right) \,,$$

where $\|\mathbf{q}_t\|$ represents the number of participants vocalizing at time t, and $\|\mathbf{q}_t \cdot \mathbf{q}_{t+1}\|$ represents the number of participants who were vocalizing at time t and who continue to vocalize at time $t+1$. Equation 2 introduces some additional notational shorthand: $n_i \equiv \|\mathbf{S}_i\|$ and $n_j \equiv \|\mathbf{S}_j\|$ are the number of vocally active participants in states $\mathbf{S}_i$ and $\mathbf{S}_j$, respectively, and $o_{ij} \equiv \|\mathbf{S}_i \cdot \mathbf{S}_j\| \leq \min(n_i, n_j)$ is the number of same participants which are vocally active in both $\mathbf{S}_i$ and $\mathbf{S}_j$.

The first factor in Equation 2 represents a time-independent, conversation-independent, and participant-independent model of transition among various degrees of multiparticipant overlap at times t and $t+1$. We refer to this factor as the Extended Degree of Overlap (EDO) model. In particular, we claim that the probability of transition between two specific states is proportional to the probability of transition between the degrees of simultaneous vocalization in each of them. Furthermore, the term $\|\mathbf{q}_t \cdot \mathbf{q}_{t+1}\|$ accounts for participant state continuity; it allows the probability of the transition $\{A, B\} \longrightarrow \{A, C\}$ to differ from that of $\{A, B\} \longrightarrow \{C, D\}$, which agrees with intuition. Figure 2 shows the total number of unique transitions in the EDO space; for reasons of figure readability, we limit the maximum degree of participant overlap to 2.

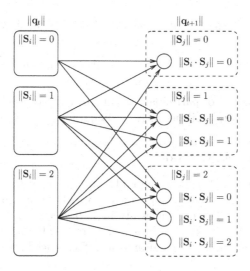

Fig. 2. Unique transition probabilities in the EDO model space with at most 2 simultaneously vocalizing participants

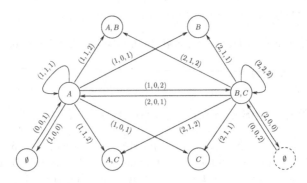

Fig. 3. The 7-state $\mathbf{S}_i$ space for a 3-participant conversation, showing the mapping of (n_i, o_{ij}, n_j) transition probabilities from the EDO space. The all-silent state $\emptyset$ is duplicated for readability; we also show the transitions from only one single one-participant state ($\{A\}$), and from only one single two-participant state. The single three-participant state is not shown.

The second factor in Equation 2 accounts for the multiplicity of specific next $\mathbf{S}_j$ states that are licensed by a particular EDO state transition (n_i, o_{ij}, n_j). We illustrate this in Figure 3. As an example, the transitions $\{A\} \longrightarrow \{A, B\}$ and $\{A\} \longrightarrow \{A, C\}$ are both of $(n_i = 1, o_{ij} = 1, n_j = 2)$ EDO transition type, and they must divide the EDO transition mass between them (for $K = 3$ participants; for $K > 3$ participants, there are additional next state candidates). Because we are constructing a participant-independent model, we assume a uniform distribution over such candidate next states,

$$P\left(\mathbf{q}_{t+1} = \mathbf{S}_j \mid \|\mathbf{q}_{t+1}\| = n_j,\, \|\mathbf{q}_{t+1} \cdot \mathbf{q}_t\| = o_{ij},\, \|\mathbf{q}_t\| = n_i\right) = \tag{3}$$

$$\left(\frac{n_i!}{o_{ij}!\,(n_i - o_{ij})!} \cdot \frac{(K - n_i)!}{(n_j - o_{ij})!\,(K - n_i - n_j + o_{ij})!}\right)^{-1},$$

where K is the number of participants in the test meeting. Equation 3 ensures that the conditional probabilities in Equation 2, for $1 \leq j \leq N$, sum to one.

4.2   Training the EDO Model

To train the EDO model, we use the multi-participant utterance-level segmentation (.mar) from the ISL Meeting Corpus [3], where the number of meetings is $R = 18$. As in [10], the references are first discretized into a time-sequence of states $\mathbf{q}_t^r$; we illustrate this process in Figure 4. The model parameters are then estimated by accumulating bigram counts from the observed time-sequence, according to

$$P\left(\|\mathbf{q}_{t+1}\| = n_j\,,\,\|\mathbf{q}_t \cdot \mathbf{q}_{t+1}\| = o_{ij}\,\big|\,\|\mathbf{q}_t\| = n_i\right) \quad = \qquad\qquad (4)$$

$$\frac{\displaystyle\sum_{\substack{r=1 \\ n_i+n_j-o_{ij}<K}}^{R}\sum_{t=1}^{T_r-1} \delta\left(\|\mathbf{q}_t^r\|, n_i\right)\delta\left(\|\mathbf{q}_t^r \cdot \mathbf{q}_{t+1}^r\|, o_{ij}\right)\delta\left(\|\mathbf{q}_{t+1}^r\|, n_j\right)}{\displaystyle\sum_{\substack{r=1 \\ n_i+n_j-o_{ij}<K}}^{R}\sum_{t=1}^{T_r-1} \delta\left(\|\mathbf{q}_t^r\|, n_i\right)}\,,$$

where $\delta\left(\cdot,\cdot\right)$ is the Kronecker delta, and r indexes training meetings. K is the number of participants in the test meeting, and is given by the number of IHM channels to segment; its appearance in Equation 4 is due to the fact that the EDO model must be recompiled each time K changes. This is because transitions may occur in the training material which are not possible in a particular test meeting: for example, a transition of type $(n_i = 2, o_{ij} = 0, n_j = 2)$, such as $\{A, B\} \longrightarrow \{C, D\}$, is not possible for a test meeting of $K = 3$ participants.

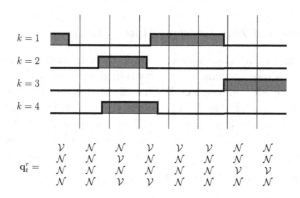

Fig. 4. Assignment of discrete multi-participant values for consecutive frames of $\mathbf{q}^r$ from a reference segmentation. A frame is assigned a value $\mathcal{V}$ for participant k if k vocalizes for at least 50% of the frame duration; otherwise, $\mathcal{N}$ is assigned.

The estimation of the first term of Equation 2, as proposed in Equation 4, is participant-independent since, at each time t, only the *number* of currently vocalizing participants is inspected, rather than specific participants (as indexed by k in $\mathbf{q}_t$). However, because the amount of vocalization overlap may vary across meetings or conversation types, the model is biased towards the interactions patterns observed in the training data; we have addressed this issue by training on a sizable corpus of meetings. There may be significant scope for selecting training material based on anticipated vocal interaction style. The model is also independent of the number of participants in each training or test meeting. Studies of overlap occurrence in meetings [4] do not report a strong correlation with participant number; the increased potential for overlap due to larger group size appears to not be realized in general [7].

5   Experiments

We assess the performance of our algorithms by directly comparing the WERs as was done in [1][6]. WERs were produced using our multi-pass NIST RT-06s Speech-to-Text submission system [8]; however, in the current work, we show only the first-pass MFCC front end WERs, obtained with our RT-06s development language model. We note that an optimistic aim of an automatic segmenter is to produce WERs achievable with human-produced reference segmentation.

The data used in the described experiments consist of two datasets from the NIST RT-05s and RT-06s evaluations. Our development set, rt05s_eval* (referred to as *confDEV* in [8]), is the complete rt05s_eval set less one anomalous meeting (with a participant on speakerphone). We use the complete rt06s_eval as held out unseen data for final evaluation purposes.

The baseline segmenter in these experiments is that used in our NIST RT-06s submission, which differs from the current system in 4 ways. In this section we evaluate these four modifications, and present experiments which explore the impact that vocal interaction modeling has on ASR performance.

5.1   Elimination of Zero-Crossing Rate (ZCR)

The first delta from our RT-06s submission is the elimination of the zero crossing rate feature, whose implementation contained an error and which, following correction, was shown not to affect WERs. Since this modification reduces the feature vector size from $2K$ to K, we have also retuned the acoustic model factors $\{\lambda_G, \lambda_R, \lambda_S\}$ on the development set. The negligible effect of this change to the WER, alongside the performance of the RT06s baseline, is shown in Table 1.

5.2   Frame Step/Size Reduction (F.100)

In a second experiment, we reduced the frame size and step from 0.110s to 0.100s. Since these parameters affect the smoothing pass, we have also modified the latter to consist of: (1) bridging gaps shorter than 0.45s; (2) eliminating spurts shorter than 0.25s; and (3) prepadding and postpadding all segments with 0.15s and 0.2s, respectively. The original smoothing consisted of 5 postprocessing passes: (1) bridging gaps shorter than 0.5s; (2) eliminating spurts shorter than 0.2s; (3) prepadding and postpadding all segments with 0.1s and 0.3s, respectively; (4) bridging remaining gaps shorter than 0.4s; and (5) eliminating remaining spurts shorter than 0.8s. As in the first experiment, these parameters were tuned to minimize WER on our development set. Table 1 shows that these two changes reduce substitutions and deletions on the development set, without increasing insertions.

5.3   Data Selection for Training the All-Silent State (ILA.0)

A third reduction in the rt05s_eval* set WER was achieved by noting that the ILA algorithm is characterized by high precision but significantly lower recall

Table 1. First-pass ASR substitutions (sub), insertions (ins), deletions (del), and overall WER before rescoring, and overall WER after rescoring in the first pass (WER'). Detailed ASR errors prior to rescoring are shown because they correlate with frame-level miss and false alarm rates (not shown) better than do post-rescoring errors. Results are for our development set rt05s_eval*; best automatic and manual performance shown in bold.

Segmentation	sub	del	ins	WER	WER'
RT06s baseline	22.5	11.9	4.8	39.2	37.0
− ZCR	21.1	13.7	4.0	38.8	36.9
+ F.100	20.7	12.8	4.0	37.4	35.2
+ ILA.0	21.2	10.8	4.6	36.6	34.2
+ MULT	21.1	11.1	4.3	**36.5**	**34.1**
maxOV.4	21.1	11.1	4.3	36.5	34.1
maxOV.3	21.1	11.2	4.3	36.5	34.1
maxOV.2	21.0	11.5	4.3	36.8	34.4
MIP	21.3	11.5	4.4	37.2	34.9
manual refs	24.4	8.3	4.8	**37.5**	**34.4**

[9]. This suggests that a large number of frames identified by the ILA as silence may in fact be missed vocal activity. To test this hypothesis, we chose to use only 50% of the ILA-identified silence frames for training the all-silent state model S_0. These are selected by picking the bottom two quartiles in terms of average per-channel log-energy, over all channels. As Table 1 shows, this leads to a significant reduction in deletions, and produces an overall WER which is lower than that produced using manual segmentation.

5.4 Sharing Probability Mass Among Candidate Next States (MULT)

The last delta between our RT-06s submission segmenter and the current system is the implementation of Equation 3. In the baseline system, this factor was ignored in Equation 2. This resulted in more frequent insertions, since the probability of transitioning to states with a high degree of overlap was not normalized by their multiplicity. This modification reduces the WER further below that obtained with manual segmentation.

5.5 Robustness and Generalization

In total, the four modifications described above and shown in Table 1 reduce the WER in the first pass from 37.0% to 34.1%, which surpasses ASR performance achieved with manual segmentation.

In Table 2, we show the performance of our segmentation system individually for each meeting in rt05s_eval. As mentioned above, the rt05s_eval set is identical to our development set, plus the meeting identified as NIST1. As can be seen, the performance of the final system exceeds that of the baseline for every

Table 2. First-pass WERs after rescoring, for individual meetings in rt05s_eval

Segm.	AMI1	AMI2	CMU1	CMU2	ICSI1	ICSI2	NIST1	NIST2	VT1	VT2	all
RT06s	33.7	47.4	36.8	37.8	34.5	27.6	119.8	37.9	37.7	40.8	45.6
− ZCR	33.8	38.8	37.6	34.5	43.5	27.1	91.1	40.9	34.5	41.9	42.5
+ F.100	33.6	36.3	33.1	34.0	42.3	27.1	91.7	39.5	33.7	38.7	41.1
+ ILA.0	34.0	36.6	32.9	33.9	34.4	27.0	94.8	37.7	34.5	38.4	40.5
+ MULT	33.3	35.7	33.3	33.5	33.0	27.2	83.1	38.3	34.0	40.4	39.2
maxOV.4	33.3	35.7	33.3	33.5	32.9	27.2	84.0	38.3	34.0	40.4	39.3
maxOV.3	33.3	35.8	33.3	33.5	33.0	27.3	81.0	38.3	34.0	40.4	39.0
maxOV.2	33.5	36.1	34.1	33.8	33.6	27.8	66.4	38.7	34.0	39.8	37.8
MIP	33.6	36.5	34.8	33.6	35.2	26.9	69.3	38.8	36.0	40.5	38.5
manual	34.7	39.3	32.9	31.3	25.8	25.3	51.2	44.0	34.3	44.8	36.1

Table 3. First-pass WERs after rescoring, for individual meetings in rt06s_eval

Segm.	CMU1	CMU2	EDI1	EDI2	NIST1	NIST2	TNO1	VT1	VT2	all
RT06s	36.9	45.1	31.6	33.3	48.1	51.8	42.9	47.8	39.4	42.1
− ZCR	37.1	45.2	35.9	41.1	43.1	49.5	46.9	45.2	37.2	42.6
+ F.100	36.1	45.5	36.3	35.8	43.8	49.7	46.6	44.3	36.0	41.8
+ ILA.0	55.0	42.6	34.6	35.3	42.8	43.5	41.2	44.7	37.0	42.5
+ MULT	36.5	42.9	35.2	46.0	40.9	43.6	40.6	43.7	36.4	40.8
maxOV.4	36.5	42.9	35.0	35.6	40.9	43.6	40.8	43.6	36.0	39.6
maxOV.3	36.5	42.9	35.0	35.6	40.8	43.6	40.8	43.8	36.0	39.6
maxOV.2	36.6	43.1	35.5	35.6	41.0	43.8	40.9	43.4	36.3	39.8
MIP	36.8	43.4	35.4	36.1	41.7	43.6	40.9	44.3	37.6	40.1
manual	37.2	40.0	34.7	32.2	39.7	35.6	41.7	39.3	33.9	37.4

meeting except NIST2. For five meetings (AMI1, AMI2, NIST2, VT1 and VT2), performance is better with automatic than with human-generated segmentation.

We show a similar analysis in Table 3 for the rt06s_eval set. Cumulatively, our post-evaluation modifications improve performance on all but the two EDI meetings. These two meetings, together with TNO1, appear to have benefited from the faulty ZCR feature, and WERs for them never fully recover once that feature is eliminated. For two of the meetings, CMU1 and TNO1, WERs with automatic segmentation are lower than those with manual segmentation.

5.6 Impact of Modeling Vocal Interaction

Finally, we show results from several experiments in which we explore the impact of modeling vocal interaction on ASR performance. In the first, we limit the state space to states of at most 4 (maxOV.4), at most 3 (maxOV.3), and at most 2 (maxOV.2) simultaneously vocalizing participants. The results on our development set are shown in Table 1; those on the complete rt05s_eval and rt06s_eval sets are shown in Tables 2 and 3, respectively.

As can be seen, limiting the maximal degree of overlap always leads to more deletion errors, although the effect asymptotes after 4-participant overlap is included. This partly corroborates the observations on overlap in [4], namely that more-than-3-participant overlap is extremely rare. However, we note that for the NIST1 meeting in rt05s_eval, which contained a participant without a microphone and suffered from a large number of ASR insertion errors as a result, limiting the maximal degree of overlap effectively reduces the insertions. This effect more than compensates for the slightly increased deletions in the remaining meetings in that set, such that the overall WER is significantly lower.

We also explore the ASR performance which would be achieved with the current segmentation system if the transition model probabilities were provided not by our vocal interaction model but by a model which treats participants in a mutually independent manner, as in other state-of-the-art meeting segmenters [1][6]. In the context of our system, such a model would have the form

$$P\left(\mathbf{q}_{t+1} = \mathbf{S}_j \mid \mathbf{q}_t = \mathbf{S}_i\right) = \prod_{k=1}^{K} P\left(\mathbf{q}_{t+1}\left[k\right] = \mathbf{S}_j\left[k\right] \mid \mathbf{q}_t\left[k\right] = \mathbf{S}_i\left[k\right]\right) . \quad (5)$$

ASR results using this model are given in Tables 1, 2, and 3 as MIP. It shows systematically worse performance; on our development set, the WER difference is 0.8% absolute, while that on the entire rt05s_eval is 0.7% absolute. On unseen data, the mutually independent participant model leads to a WER which is 0.5% absolute higher. We note that this is a conservative estimate of the difference; a fair estimate in the context of our system would require acoustic models for all possible overlap states, whereas our acoustic model training procedure typically produces models for at most 4-participant overlap. Furthermore, our full-covariance acoustic models treat participants jointly [10].

6   Conclusions

We have described the automatic segmentation system used in our NIST RT-06s Speech-to-Text Evaluation submission, together with several improvements. The system implements a novel approach to segmenting multi-channel, multi-speaker meeting recordings, in particular in its use of multi-participant acoustic and transition models. In its current state, the system outperforms human segmentation in first-pass ASR performance on our development set. The performance on the complete rt05s_eval and rt06s_eval sets leads to first-pass WERs which are 1.6%-2.2% absolute higher than with human segmentation, comparing favorably with other state-of-the-art systems [1][6].

Acknowledgments

We would like to thank Mari Ostendorf for her suggestions and help in clarifying and presenting the ideas in this paper. This work was partly supported by the European Union under the integrated project CHIL (IST-506909), Computers in the Human Interaction Loop (http://chil.server.de).

References

1. Boakye, K., Stolcke, A.: Improved speech activity detection using cross-channel features for recognition of multiparty meetings. In: Proc. of INTERSPEECH 2006, Pittsburgh PA, USA, pp. 1962–1965 (2006)
2. Brady, P.: A model for generating on-off speech patterns in two-way conversation. Bell Systems Technical Journal 48(7), 2445–2472 (1969)
3. Burger, S., MacLaren, V., Yu, H.: The ISL Meeting Corpus: The impact of meeting type on speech style. In: Proc. of ICSLP 2002, Denver CO, USA, pp. 301–304 (2002)
4. Çetin, Ö., Shriberg, E.: Overlap in meetings: ASR effects and analysis by dialog factors, speakers, and collection site. In: Renals, S., Bengio, S., Fiscus, J.G. (eds.) MLMI 2006. LNCS, vol. 4299, pp. 200–211. Springer, Heidelberg (2006)
5. Dabbs Jr., J., Ruback, R.: Dimensions of group process: Amount and structure of vocal interaction. Advances in Experimental Social Psychology 20, 123–169 (1987)
6. Dines, J., Vepa, J., Hain, T.: The segmentation of multi-channel meeting recordings for automatic speech recognition. In: Proc. of INTERSPEECH 2006, Pittsburgh PA, USA, pp. 1213–1216 (2006)
7. Fay, N., Garrod, S., Carletta, J.: Group discussion as interactive dialogue or serial monologue: The influence of group size. Psychological Science 11(6), 487–492 (2000)
8. Fügen, C., Ikbal, S., Kraft, F., Kumatani, K., Laskowski, K., McDonough, J., Ostendorf, M., Stüker, S., Wölfel, M.: The ISL RT-06S Speech-to-Text System. In: Renals, S., Bengio, S., Fiscus, J.G. (eds.) MLMI 2006. LNCS, vol. 4299, pp. 407–418. Springer, Heidelberg (2006)
9. Laskowski, K., Jin, Q., Schultz, T.: Crosscorrelation-based multispeaker speech activity detection. In: Proc. of ISLP 2004, Jeju Island, South Korea, pp. 973–976 (2004)
10. Laskowski, K., Schultz, T.: Unsupervised learning of overlapped speech model parameters for multichannel speech activity detection in meetings. In: Proc. of ICASSP 2006, Toulouse, France, vol. I, pp. 993–996 (2006)
11. Laskowski, K., Schultz, T.: A geometric interpretation of normalized maximum crosscorrelation for vocal activity detection in meetings. In: Proc. of HLT-NAACL, Short Papers, Rochester NY, USA, pp. 89–92 (2007)
12. Pfau, T., Ellis, D., Stolcke, A.: Multispeaker speech activity detection for the ICSI Meeting Recorder. In: Proc. of ASRU 2001, Madonna di Campiglio, Italy, pp. 107–110 (2001)
13. Sacks, H., Schegloff, E., Jefferson, G.: A simplest sematics for the organization of turn-taking for conversation. Language 50(4), 696–735 (1974)
14. Wrigley, S., Brown, G., Wan, V., Renals, S.: Feature selection for the classification of crosstalk in multi-channel audio. In: Proc. of EUROSPEECH 2003, Geneva, Switzerland, pp. 469–472 (2003)

Binaural Speech Separation Using Recurrent Timing Neural Networks for Joint F0-Localisation Estimation

Stuart N. Wrigley and Guy J. Brown

Department of Computer Science, University of Sheffield
211 Portobello Street, Sheffield S1 4DP, United Kingdom
s.wrigley@dcs.shef.ac.uk, g.brown@dcs.shef.ac.uk

Abstract. A speech separation system is described in which sources are represented in a joint interaural time difference-fundamental frequency (ITD-F0) cue space. Traditionally, recurrent timing neural networks (RTNNs) have been used only to extract periodicity information; in this study, this type of network is extended in two ways. Firstly, a coincidence detector layer is introduced, each node of which is tuned to a particular ITD; secondly, the RTNN is extended to become two-dimensional to allow periodicity analysis to be performed at each best-ITD. Thus, one axis of the RTNN represents F0 and the other ITD allowing sources to be segregated on the basis of their separation in ITD-F0 space. Source segregation is performed within individual frequency channels without recourse to across-channel estimates of F0 or ITD that are commonly used in auditory scene analysis approaches. The system is evaluated on spatialised speech signals using energy-based metrics and automatic speech recognition.

1 Introduction

When listening in natural environments, the ear is bombarded with energy from multiple sound sources. Despite these sounds being mixed together, the human auditory system has the ability to analyse and extract cognitive representations for the individual sounds that are present—possibly performing this task simultaneously for multiple sources. It has been proposed that the acoustic signal is subjected to an *auditory scene analysis* in which a number of cues are extracted and used to segregate sounds on the basis of them 'belonging' to same physical source [1]. Such cues include common periodicity, common onset/offset, proximity in frequency, etc.

Human speech perception is robust even in very challenging acoustic environments; conversely, automatic speech recognition (ASR) systems can be susceptible to relatively small changes in the background acoustics. For many years, there has been interest in developing computational models of auditory scene analysis (CASA; see [2] for a review) and one use of such models is as an aide to ASR systems. In this paper, we present a novel technique of computing a joint

A. Popescu-Belis, S. Renals, and H. Bourlard (Eds.): MLMI 2007, LNCS 4892, pp. 271–282, 2007.
© Springer-Verlag Berlin Heidelberg 2007

harmonicity-location cue in a neurobiologically plausible manner which can be used to segregate concurrent talkers and produce a mask for use with 'missing data' automatic speech recognisers [3].

The remainder of this article is organized as follows. The next section describes the two grouping cues that will be used by our system. Section 3 provides a brief overview of the competing approaches to neural representations of sounds. Following this, recurrent timing neural networks (RTNNs) are described in detail. This is followed by the implementation details of the auditory front end and the way in which this is coupled to an array of RTNNs. We present a number of evaluation techniques which have been used to assess the system and describe their results. We conclude with a discussion of the presented work and directions for future work.

2   Grouping Cues

2.1   Harmonicity

Fundamental frequency (F0) is a potent grouping cue. When listening, harmonically related components tend to form perceptual wholes (streams), whereas differences in F0 promote segregation. For example, Brokx and Nooteboom found that listeners were better able to identify two simultaneous speech utterances if they had different F0s [4].

Further support for the role of F0 in grouping comes from a number of studies which have investigated the perception of 'double vowels'. In this paradigm, a pair of steady-state, synthetic vowels are presented simultaneously, with identical onset and offset, and subjects are required to identify both vowels. Scheffers [5] found that listeners were able to identify both simultaneous vowels more accurately when they were on different F0s than when they were on the same F0. From these studies, it was proposed that a F0-guided segregation strategy is used to separate, and subsequently identify, simultaneous sounds.

Despite the fact that listeners' recognition does improve with increasing F0, doubt has been cast upon the proposed F0-guided segregation strategy. Bird and Darwin [6] investigated the mechanisms by which the auditory system exploits F0 differences in separating two sentence-length utterances. They used a stimulus in which the low-pass part of the target sentence had the same F0 as the high-pass part of the interfering sentence. The remaining parts shared the same variable F0. If the auditory system relies on global mechanisms, performance would be impaired due to inappropriate grouping of low- and high-pass parts. It was found that listener performance on the band swapped stimuli was the same as on the unmanipulated stimuli up to 2 semitones. Thus, across-frequency grouping of components across the low- and high-frequency regions only occurred for F0 differences of 5 semitones and above, but not 2 semitones and below.

2.2   Location

Listeners can also take advantage of the differing signals reaching the two ears to determine the direction of a sound source [7]. Provided a sound is not in the median plane[1], sound energy will reach the closer ear slightly before the further ear and also with a slightly higher intensity. These two cues are referred to as *interaural time difference* (ITD) and *interaural intensity difference* (IID), the latter caused by 'shadowing' due to the head. ITD will be the focus in this study. ITDs range from 0 s for a sound directly in front of the listener's head (i.e., at an azimuth of $\pm 0°$) to about 690 μs for a sound directly opposite one of the ears (i.e., at an azimuth of $\pm 90°$).

In general, the constituent energies of a sound originating from the same location will share approximately the same ITD (we note, however, that ITD coherence is eroded in reverberant conditions; see [2, Chap. 7]). Thus, across-frequency grouping by ITD ought to provide a powerful mechanism for segregating multiple voices. Indeed, across-frequency grouping by ITD has been employed by computational models of voice separation (e.g., [8,9]).

However, analogous to F0-based segregation, there is also evidence that across-frequency grouping does not occur for interaural time difference (ITD). A number of studies have drawn across-frequency grouping by ITD into question; Edmonds and Culling [10] studied this using target and interferer pairings each of which had been low- and high-pass filtered. Even when the low-pass portion of the target and the high-pass portion of the interferer were placed at the same ITD and the remaining portions placed at a different ITD, listeners performed as well as when both target portions were presented at a consistent ITD. When both target and interferer are placed at the same ITD, performance was significantly reduced. This suggests that the auditory system exploits differences in ITD independently within each frequency channel.

3   Neural Mechanisms

The precise mechanism by which the auditory system can exploit different grouping cues (the 'neural code') remains unclear. Taking the example of harmonicity, theories of pitch perception can be considered to lie on a continuum with 'place code' models and 'temporal code' models at the extremities. The place code states that the pitch of a periodic sound corresponds to the position of maximum excitation in some tonotopically organised site in the brain. In contrast, temporal models of pitch perception use the temporal fine structure of the auditory nerve firings to determine the pitch.

A class of neural networks called *timing nets* have been suggested as a means of explaining how the auditory system uses temporally-coded input to produce meaningful outputs [11,12]. Such networks consist of coincidence detectors and delay lines and can be considered to encapsulate a range of architectures which employ analyses of interspike intervals (e.g., auto-correlation and cross-correlation). A specific form of timing nets called recurrent timing neural networks (RTNNs) requires only one spike pattern as input and has been successfully used for periodicity analysis [12]. It should be emphasised, however, that the stimuli used in

---

[1] A vertical plane passing through the head such that all points on the plane are equidistant from both ears.

[12] consisted of synthetic, stationary F0s. In the study presented here, we extend this work to operate on natural speech and extend the network architecture such that interaural time delay is also represented within the same network. This novel architecture allows concurrent speech to be separated on the basis of a joint F0-location cue without need for across-channel grouping: all processing is strictly within-channel.

4   Recurrent Timing Neural Networks

An RTNN consists of a bank of coincidence detectors, all operating on the same stimulus; Fig. 1(b). Each node of the network has a recurrent input exhibiting a slightly different delay; Fig. 1(a). The pattern circulating in the recurrent delay loop re-emerges after τ milliseconds; this is then compared with the stimulus arriving at the node; if a coincidence is detected, the amplitude of the delay loop input is increased by a certain factor. Regardless of the detection of a coincidence, an attenuated version of the incoming signal is fed into the delay line: without this, there would be no circulating signal to produce coincidences. Thus, stimulus periodicities equal to a node's recurrent delay will be emphasised by that node. Over time, repeating temporal patterns are enhanced relative to the rest of the stimulus. Furthermore, multiple repeating patterns with different periodicities can be detected and encoded by such networks. Cariani showed that such a relatively simple network was able to successfully separate up to three concurrent synthetic vowels [12].

In this study, we wish to represent both pitch information and ITD information in the same feature space. To achieve this, we make two important alterations to

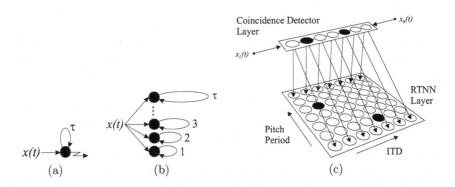

Fig. 1. (a) Coincidence detector with recurrent delay loop. (b) A group of coincidence detectors with recurrent delay loops of increasing length form a recurrent timing neural network (RTNN). Note that all nodes in the RTNN receive the same input. (c) 2D RTNN (bottom layer) with extra coincidence detector layer (top) allowing joint estimation of F0 and ITD. Downward connections are only shown for the front and back rows. Recurrent delay loops for the RTNN layer are omitted for clarity. $x_L(t)$ and $x_R(t)$ represent signals from the left and right ears respectively. Solid nodes represent activated coincidence detectors.

the architecture shown in Fig. 1(b). In order to compute the ITD feature, a one-dimensional row of coincidence detectors coupled by a delay line is introduced. In our system, 41 nodes are used, each coupled by a delay of equal duration (1 sample at 20 kHz or 50 μs). The end nodes of this row each receive one of the individual ear signals as shown in Fig. 1(c). Each signal propagates down the row; in essence, this performs a cross-correlation analysis equivalent to Jeffress' neural coincidence model of sound localisation [13]. Hence, the ITD of sounds to the right of the head are represented by coincidences in the left-hand half of the delay line (since they reach the right ear first and gain a headstart down the delay line travelling toward the left-hand edge) and vice versa.

Functionally, the delay line has the effect of performing an initial stage of source separation: activities due to spatially distinct sound sources will be emitted by their corresponding delay line node. At any one time, all nodes in the delay line will be emitting activity of some form (although only a small number will be responding strongly to sources at their best-ITD). In order to perform periodicity analysis on the output of each of these nodes, the one-dimensional network architecture shown in Fig. 1(b) is replicated to create a two-dimensional network in which each column performs periodicity analysis on the output of a single ITD delay line node; see Fig. 1(c). The activity of the two-dimensional layer, therefore, is a map with ITD on one axis and pitch on the other. Continuing the example shown in Fig. 1(c), the two-dimensional network is showing that the source nearest the right side of the head has a large pitch period, while the source towards the left side of the head has a small pitch period.

The ability to represent both F0 and ITD on the same feature space, allows the model to avoid a common problem in CASA: when multiple concurrent sources are present, how is the correct ITD associated with the correct F0? The two features are commonly computed in distinct feature spaces. In this model, they are automatically associated. Furthermore, it is easier to separate multiple sources in this feature space since it is unlikely that two sources will exhibit the same pitch and location simultaneously, thus being represented in different areas. Indeed, given a static separation of the sources, there is no need for explicit tracking of F0 or location: we simply connect the closest activity regions over time. A further advantage is that source separation can proceed within-channel without reference to a dominant F0 or dominant ITD estimate as required in an across-frequency grouping technique. Provided there is some separation in one or both of the cues, two activity regions (in the case of two simultaneous talkers) can be extracted and assigned to different sources.

5   The Model

5.1   Auditory Periphery

A bank of 20 gammatone filters [14] with overlapping passbands simulate the frequency analysis performed by the basilar membrane. Their centre frequencies range from 100 Hz to 8 kHz and are equally spaced on the equivalent rectangular bandwidth (ERB) scale [15]. A gammatone filter of order n and centre frequency f Hz is defined as:

$$gt(t) = t^{n-1} \exp(-2\pi bt) \cos(2\pi ft + \phi) H(t) \tag{1}$$

where f is the centre frequency, ϕ is phase, b is related to bandwidth, and $H(t)$ is the unit step (Heaviside) function defined as $H(t) = 1$ if $t \geq 0$, $H(t) = 0$ otherwise. Here, we use fourth-order gammatone filters. Since later stages of the model only require periodicity information, the auditory nerve response is approximated by half-wave rectifying and cube root compressing the output of each filter.

5.2   RTNN

The RTNN layer consists of a grid of independent (i.e., unconnected) coincidence detectors with an input from the ITD estimation layer (described above) and a recurrent delay loop. For a node with a recurrent delay loop duration of τ whose input $x_\theta(t)$ is received from the ITD node tuned to an interaural delay of θ, the update rule is:

$$C(t) = \alpha x_\theta(t) + \beta x_\theta(t) C(t - \tau) . \tag{2}$$

The output of the node (and, hence, also about to enter the recurrent delay loop) is denoted by $C(t)$; $C(t-\tau)$ is the response which is just emerging from the delay loop. To ensure some form of signal is always circulating in the delay loop to allow coincidences to occur, α acts as an attenuator for the incoming signal. Note that α is sufficiently small so as not to dominate the node's response ($\alpha = 0.2$). Should a coincidence occur, the weight β determines the rate of adjustment and is dependent on τ such that coincidences at low pitches are de-emphasized [12]. Here, β increases linearly from 3 at the smallest recurrent delay loop length to 10 at the largest.

In order to perform the joint pitch-ITD analysis on each auditory nerve centre frequency, the model employs 20 independent networks (of the form shown in Fig. 1(c)), one for each frequency channel. Network activity is captured using a sliding temporal window in which activity over the duration of the window is averaged. In our model, we use a window size of 25 ms and a temporal shift of 5 ms. Therefore, for every frequency channel, a sequence of two-dimensional activity plots is built up over time.

Ultimately, the system should allow concurrent speakers to be separated and be transcribed by an automatic speech recognition system. Thus, it is necessary to know, at any time-frequency point during the signal, whether that point is dominated by the target speech or by some form of interference. The network activity plots (one generated every 5 ms per frequency channel) can be used to make an estimate of talker activity at a particular time-frequency point: a highly active node relative to the rest of the network indicates that the source at that F0-ITD combination is active. Specifically, a time-frequency binary mask for the target talker is created from the RTNN output. A time-frequency mask unit is set to 1 if the target talker's activity was greater than the target's mean activity for that frequency channel, otherwise it is set to 0. Talker activity can be grouped across time frames by associating the closest active nodes in F0-ITD space (assuming the two talkers don't momentarily have the same ITD *and* F0).

6   Evaluation

The separation technique described above was evaluated on a number of speech mixtures drawn from the TIdigits Studio Quality Speaker-Independent Connected-Digit Corpus [16]. The sampling rate for the corpus is 20 kHz.

In order to investigate the influence of spatial separation on the ability of the system to successfully segregate concurrent speakers, three different target+interferer spatialisations were used: -40°+40°, -20°+20° and -10°+10°. For each scenario, 100 randomly selected utterance pairs were created, all of which were from male talkers. To avoid duration mismatches due to differing speaking rates across subjects, the target utterance consisted of five digits and the interferer utterance consisted of seven digits. Furthermore, the target was always on the left of the azimuth midline. The signals were spatialised by convolving them with head related transfer functions (HRTFs) measured from a KEMAR artificial head in an anechoic environment [17]. The two speech signals where then combined with a signal-to-noise ratio (SNR) of 0 dB. The SNR was calculated using the original, monaural, signals prior to spatialisation.

Three forms of evaluation were employed: assessment of the amount of target energy lost (P_{EL}) and interferer energy remaining (P_{NR}) in the mask [18, p. 1146]; target speaker SNR improvement; ASR performance improvement.

All three techniques require the use of an a priori binary mask (an 'optimal' mask). The a priori binary mask is formed by placing a 1 in any time-frequency units where the energy in the mixed signal is within 1 dB of the energy in the clean target speech (the regions which are dominated by target speech), otherwise they are set to 0. In other words, an a priori binary mask uses information about regions of uncorrupted speech within the mixture.

6.1   Signal Energy

In order to assess the quality of segregation based upon an energy metric, it is necessary to obtain a number of time-domain signals. These signals are derived from a resynthesis process which uses a binary mask to determine which time-frequency portions are required and which are to be discarded. The gammatone filter outputs for each frequency channel are divided into frames of size equal to the binary mask resolution. Each signal frame is then weighted by the value of the binary mask at that time-frequency point. Individual channels are recovered using the overlap-and-add method and these are summed across frequencies to yield a resynthesized signal. The percentage of target speech excluded from the segregated speech (P_{EL}), and the percentage of interferer included (P_{NR}) are defined to be [18, p. 1146]:

$$P_{EL} = \frac{\sum_n e_1^2(n)}{\sum_n I^2(n)} , \tag{3}$$

$$P_{\mathrm{NR}} = \frac{\sum_n e_2^2(n)}{\sum_n O^2(n)} \ . \tag{4}$$

The clean target signal which has been resynthesized using the a priori binary mask is denoted by $I(n)$. $O(n)$ is the clean target signal which has been resynthesized using the RTNN produced mask (the actual separated signal produced by our system). $e_1(n)$ is the clean target signal which has been resynthesized using a mask in which 1s are present at all time-frequency points which are 1 in the a priori binary mask and 0 in the RTNN produced mask (the portions of the signal which ought to be present but are missing in the system's separated signal). $e_2(n)$ is the opposite of this; in other words, $e_2(n)$ is the clean target signal which has been resynthesized using a mask in which 1s are present at all time-frequency points which are 1 in the RTNN produced mask and 0 in the a priori binary mask (the portions of the signal which are present but should not be: remaining interferer).

In addition to resynthesizing the target $O(n)$, the interferer signal is also resynthesized using the RTNN generated target mask. This allows the calculation of SNR (an easily understood metric) before and after processing.

6.2   Automatic Speech Recognition

The third evaluation technique involves using an ASR system which can exploit the 'missing data' technique [3]. The task of ASR can be defined as the assignment of an acoustic observation x to a class of speech sound C. However, some components of x may be unreliable or missing due to an interfering sound source. In this situation, the likelihood of the acoustic model $f(x|C)$ cannot be established in the usual manner. To overcome this problem, the missing data approach partitions x into reliable and unreliable components, x_r and x_u. The reliable components x_r (whose values are known) are directly available to the classifier whereas the unreliable components x_u are uncertain.

More specifically, the marginal distribution $f(x_r|C)$ is used directly and the likelihood $f(x_u|C)$ is estimated by integrating over bounds (i.e., between zero and the observed energy) [3].

The missing data approach requires a process which identifies the reliable and unreliable components, x_r and x_u. Here, we use the RTNN time-frequency binary mask to indicate whether the acoustic evidence in each time-frequency region is reliable; units assigned 1 in the binary mask define the reliable areas of target speech whereas units assigned 0 represent unreliable regions.

The features used by the recogniser in this study are known as *auditory rate maps*. They are computed by calculating the instantaneous Hilbert envelope of each gammatone filter response [19]. This is smoothed by a low-pass filter with an 8 ms time constant, sampled at 5 ms intervals (to match the RTNN binary mask resolution), and cube root compressed to give a pseudo-spectrogram representation of auditory firing rate (Fig. 2).

Auditory rate maps were obtained for the training section of the corpus, and were used to train 12 word-level HMMs (a silence model, 'oh', 'zero' and '1' to '9') each consisting of 18 no-skip, straight-through states with observations modelled

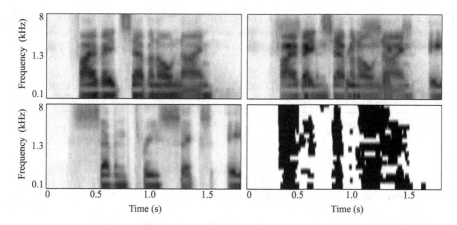

Fig. 2. The upper left panel shows the ratemap for the target utterance ('587o9') and the interfering utterance ('736883o') ratemap is shown below it. Darker areas indicate higher energy; the interfering ratemap has been truncated to match the duration of the target. The upper right panel shows the ratemap of the two utterances which have been spatialised to -40° (target) and 40° (interferer) and mixed at a monaural SNR of 0 dB. The bottom right panel shows the RTNN missing data mask for the target utterance—black pixels denote time-frequency regions dominated by the target utterance.

by a 12 component diagonal Gaussian mixture. Training was performed using HTK [20] and testing used Barker's Computational Auditory Scene Analysis Toolkit (CTK)[2].

6.3 Results

The results from our model are shown in Table 1; each value represents the average performance of 100 target and interferer utterance pairs for the three different spatial separations. In addition, the average performance across all spatial separations is included. The data in the table is split into three main classes: P_{EL} and P_{NR}, SNR improvement and ASR improvement. For comparison, a priori performance is also shown for SNR and ASR approaches. These values are calculated for the left ear (the ear closest to the target). Although the speech signals were mixed at 0 dB relative to the monaural signals, the actual SNR at the left ear for the spatialised signals will depend on the spatial separation of the two talkers (hence its inclusion in the table).

On average, the RTNN system removes over 91% of the interfering utterance with this figure rising to 94% at the most favourable separation. In addition to this, the amount of energy incorrectly removed from the target P_{EL} is approximately 11%. The SNR metric shows a significant improvement at all interferer positions; on average, our model exhibits an improvement factor of 3.7 when compared to the SNR before separation. Furthermore, it can be observed that

---

[2] Available from http://www.dcs.shef.ac.uk/~jon/ctk.html

Table 1. Separation performance for concurrent speech at different interferer azimuth positions in degrees; 'pre' denotes performance before processing; 'RTNN' denotes performance after processing and 'a priori' denotes 'optimal' performance. ASR accuracy is (100% - word error rate).

	$\pm 10°$	$\pm 20°$	$\pm 40°$	AVERAGE
SNR (dB) pre	1.64	3.13	5.19	3.32
SNR (dB) RTNN	10.03	11.55	14.49	12.02
SNR (dB) a priori	12.35	13.27	15.01	13.54
Mean P_{EL} (%)	10.62	12.74	10.22	11.19
Mean P_{NR} (%)	9.99	8.42	6.02	8.14
ASR Acc. (%) pre	15.00	22.20	28.20	21.80
ASR Acc. (%) RTNN	71.60	74.60	83.40	76.53
ASR Acc. (%) a priori	93.40	94.00	94.60	94.00

SNR performance approaches the a priori values at wider separations. Such SNR performance is supported by the low values for P_{NR} which indicate good levels of interferer rejection and relatively little target loss.

Importantly, the missing data ASR paradigm is tolerant to this relatively low level of target energy loss as indicated by the ASR accuracy performance. Indeed, the missing data ASR performance remains relatively robust when compared to the baseline system which used a unity mask (all time-frequency units assigned 1). We note that ASR performance also approaches the a priori values at wider angular separations. Furthermore, we predict that an increased sampling rate would produce improvements in performance for both SNR and ASR at smaller separations due to the higher resolution of the ITD sensitive layer (see below).

7   Conclusions

A number of grouping cues play important roles in the auditory system's ability to segregate competing sounds. Two of these cues are harmonicity and location. The neural coding strategy by which such cues are represented is the subject of continued debate. A type of neural network called recurrent timing neural networks has been proposed as a means of explaining how the auditory system uses temporally-coded input to produce meaningful outputs [11,12]. Such networks have been used successfully in previous studies to separate multiple concurrent synthetic vowels using periodicity information.

In this study we extended such one-dimensional networks to allow the architecture to represent sounds in a joint F0-ITD cue space. The system was evaluated using a much more challenging paradigm than the synthetic static vowels used in previous RTNN studies [11,12]. Here, the scenarios consist of concurrent real speech mixed at an SNR of 0 dB. Unlike stationary vowels, each constituent signal contains fluctuating F0s and sections of unvoiced speech are common.

The results shown in Table 1 indicate high levels of interferer rejection with low levels of incorrectly removed target speech. The system retained an average of 88.81% of target speech energy and removed over 91% of the interferer. SNR values for the separated target speech also indicate good separation, and informal listening tests found that target speech extracted by the system was of good quality. SNR performance reported here (10.03 dB at the smallest separation) also compares well with those of [9], although direct comparison is difficult due to differing stimuli and spatial separations. The energy-based mechanism allowing unvoiced segments to be represented in the RTNN binary mask successfully included the utterances' fricatives. We note that the target signals commonly used in such evaluations tend to be voiced throughout (e.g., [18]), thus avoiding the problem of unvoiced energy.

The relatively wide spatial separations employed here were by necessity of the sampling rate of the speech corpus: at 20 kHz an ITD of one sample is equivalent to an angular separation of approximately 5.4°. Thus, the smallest separation used here corresponds to an ITD of just 3.7 samples. A means of addressing this issue is to upsample the corpus to a higher sampling rate of, for example, 48 kHz. However, this has the effect of significantly increasing the size of the RTNN and thus the computational load—a topic of future work. We will also test the system on a larger range of SNRs and larger set of interferer positions.

Furthermore, an assumption made by the system is that the target is always on the left side of the head. At each frequency, the activity of a source is represented in F0-ITD space. Over the duration of the signal, the position of this source will fluctuate with pitch and location, hence, creating a trace through the 3-dimensional space of F0, ITD and time. This presents a permutation ambiguity problem similar to that encountered in frequency-domain blind source separation approaches [21], which could be solved by combining channels that have a similar temporal structure. Alternatively, an attentional process could be employed which would select one of the sources to be the target based upon some a priori knowledge of the target and track it across time (e.g., [22]).

Acknowledgments. This work was partly supported by the European Union 6th FWP IST Integrated Project AMI (Augmented Multi-party Interaction, FP6-506811).

References

1. Bregman, A.S.: Auditory Scene Analysis. The Perceptual Organization of Sound. MIT Press, Cambridge (1990)
2. Wang, D., Brown, G.J. (eds.): Computational Auditory Scene Analysis: Principles, Algorithms and Applications. IEEE Press / Wiley-Interscience (2006)
3. Cooke, M., Green, P., Josifovski, L., Vizinho, A.: Robust automatic speech recognition with missing and unreliable acoustic data. Speech Commun. 34(3), 267–285 (2001)

4. Brokx, J.P.L., Nooteboom, S.G.: Intonation and the perceptual separation of simultaneous voices. J. Phonetics 10, 23–36 (1982)
5. Scheffers, M.T.M.: Sifting Vowels: Auditory Pitch Analysis and Sound Segregation. PhD thesis, Groningen University, The Netherlands (1983)
6. Bird, J., Darwin, C.J.: Effects of a difference in fundamental frequency in separating two sentences. In: Palmer, A.R., Rees, A., Summerfield, A.Q., Meddis, R. (eds.) Psychophysical and physiological advances in hearing, Whurr, pp. 263–269 (1997)
7. Blauert, J.: Spatial Hearing — The Psychophysics of Human Sound Localization. MIT Press, Cambridge (1997)
8. Lyon, R.F.: A computational model of binaural localization and separation. In: Proc. Intl. Conf. on Acoustics, Speech and Signal Processing (ICASSP), pp. 1148–1151 (1983)
9. Roman, N., Wang, D., Brown, G.J.: Speech segregation based on sound localization. J. Acoust. Soc. Am. 114, 2236–2252 (2003)
10. Edmonds, B.A., Culling, J.F.: The spatial unmasking of speech: Evidence for within-channel processing of interaural time delay. J. Acoust. Soc. Am. 117, 3069–3078 (2005)
11. Cariani, P.A.: Neural timing nets. Neural Networks 14, 737–753 (2001)
12. Cariani, P.A.: Recurrent timing nets for auditory scene analysis. In: Proc. Intl. Conf. on Neural Networks (IJCNN) (2003)
13. Jeffress, L.A.: A place theory of sound localization. J. Comp. Physiol. Psychol. 41, 35–39 (1948)
14. Patterson, R.D., Nimmo-Smith, I., Holdsworth, J., Rice, P.: An efficient auditory filterbank based on the gammatone function. Technical Report 2341, Applied Psychology Unit, University of Cambridge, UK (1988)
15. Glasberg, B.R., Moore, B.C.J.: Derivation of auditory filter shapes from notched-noise data. Hearing Res. 47, 103–138 (1990)
16. Leonard, R.G.: A database for speaker-independent digit recognition. In: Proc. Intl. Conf. on Acoustics, Speech, and Signal Processing (ICASSP). vol. 3 (1984)
17. Gardner, W.G., Martin, K.D.: HRTF measurements of a KEMAR. J. Acoust. Soc. Am. 97(6), 3907–3908 (1995)
18. Hu, G., Wang, D.: Monaural speech segregation based on pitch tracking and amplitude modulation. Neural Networks 15(5), 1135–1150 (2004)
19. Cooke, M.P.: Modelling auditory processing and organisation. Cambridge University Press, Cambridge (1991/1993)
20. Young, S., Evermann, G., Gales, M., Hain, T., Kershaw, D., Moore, G., Odell, J., Ollason, D., Povey, D., Valtchev, V., Woodland, P.: The HTK Book (for HTK Version 3.3). Cambridge University Engineering Department (2005)
21. Murata, N., Ikeda, S., Ziehe, A.: An approach to blind source separation based on temporal structure of speech signals. Neurocomputing 41, 1–24 (2001)
22. Wrigley, S.N., Brown, G.J.: A computational model of auditory selective attention. IEEE Trans. Neural Networks 15(5), 1151–1163 (2004)

To Separate Speech
A System for Recognizing Simultaneous Speech

John McDonough[1,3], Kenichi Kumatani[2,3], Tobias Gehrig[4],
Emilian Stoimenov[4], Uwe Mayer[4], Stefan Schacht[1], Matthias Wölfel[4],
and Dietrich Klakow[1]

[1] Spoken Language Systems, Saarland University, Saarbrücken, Germany
[2] IDIAP Research Institute, Martigny, Switzerland
[3] Institute for Intelligent Sensor-Actuator Systems, University of Karlsruhe, Germany
[4] Institute for Theoretical Computer Science, University of Karlsruhe, Germany

Abstract. The *PASCAL Speech Separation Challenge* (SSC) is based
on a corpus of sentences from the Wall Street Journal task read by two
speakers simultaneously and captured with two circular eight-channel
microphone arrays. This work describes our system for the recognition
of such *simultaneous* speech. Our system has four principal components:
A person tracker returns the locations of both active speakers, as well
as segmentation information for each utterance, which are often of unequal length; two beamformers in *generalized sidelobe canceller* (GSC)
configuration separate the simultaneous speech by setting their active
weight vectors according to a minimum mutual information (MMI) criterion; a postfilter and binary mask operating on the outputs of the
beamformers further enhance the separated speech; and finally an *automatic speech recognition* (ASR) engine based on a *weighted finite-state
transducer* (WFST) returns the most likely word hypotheses for the separated streams. In addition to optimizing each of these components, we
investigated the effect of the filter bank design used to perform subband
analysis and synthesis during beamforming. On the SSC development
data, our system achieved a word error rate of 39.6%.

1 Introduction

The *PASCAL Speech Separation Challenge* (SSC) is based on a corpus of sentences from the Wall Street Journal (WSJ) task read by two speakers simultaneously and captured with two circular eight-channel microphone arrays. This
work describes our system for the automatic recognition of such *simultaneous*
speech. Our system has four principal componenents: A person tracker returns
the locations of both active speakers, as well as segmentation information for
each utterance, which are often of unequal length; two beamformers in *generalized sidelobe canceller* (GSC) configuration separate the simultaneous speech
by setting their active weight vectors according to a minimum mutual information (MMI) criterion; a postfilter and binary mask operating on the outputs

A. Popescu-Belis, S. Renals, and H. Bourlard (Eds.): MLMI 2007, LNCS 4892, pp. 283–294, 2007.

of the beamformers further enhance the separated speech; and finally an *automatic speech recognition* (ASR) engine based on a *weighted finite-state transducer* (WFST) returns the most likely word hypotheses for the separated streams.

Our speaker tracking system was previously described in [1]. It is based on a *joint probabilistic data association filter* (JPDAF). The JPDAF is capable of tracking multiple targets simultaneously and consists of multiple *Kalman filters*, once for each target to be tracked [2, §6.4]. When new observations become available, they are associated either with an active target or with the *clutter model*, which models spurious acoustic events, through the calculation of posterior probabilities. After the association step, the position of each target can be updated independently through suitably modified Kalman filter update formulae.

In acoustic beamforming, it is typically assumed that the position of the speaker is estimated by a speaker localization system. A conventional beamformer in GSC configuration is structured such that the direct signal from the speaker is undistorted [3, §6.7.3]. Subject to this *distortionless constraint*, the total output power of the beamformer is minimized through the appropriate adjustment of an active weight vector, which effectively places a null on any source of interference, but can also lead to an undesirable *signal cancellation*. To avoid the latter, the adaptation of the active weight vectors is typically halted whenever the desired source is active.

For the speech separation task, we implemented two beamformers in GSC configuration, where one GSC was directed at each source. We then jointly adjusted the active weight vectors of the GSCs so as to provide output streams with minimum mutual information. Better speech separation was achieved through the use of non-Gaussian pdfs for calculating mutual information. In our initial experiments on the SSC development data, a simple delay-and-sum beamformer achieved a word error rate (WER) of 70.4%. The MMI beamformer under a Gaussian assumption achieved 55.2% WER which was further reduced to 52.0% with a K_0 pdf, whereas the WER for data recorded with close-talking microphone was 21.6%.

We also used novel techniques to represent a *full* WSJ trigram language model with 1,639,687 bigrams and 2,684,151 trigrams as a statically-expanded WFST for decoding the separated streams. The final decoding graph constructed from this trigram contained nearly 50 million states and over 100 million transitions. We were able to construct such a large decoding graph by introducing an additional symbol into the language model to explicitly model transitions to the back-off node and thereby make the language model transducer *sequential*. Because the components to be composed together to create the final decoding graph were likewise sequential, we were able to forego the last determinization step, which is usually the most demanding operation in terms of computation and main memory requirements. The use of the full trigram provided a reduction in WER from 52.5% to 47.7%.

In a final set of experiments, we used four different filter bank designs to perform subband analysis and synthesis. We also tested different postfiltering configurations, and applied binary masking to the postfiltered streams. Our best current result on the SSC development data is 39.6% WER. Our best result on the SSC 2007 evaluation set was 46.9% WER.

The balance of this work is organized as follows. In Section 2, we review the definition of mutual information, and demonstrate that, under a Gaussian assumption, the mutual information of two complex random variables is a simple function of their cross-correlation coefficient. We then discuss our MMI beamforming criterion and present the framework needed to apply minimum mutual information beamforming when the Gaussian assumption is relaxed. In Section 3 we describe sequence of operations used to optmize the search space for automatic recognition on the separated streams of speech. We also present the sizes of the language models and decoding graphs used for our experiments. In Section 4, we present the results of far-field automatic speech recognition experiments conducted on data from the PASCAL Speech Separation Challenge. Finally, in Section 5, we present our conclusions and plans for future work.

2   Beamforming

Consider two r.v.s Y_1 and Y_2. By definition, the *mutual information* [4] between Y_1 and Y_2 can be expressed as

$$I(Y_1, Y_2) = \mathcal{E}\left\{\log \frac{p(Y_1, Y_2)}{p(Y_1)p(Y_2)}\right\} \tag{1}$$

where $\mathcal{E}\{\}$ indicates the ensemble expectation.

The univariate Gaussian pdf for complex r.v.s Y_i can be expressed as

$$p(Y_i) = \frac{1}{\pi \sigma_i^2} \exp\left(-|Y_i|^2/\sigma_i^2\right) \tag{2}$$

where $\sigma_i^2 = \mathcal{E}\{Y_i Y_i^*\}$ is the variance of Y_i. Let us define the zero-mean complex random vector $\mathbf{Y} = \begin{bmatrix} Y_1 & Y_2 \end{bmatrix}^T$ and the *covariance matrix*.

$$\Sigma_Y = \mathcal{E}\{\mathbf{Y}\mathbf{Y}^H\} = \begin{bmatrix} \sigma_1^2 & \sigma_1\sigma_2\rho_{12} \\ \sigma_1\sigma_2\rho_{12} & \sigma_2^2 \end{bmatrix} \tag{3}$$

where $\rho_{12} = \epsilon_{12}/\sigma_1\sigma_2$ and $\epsilon_{12} = \mathcal{E}\{Y_1 Y_2^*\}$. The bivariate Gaussian pdf for complex r.v.s is given by

$$p(Y_1, Y_2) = \frac{1}{\pi^2 |\Sigma_Y|} \exp\left(-\mathbf{Y}^T \Sigma_Y^{-1} \mathbf{Y}\right) \tag{4}$$

It follows that the mutual information (1) for jointly Gaussian complex r.v.s can be expressed as [5]

$$I(Y_1, Y_2) = -\log\left(1 - |\rho_{12}|^2\right) \tag{5}$$

From (5), it is clear that minimizing the mutual information between two zero-mean Gaussian r.v.s is equivalent to minimizing the magnitude of their *cross correlation coefficient* ρ_{12}, and that $I(Y_1, Y_2) = 0$ if and only if $|\rho_{12}| = 0$.

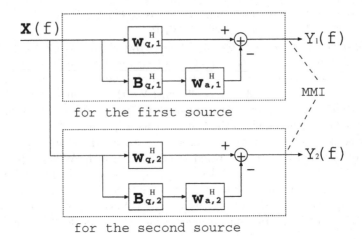

for the first source

for the second source

Fig. 1. A beamformer in GSC configuration

Consider two subband beamformers in GSC configuration as shown in Figure 1. The output of the i-th beamformer for a given
subband can be expressed as,

$$Y_i = (\mathbf{w}_{q,i} - \mathbf{B}_i \mathbf{w}_{a,i})^H \mathbf{X} \tag{6}$$

where $\mathbf{w}_{q,i}$ is the *quiescent weight vector* for the i-th source, $\mathbf{B}_i$ is the *blocking matrix*, $\mathbf{w}_{a,i}$ is the *active weight vector*, and $\mathbf{X}$ is the input subband *snapshot vector*. In keeping with the GSC formalism, $\mathbf{w}_{q,i}$ is chosen to preserve a signal from the *look direction* [3, §6.3]. The blocking matrix $\mathbf{B}_i$ is chosen such that $\mathbf{B}_i^H \mathbf{w}_{q,i} = \mathbf{0}$. The active weight vector $\mathbf{w}_{a,i}$ is typically chosen to maximize the signal-to-noise ratio (SNR). Here, however, we develop an optimization procedure to find that $\mathbf{w}_{a,i}$ which *minimizes* the mutual information $I(Y_1, Y_2)$ where Y_1 and Y_2 are the outputs of the two beamformers. Minimizing a mutual information criterion yields a weight vector $\mathbf{w}_{a,i}$ capable of canceling interference that leaks through the sidelobes without the signal cancellation problems encountered in conventional beamforming. The details of the estimation of the optimal active weights $\mathbf{w}_{a,i}$ under the MMI criterion (5) as well as the application of a *regularization term* are described in Kumatani *et al* [6].

Beamforming in the subband domain has the considerable advantage that the active sensor weights can be optimized for each subband independently, which provides a tremendous computational savings. The subband analysis and resynthesis can be performed with a *perfect reconstruction* (PR) filter bank such as the popular *cosine modulated filter bank* [7, §8]. As this PR filter bank is based on assumptions that are not satisfied in beamforming and adptive filtering applications, however, there are other designs that are better suited for such applications. In Section 4 we present the results of ASR experiments comparing the effectiveness of frequency domain beamforming with subband beamforming

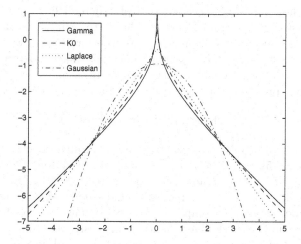

Fig. 2. Plot of the log-likelihood of the super-Gaussian and Gaussian pdfs

based on the PR design, the design proposed by De Haan *et al* [8], and a further novel design technique. We also compare the performance of subband beamformers based on the filter designs with frequency domain beamforming based on a simple FFT.

A plot of the log-likelihood of the Gaussian and three super-Gaussian *real* univariate pdfs considered here is provided in Figure 2.

From the figure, it is clear that the Laplace, K_0 and Γ densities exhibit the "spikey" and "heavy-tailed" characteristics that are typical of super-Gaussian pdfs. This implies that they have a sharp concentration of probability mass at the mean, relatively little probability mass as compared with the Gaussian at intermediate values of the argument, and a relatively large amount of probability mass in the tail; i.e., far from the mean. As explained in [6], univariate and bivariate forms of the complex Laplace, K_0 and Γ pdfs can be derived using the theory of *Meijer G-functions* [9].

3 Search Space Optimization

As originally proposed by Mohri *et al* [10,11], a *weighted finite-state transducer* (WFST) that translates phone sequences into word sequences can be obtained by forming the *composition* $L \circ G$, where L is a *lexicon* which translates the phonetic transcription of a word to the word itself, and G is a *grammar* or *language model* which assigns to valid sequences of words a weight consisting of the negative log probability of this sequence. In the original formulation of Mohri and Riley [12], phonetic context is modeled by the series of compositions $H \circ C \circ L \circ G$, where H is a transducer converting sequences of Gaussian mixture models (GMMs) to sequences of polyphones, and C is a transducer that converts these polyphone sequences to corresponding sequences of phones.

In [13], we showed how the necessity of explicitly modeling C could be circumvented by constructing a transducer HC that maps directly from sequences of GMM names to sequences of phones. In more recent work [14], we demonstrated that HC can be incrementally expanded and immediately determinized. Such an incremental procedure enables a much larger decision tree to be modeled as a WFST. In our previous work, we constructed a recognition network based on

$$\text{min push det}(\text{min det } HC \circ \text{det}(L \circ G)) \qquad (7)$$

where det, push, and min represent the WFST equivalence transformations, *determinization* [15], *weight pushing* [16], and *minimization* [17]. The sequence represented by (7) is the "standard" build procedure [11]. By far, the most memory and time intensive portion of this build sequence is the determinization after HC has been statically composed with $L \circ G$. Hence, we sought to construct a larger recognition network by eliminating this costly determinization.

In the context of WFSTs, ϵ–symbols represent that null symbol that consumes no input or produces no ouput. A *sequential transducer* is deterministic on the input side and has no ϵ–symbols as input. A well-known theorem from Mohri [15] states that the composition of two sequential transducers is sequential. As typically constructed, the grammar G is *not* sequential, as ϵ–symbols are used to allow transitions to nodes modeling *back-off* probabilities, which in turn implies $L \circ G$ is not sequential. We remedied this problem by replacing the ϵ–symbols in G with an explicit *back-off symbol* %, much the way word end markers are introduced to disambiguate homophones [11] thereby allowing $L \circ G$ to be determinized. Similarly, a back-off self-loop was added to the end of each word in L, and to the end of each three-state sequence in HC. These changes were sufficient to make $L \circ G$ sequential. As HC was already sequential, we were able to entirely forego the determinization after composing HC and $L \circ G$. Adding the back-off symbols had an additional salutary effect in that $L \circ G$, and hence the final recognition network, became much smaller, which provided for the use of a still larger language model G.

The sizes of the shrunken and full bigram trigram language models along with the decoding graphs built from them and used for the speech recognition experiments reported in Section 4 are given in Table 1. We performed our initial experiments with the a decoding graph built based on (7) *without* explicit back-off symbols. Thereafter, we built decoding graphs with the full bigram, then with shrunken and full trigrams using the new build technique *with* explicit back-off symbols in the LM. It is worth noting that the decoding graph built from the full bigram with the back-off symbols actually has fewer nodes and transitions than the decoding graph built from the shrunken bigram without back-off symbols. Moreover, as we were able to eliminate the costly determinization after composing HC and $L \circ G$, we were able build a decoding graph from the full WSJ trigram with nearly 50 million states and over 100 million transitions.

Table 1. Dimensions of the various language models and the decoding graphs built from them

Language	G		$HC \circ L \circ G$	
Model	Bigrams	Trigrams	Nodes	Arcs
Shrunken Bigram	323,703	0	4,974,987	16,672,798
Full Bigram	835,688	0	4,366,485	10,639,728
Shrunken Trigram	431,131	435,420	14,187,005	32,533,593
Full Trigram	1,639,687	2,684,151	49,082,515	114,304,406

4 Experiments

We performed far-field automatic speech recognition experiments on development data from the *PASCAL Speech Separation Challenge* (SSC) [18]. The data contain recordings of five pairs of speakers and each pair of speakers reads approximately 30 sentences taken from the 5,000 word vocabulary Wall Street Journal (WSJ) task. The data were recorded with two circular, eight-channel microphone arrays. The diameter of each array was 20 cm, and the sampling rate of the recordings was 16 kHz. The database also contains speech recorded with close talking microphones (CTM). This is a challenging task for source separation algorithms given that the room is reverberant and some recordings include significant amounts of background noise. In addition, as the recorded data is real and not artificially convoluted with measured room impulse responses, the position of the speaker's head as well as the speaking volume varies.

After beamforming, the feature extraction of our ASR system was based on cepstral features estimated with a warped *minimum variance distortionless response* [19] (MVDR) spectral envelope of model order 30. We concatenated 15 cepstral features, each of length 20, then applied linear discriminant analysis (LDA) [20, §10] and a *semi-tied covariance* (STC) [21] transform to obtain final features of length 42 for speech recognition.

4.1 Beamforming Experiments

The training data used for our initial beamforming experiments were taken from the ICSI, NIST, and CMU meeting corpora, as well as the Transenglish Database (TED) corpus, for a total of 100 hours of training material. In addition to these corpora, approximately 12 hours of speech from the WSJCAM0 corpus [22] was used for HMM training in order to cover the British accents for the speakers [18]. Acoustic models estimated with two different HMM training schemes were used for the several decoding passes: conventional maximum likelihood (ML) HMM training [23, §12] and speaker-adapted training under a ML criterion (ML-SAT) [24]. Our baseline system was fully continuous with 3,500 codebooks and a total of 180,656 Gaussian components.

We performed four passes of decoding on the waveforms obtained with each of the beamforming algorithms. Parameters for speaker adaptation were estimated

Table 2. Word error rates for every beamforming algorithm after every decoding passes

Beamforming	Pass (%WER)			
Algorithm	1	2	3	4
Delay & Sum	85.1	77.6	72.5	70.4
MMI: Gaussian	79.7	65.6	57.9	55.2
MMI: Laplace	81.1	67.9	59.3	53.8
MMI: K_0	78.0	62.6	54.1	52.0
MMI: Γ	80.3	63.0	56.2	53.8
CTM	37.1	24.8	23.0	21.6

using the word lattices generated during the prior pass [25]. A description of the individual decoding passes follows:

1. Decode with the unadapted, conventional acoustic model and bigram language model (LM).
2. Estimate vocal tract length normalization (VTLN) [26] parameters and constrained maximum likelihood linear regression parameters (CMLLR) [27] for each speaker, then redecode with the conventional acoustic model and bigram LM.
3. Estimate VTLN, CMLLR, and maximum likelihood linear regression (MLLR) [28] parameters for each speaker, then redecode with the conventional model and bigram LM.
4. Estimate VTLN, CMLLR, MLLR parameters, then redecode with the ML-SAT model and bigram LM.

Table 2 shows the word error rate (WER) for every beamforming algorithm and speech recorded with the CTM after every decoding pass on the SSC development data. These results were obtained with subband-domain beamforming where subband analysis and synthesis was performed with the perfect reconstruction cosine modulated filter bank described in [7, §8]. After the fourth pass, the delay-and-sum beamformer has the worst recognition performance of 70.4% WER. The MMI beamformer with a Gaussian achieved a WER of 55.2%. The best performance of 52.0% WER was achieved with the MMI beamformer by assuming the subband samples are distributed according to the K_0 pdf.

4.2 Language Modeling Experiments

To test the effect of language modeling improvements, we trained a triphone acoustic model on 30 hours of American WSJ data, and the 12 hours of Cambridge WSJ data. For the language modeling experiments, we used the same acoustic features and same sequence of decoding passes as in the prior section. The word error rate reduction achieved through larger language models are shown in Table 3. The most dramatic reduction in WER was achieved by replacing the bigram LMs with the shrunken trigram. As shown in Table 1, the shrunken trigram produced a decoding graph that was still small enough to run on our standard 32-bit workstations. The full trigram, on the other hand, could

Table 3. ASR results on the SSC development data

Language Model/Pass	Pass (%WER)			
	1	2	3	4
shrunken bigram	85.7	64.8	53.8	52.5
full bigram		65.5	53.8	52.4
shrunken trigram	86.1	61.7	49.8	47.7
full trigram	88.3	61.0	48.9	47.0

not be used on the 32-bit machines used for the experiments reported here. On a work station with a 64-bit operating system, more than 7 Gb of RAM were required merely to load the decoding graph built from the full trigram, and the entire task image of the recognition job was approximately 8 Gb. Moreover, the reduction in WER with respect to the shrunken trigram that was achieved by the full trigram was less than one percent absolute. Hence, we used the decoding graph built from the shrunken trigram to decode the evaluation data, as that system was much more tractable.

The results of further experiments with these language models, as well as a description of a technique for dynamically composing the HC and $L \circ G$ components, and thereby radically reducing the enormous amount of random access memory required by the full trigram, are given in [29].

4.3 Filter Bank Experiments

As explained in Section 2, our MMI beamformer operates in the frequency or subband domain. Hence, the digital filter bank used for subband analysis and resynthesis is an important component of the speech separation system. We investigated four different filter bank designs, including:

1. The *cosine modulated filter bank* described by Vaidyanathan [7, §8], which yields *perfect reconstruction* (PR) under optimal conditions. In such a filter bank, PR is achieved through *aliasing cancellation*, wherein the aliasing that is perforce present in one subband is cancelled by the aliasing in all others. Aliasing cancellation breaks down if arbitrary complex factors are applied to the subband samples. For this reason, such a PR filter bank is not optimal for beamforming or adaptive filtering applications.
2. An DFT filter bank based on overlap-add.
3. The modulated filter bank proposed by De Haan *et al* [8], wherein separate analysis and synthesis prototypes are designed to minimize an error criterion consisting of a weighted combination of the total spectral response error and the aliasing distortion. This design is dependent on the use of oversampling to reduce aliasing error.
4. A novel cosine modulated design which differs from the De Haan filter bank in that a *Nyquist(M)* constraint [7, §4] is imposed on the prototype in order to ensure that the total response error vanishes. Thereafter the remaining components of the prototype are chosen to minimize aliasing error, as with

Table 4. ASR results on the SSC development data

Filter Bank	Pass (%WER)			
	1	2	3	4
PR	87.7	65.2	54.0	50.7
PR + postfilter + binary mask	87.1	66.6	55.7	52.5
FFT	88.5	71.1	58.8	55.5
De Haan	88.7	68.2	56.1	53.3
De Haan + postfilter + binary mask	82.7	57.7	42.7	39.6
Nyquist(M) + postfilter + binary mask	84.8	58.0	43.4	40.9

the De Haan design. The Nyquist(M) design similarly uses oversampling to reduce aliasing distortion.

The word error rates (WERs) obtained with the four filter banks on the SSC development data are shown in Table 4. For these experiments, the Gaussian pdf was used exclusively. We also investigated the effect of applying a Zelinski post filter [30] to the output of the beamformer in the subband domain, as well as the binary mask [1] described in [31]. The results indicate that the performnace of PR filter bank is actually quite competitive if no postfiltering nor binary masking is applied to the output of the beamformer. For the PR design, performance *degrades* from 50.7% WER to 52.5% when such postfiltering and masking is applied, which is not surprising given that both will tend to destroy the aliasing cancellation on which this design is based. When postfiltering and masking is applied to either the De Haan or the Nyquist(M) designs, performance is greatly enhanced. With the De Haan design adding postfiltering and masking reduced WER from 53.5% to 39.6%. With postfiltering and masking the Nyquist(M) design achieved very similar performance of 40.9%. For both the De Haan and Nyquist(M) designs, an oversampling factor of eight was used. The simple FFT achieved significantly worse performance than all of the subband filter banks.

5   Conclusions and Future Work

In this work, we have described our system for the automatic recognition of simultaneous speech. Our system consisted of three principal components: A person tracker returns the locations of both active speakers, as well as segmentation information for each utterance, which are often of unequal length; two beamformers in GSC configuration separate the simultaneous speech by setting their active weight vectors according to a minimum mutual information (MMI) criterion; a postfilter and binary mask operating on the outputs of the beamformers further enhance the separated speech; and finally an ASR engine based

---

[1] We learned of the binary masking technique only by attending MLMI and listening to Iain McCowan's presentation about the SSC system developed by him and his collaborators. Our experiments with the binary mask were conducted *after* the SSC deadline.

on a WFST returns the most likely word hypotheses for the separated streams. In addition to developing and optimizing each of these three components, we have also proposed a novel filter bank design in this work that, when used for subband beamforming, provided ASR performance comparable or superior to any design that has previously appeared in the literature. Our final results on the SSC development data were 39.6% WER. On the SSC evaluation data, our system achieved a WER of 46.2%.

In future, we plan to continue our investigations into the use of super-Gaussians pdfs for MMI beamforming. This will entail systematically searching the entire class of multi-dimensional super-Gaussians pdfs that can be represented with the aid of the Meijer-G function. We will also develop an online or LMS-style algorithm for updating the active weight vectors of the GSCs during MMI beamforming. Finally, we hope to investigate other optimization criteria such the *negentropy* metric typically used in the field of independent component analysis [4].

References

1. Gehrig, T., Klee, U., McDonough, J., Ikbal, S., Wölfel, M., Fügen, C.: Tracking and beamforming for multiple simultaneous speakers with probabilistic data association filters. In: Proc. Interspeech, pp. 2594–2597 (2006)
2. Bar-Shalom, Y., Fortmann, T.E.: Tracking and Data Association. Academic Press, San Diego (1988)
3. Van Trees, H.L.: Optimum Array Processing. Wiley-Interscience, Chichester (2002)
4. Hyvärinen, A., Oja, E.: Independent component analysis: Algorithms and applications. Neural Networks 13, 411–430 (2000)
5. McDonough, J., Kumatani, K.: Minimum mutual information beamforming. Technical Report 107, Interactive Systems Lab, Universität Karlsruhe (August 2006)
6. Kumatani, K., Gehrig, T., Mayer, U., Stoimenov, E., McDonough, J., Wölfel, M.: Adaptive beamforming with a minimum mutual information criterion. IEEE Trans. Audio Speech and Lang. Proc. (to appear)
7. Vaidyanathan, P.P.: Multirate Systems and Filter Banks. Prentice-Hall, Englewood Cliffs (1993)
8. de Haan, J.M., Grbic, N., Claesson, I., Nordholm, S.E.: Filter bank design for subband adaptive microphone arrays. IEEE Trans. Speech and Audio Proc. 11(1), 14–23 (2003)
9. Brehm, H., Stammler, W.: Description and generation of spherically invariant speech-model signals. Signal Processing 12, 119–141 (1987)
10. Mohri, M., Riley, M., Hindle, D., Ljolje, A., Periera, F.: Full expansion of context-dependent networks in large vocabulary speech recognition. In: Proc. ICASSP, Seattle, vol. II, pp. 665–668 (1998)
11. Mohri, M., Pereira, F., Riley, M.: Weighted finite-state transducers in speech recognition. Computer Speech and Language 16, 69–88 (2002)
12. Mohri, M., Riley, M.: Network optimizations for large vocabulary speech recognition. Speech Communication 25(3) (1998)
13. Stoimenov, E., McDonough, J.: Modeling polyphone context with weighted finite-state transducers. In: Proc. ICASSP (2006)
14. Stoimenov, E., McDonough, J.: Memory efficient modeling of polyphone context with weighted finite-state transducers. In: Proc. Interspeech (2007)

15. Mohri, M.: Finite-state transducers in language and speech processing. Computational Linguistics 23(2) (1997)
16. Mohri, M., Riley, M.: A weight pushing algorithm for large vocabulary speech recognition. In: Proc. ASRU, Aarlborg, Denmark, September 2001, pp. 1603–1606 (2001)
17. Mohri, M.: Minimization algorithms for sequential transducers. Theoretical Computer Science 234(1–2), 177–201 (2000)
18. Lincoln, M., McCowan, I., Vepa, J., Maganti, H.: The multi-channel wall street journal audio visual corpus (mc-wsj-av): specification and initial experiments. In: Proc. ASRU, pp. 357–362 (November 2005)
19. Wölfel, M., McDonough, J.: Minimum variance distortionless response spectral estimation, review and refinements. IEEE Signal Processing Magazine 22(5), 117–126 (2005)
20. Fukunaga, K.: Introduction to Statistical Pattern Recognition. Academic Press, New York (1990)
21. Gales, M.J.F.: Semi-tied covariance matrices. In: Proc. ICASSP (1998)
22. Fransen, J., Pye, D., Robinson, T., Woodland, P., Young, S.: Wsjcam0 corpus and recording description. Technical Report CUED/F-INFENG/TR.192, Cambridge University Engineering Department (CUED) Speech Group (September 1994)
23. Deller, J., Hansen, J., Proakis, J.: Discrete-Time Processing of Speech Signals. Macmillan Publishing, New York (1993)
24. Anastasakos, T., McDonough, J., Schwarz, R., Makhoul, J.: A compact model for speaker-adaptive training. In: Proc. ICSLP, pp. 1137–1140 (1996)
25. Uebel, L., Woodland, P.: Improvements in linear transform based speaker adaptation. In: Proc. ICASSP (2001)
26. Wölfel, M.: Mel-Frequenzanpassung der Minimum Varianz Distortionless Response Einhüllenden. In: Proc. of ESSV (2003)
27. Gales, M.J.F.: Maximum likelihood linear transformations for HMM-based speech recognition. Computer Speech and Language 12 (1998)
28. Leggetter, C.J., Woodland, P.C.: Maximum likelihood linear regression for speaker adaptation of continuous density hidden markov models. Computer Speech and Language 9, 171–185 (1995)
29. McDonough, J., Stoimenov, E., Klakow, D.: An algorithm for fast composition of weighted finite-state transducers. In: Proc. ASRU (submitted, 2007)
30. Simmer, K.U., Bitzer, J., Marro, C.: Post-filtering techniques. In: Branstein, M., Ward, D. (eds.) Microphone Arrays, pp. 39–60. Springer, Heidelberg (2001)
31. McCowan, I., Hari-Krishna, M., Gatica-Perez, D., Moore, D., Ba, S.: Speech acquisition in meetings with an audio-visual sensor array. In: Proceedings of the IEEE International Conference on Multimedia and Expo (ICME) (July 2005)

Microphone Array Beamforming Approach to Blind Speech Separation

Ivan Himawan[1,3], Iain McCowan[1,2], and Mike Lincoln[3]

[1] Queensland University of Technology, Brisbane QLD, Australia
[2] CSIRO e-HEALTH Research Centre, Brisbane QLD, Australia
[3] Centre for Speech Technology Research, Edinburgh, United Kingdom
i.himawan@qut.edu.au, iain.mccowan@csiro.au, mlincol1@inf.ed.ac.uk

Abstract. In this paper, we present a microphone array beamforming approach to blind speech separation. Unlike previous beamforming approaches, our system does not require a-priori knowledge of the microphone placement and speaker location, making the system directly comparable other blind source separation methods which require no prior knowledge of recording conditions. Microphone location is automatically estimated using an assumed noise field model, and speaker locations are estimated using cross correlation based methods. The system is evaluated on the data provided for the PASCAL Speech Separation Challenge 2 (SSC2), achieving a word error rate of 58% on the evaluation set.

Keywords: Speech Separation, Microphone Array, Speech Recognition.

1 Introduction

Recently there has been a large effort to produce automatic speech recognition systems which can automatically transcribe multi-party conversations such as those which occur during meetings [1, 2]. Overlapping speech segments, where more than one individual is talking simultaneously, pose a serious problem for such systems, increasing the absolute word error rate between 15-30% when using close talking microphones for a large vocabulary task [3]. However most systems are optimised for a single dominant speaker, with performance degrading significantly when competing speech is present. One of the reasons for this is that, while a number of different algorithms exist for separating the speech of competing talkers, to date there has been no systematic comparison of the approaches in terms of their effects on recognition performance. Algorithms are typically evaluated in terms of signal to noise ratio on small datasets containing signals in the presence of noise or other interfering signals, and these datasets are often specific to the group conducting the research, making comparison of algorithms difficult. The PASCAL SSC2 attempts to overcome this lack of comparable results by providing a unified framework in which speech separation algorithms may be compared in terms of speech recognition word error rate (WER) on a standardised task, with a baseline recogniser. The task consists of separating the overlapping speech of two talkers simultaneously reading sentences from the

A. Popescu-Belis, S. Renals, and H. Bourlard (Eds.): MLMI 2007, LNCS 4892, pp. 295–305, 2007.

Wall Street Journal database in an instrumented meeting room. The speakers are recorded using headset and lapel microphones and two 8-element circular microphone arrays placed on a table in the centre of the meeting room. A development set consisting of 178 sentences and an evaluation set of 143 sentences are provided along with a baseline Wall Street Journal recogniser capable of recognising the separated speech.

There are two common approaches to the separation of overlapping speech: blind source separation (BSS) and microphone array processing. Blind source separation attempts to exploit the assumption of statistical independence between the overlapped signals in order to separate them, while microphone array processing provides an enhanced version of the input speech based on the location of the speakers. Microphone array processing techniques generally require a-priori knowledge of the microphone and speaker locations, while BSS algorithms typically do not. In order to emulate real world conditions in which microphone and speaker locations are unknown, and to provide a fair comparison between techniques, one of the rules of the challenge states that 'knowledge of the room layout and speaker and microphone placements may *not* be used unless it is derived automatically by the algorithm'. We have developed a microphone array beamforming technique which automatically generates the required speaker and microphone location information which is therefore directly comparable to BSS approaches.

Section 2 presents the separation algorithm and give a detailed description of each of its sub-components. Section 3 then presents and discusses experimental results for the accuracy of both the sub-components and the separation system as a whole, and Section 4 gives conclusions and further work which may improve the performance of the system.

2   System Description

The system consists of four sub-components as shown in Figure 1. Initially the microphone locations are estimated using a procedure based on the assumption that the noise field within the recording room is diffuse - this is a reasonable assumption for many practical reverberant environments such as offices and cars. An SRP-PHAT based localisation system is then used to estimate the location of the speakers in the room. The microphone and speaker location information is used in a standard superdirective beamformer to separate the speech of the

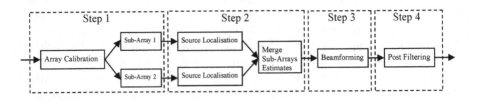

Fig. 1. System Diagram

two speakers, and finally crosstalk between the speakers is canceled by means of a masking post-filter applied to the beamformer output. The separated signals are then recognised by the ASR system for evaluation.

2.1   Array Calibration

The microphone locations are estimated using the procedure described in [4]. The coherence between two signals at points i and j, as a function of frequency f, is defined as

$$\Gamma_{ij}(f) \triangleq \frac{\phi_{ij}(f)}{\sqrt{\phi_{ii}(f)\phi_{jj}(f)}} \tag{1}$$

where ϕ_{ij} is the cross-spectral density between the signals on microphones i and j. For a diffuse noise field, the coherence function may be shown to be [5]:

$$\Gamma_{ij}^{\text{diffuse}} = \text{sinc}\left(\frac{2\pi f d_{ij}}{c}\right) \tag{2}$$

where $\text{sinc}(x) \equiv \sin(x)/x$, and d_{ij} is the distance between the spatial locations of microphones i and j.

The distance between each microphone pair, d_{ij} can be estimated by fitting Equation 2 to the measured coherence from Equation 1:

$$\varepsilon_{ij}(d) = \sum_{f=0}^{fs/2} \left| \mathcal{R}\{\Gamma_{ij}(f)\} - \text{sinc}\left(\frac{2\pi f d}{c}\right) \right|^2 \tag{3}$$

$$d_{ij} = \underset{d}{\text{argmin}}\, \varepsilon_{ij}(d) \tag{4}$$

Given these inter-microphone distance estimates, multidimensional scaling [6, 7] is then used to estimate the relative locations of the microphones.

Initial experiments on the SSC2 development data showed that the microphone calibration system was unable to precisely estimate the microphone positions when calibrating both arrays simultaneously, however microphones from the same array were successfully clustered near each other. This is due to the larger distances separating the two arrays - approximately 1.5 meters, compared to intra-array spacings of several centimetres. While the diffuse coherence model still holds for larger distances, it is difficult to obtain precise curve fitting as the width of the sinc main lobe approaches zero, leading to inaccurate estimates. To overcome this, a two pass procedure was used :

1. Find initial position of all microphones using the calibration procedure.
2. Cluster microphones into K sub-arrays using K-Means clustering. K is initialised to 1 and incremented until no microphone within a cluster is more than 5cm from any other microphone within that cluster.
3. Re-calibrate the positions of the microphones within each sub-array.

4. Since the MDS algorithm estimates locations subject to arbitrary rotation and scaling, each sub-array is aligned to the original estimates from step 1 to ensure a common basis for all microphones.

Evaluation of the microphone localisation procedure is given in Section 3.1

2.2   Source Localisation

The automatically derived microphone locations are then used in an SRP-PHAT [8] (Steered Response Power with Phase Transform) source localisation system to estimate the location of the speakers.

Steered response power (SRP) localisation relies on directing a standard beam-former, such as delay-and-sum, to a number of points of interest and evaluating the power of the output signal. A search is usually performed over the area in which the source is expected to lie, and the source position taken as the location at which the beamformer's output power is maximised.

In SRP-PHAT localisation, the PHAT weighted Generalised Cross Correlation for the signal at each pair of microphones i, j is calculated for each time frame.

$$R^{ij}_{GCC\_PHAT}(\tau) = \frac{1}{2\pi} \int_{-\infty}^{+\infty} \frac{X_i(\omega)X_j^*(\omega)}{|X_i(\omega)X_j^*(\omega)|} e^{j\omega\tau} d\omega \qquad (5)$$

The PHAT weighting has been shown to emphasise the peak in the GCC seen at the time lag corresponding to the true source location, and de-emphasise those caused by reverberation and noise.

The SRP of a filter and sum beamformer directed at location S can be shown to be equivalent to the sum of the GCC's for all pairs of microphones within the array :

$$P(S) = \sum_{i=1}^{N} \sum_{j=1}^{N} R^{ij}_{GCC-PHAT}(\delta(S, i, j)) \qquad (6)$$

$\delta(S, i, j)$ is the time delay of arrival between microphones i and j for a source located at position S

$$\delta(S, i, j) = \frac{|S - L_i| - |S - L_j|}{c} \qquad (7)$$

where L_i and L_j are the locations of microphones i and j (in this case estimated using the microphone localisation procedure) respectively and c is the speed of sound.

In a single speaker scenario, a search is performed over a number of locations S and the speaker's location is taken as the value of S which maximises $P(S)$. Since the SSC2 recordings contain two speakers, the two highest peaks are taken to be the locations of the two speakers. In order to make the localisation more ro-bust, any frames in which the second peak is below 70% of the first is discarded, since this typically implies that only a single speaker is active and that the second location is spurious. A minimum angular separation between estimates is also enforced, in order to avoid having both location estimates come from the same speaker.

2.3 Superdirective Beamforming

In this work, superdirective beamformers are used to provide spatial discrimination between the two speakers. A set of superdirective channel filters are calculated for each speaker, minimising the diffuse noise power received by the sensors under the constraint of unity gain to the desired direction. The array steering vector, which represents microphone and source spatial configurations, is constructed from automatically derived positions. Assuming near-field condition, this vector is given by [9]:

$$\mathbf{d} = [a_0 e^{-j\omega\tau_0}, a_1 e^{-j\omega\tau_1}, ..., a_{N-1} e^{-j\omega\tau_n}]^T,$$

$$a_n = \frac{d_{ref}}{d_n}, \quad \tau_n = \frac{d_n - d_{ref}}{c},$$

where d_n and d_{ref} denote the Euclidian distance between the source and the microphone n, or the reference microphone, respectively. To increase the robustness of superdirective beamforming in the presence of array perturbation, a weighted identity matrix is usually added to the noise cross-spectral matrix to satisfy the white noise gain constraint [10].

2.4 Cross-Talk Noise Cancelling Post-filter

In order to further eliminate the signal from competing speakers, a frequency domain masking post-filter is applied to each beamformer output. The post-filter is motivated by the assumption that, at each time frame, the energy spectrum of the sum of two signals can be approximated by the maxima of the individual spectra for each signal in each frequency bin. This is due to the fact that two independent signals rarely contain high energy in the same frequency bin at the same time, a property which has been exploited to develop a series of techniques for speech separation using a single microphone [11].

If $b_s(f)$ is the frequency domain output of beamformer s, $z_s(f)$ is the post-filtered output and $h_s(f)$ is the frequency response of the post-filter, then for S beamformers directed at competing sources [12],

$$z_s(f) = b_s(f)h_s(f) \tag{8}$$

where

$$h_s(f) = \begin{cases} 1 & \text{if } s = \overset{\text{argmax}}{s'} |b_{s'}|, s' = 1...S \\ 0 & \text{otherwise} \end{cases} \tag{9}$$

In the case of the SSC2 system with $S = 2$ the output of the beamformer is set to zero in a given frequency bin if the output of the second beamformer is greater in that bin.

3   Experiments and Results

3.1   Array Calibration Performance

In order to evaluate the performance of the array calibration technique described in Section 2.1, the inter-microphone distances are compared to the known ground truth microphone locations. Table 1 shows the mean, standard deviation, minimum and maximum errors of the inter-microphone distance estimates for each of the two arrays.

Table 1. Mean, standard deviation, minimum and maximum inter-microphone distance error. All measurements are in centimeters.

sub-array	$\bar{\delta}_d$	σ_d	$\min(\delta_d)$	$\max(\delta_d)$
1	0.80	0.65	0.02	2.33
2	0.46	0.31	0.02	1.15

The MDS algorithm returns the microphone locations subject to arbitrary rotation and translation. So that the location estimates may be evaluated, for each sub-array an incremental search is performed to align the estimated locations to those of the ground truth array geometry. Figure 2 shows the actual and estimated microphone positions after the search, and Table 2 gives the mean, standard deviation, minimum and maximum position errors in the microphone location estimates.

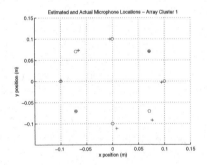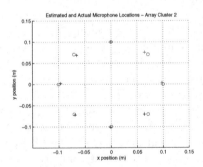

Fig. 2. The actual (circles) and estimated (pluses) microphone positions for sub-array1 and sub-array2

The inter-microphone distance and position errors for sub-array 1 are about approximately twice that of sub-array 2, however, the microphone positions estimates for both sub-arrays show good accuracy, with errors ranging from 0cm to 2.18cm.

Table 2. Results measuring accuracy of positions estimates in terms of the mean, standard deviation, minimum and maximum distance error. All measurements are in centimetres.

sub-array	$\bar{\delta}_d$	σ_d	$\min(\delta_d)$	$\max(\delta_d)$
1	0.69	0.75	0.01	2.18
2	0.42	0.28	0.00	0.82

3.2 Source Localisation Performance

To assess the performance of the source localisation system, the estimated source positions were compared with ground truth source locations obtained from the MC-WSJ-AV data collection documentation [13]. The documentation gives the location of the speaker in terms of their seating position around the meeting room table and from this, and measurements of the table size, the approximate location of the speaker can be inferred. Tables 3 and 4 show the localisation errors in terms of azimuth (degrees), elevation (degrees), and radius (meters) for both ground truth microphone locations and those estimated using the procedure described in Section 2.1. The results show that, for sub-array 1, the use of estimated microphone positions has little effect on the ability to locate both speakers, however for sub-array 2 there is a significant decrease in the number of files in which locations for both speakers are found. This may likely be attributed to the fact that sub-array 2 is typically further from the speakers than sub-array 1.

To improve the speaker localisation performance, location estimates from the two arrays are calculated independently using the technique described in 2.2 and then combined as follows - Since sub-array 1 provides better localisation performance than sub-array 2, the location corresponding to the most dominant peak in sub-array 1's GCC-Phat is taken as the location of speaker 1. For the second speaker, if estimate 2 from sub-array 1 is more than 22.5 degrees away from estimate 1 in the azimuth, then it is retained as the second location estimate, otherwise the location estimate from sub-array 2 with greatest angular azimuth separation from estimate 1 of sub-array 1 is selected. This technique attempts to avoid the problem of localising to the same source twice. Localisation error

Table 3. Accuracy of speaker localisation results on development set using actual microphone positions, in terms of azimuth (degrees), elevation (degrees), and radius (metres) error

sub-array	found both (total of 178)	speaker	azimuth		elevation		radius	
			mean	stdev	mean	stdev	mean	stdev
1	134	1	4.36	6.84	7.84	4.82	0.41	0.37
		2	4.59	5.03	8.21	5.11	0.48	0.36
2	143	1	8.66	6.57	7.42	5.67	0.35	0.28
		2	10.33	24.06	8.24	8.98	0.37	0.28

Table 4. Accuracy of speaker localisation results on development set using estimated microphone positions, in terms of azimuth (degrees), elevation (degrees), and radius (metres) error

sub-array	found both (total of 178)	speaker	azimuth mean	azimuth stdev	elevation mean	elevation stdev	radius mean	radius stdev
1	133	1	3.53	2.60	4.83	3.95	0.41	0.34
		2	5.43	12.04	9.30	9.13	0.77	0.43
2	115	1	8.16	7.53	8.15	6.37	0.32	0.31
		2	9.80	7.24	6.72	7.11	0.36	0.32

Table 5. Accuracy of merged speaker localisation results on development set using actual and estimated microphone positions, in terms of azimuth (degrees), elevation (degrees), and radius (metres) error

geometry	found both (total of 178)	speaker	azimuth mean	azimuth stdev	elevation mean	elevation stdev	radius mean	radius stdev
True	153	1	4.36	6.84	7.84	4.82	0.41	0.37
		2	4.78	5.09	7.82	5.05	0.43	0.35
Calibrated	148	1	3.53	2.60	4.83	3.95	0.41	0.34
		2	6.73	12.06	9.63	8.92	0.71	0.48

for the combined estimates are shown in Table 5 for both estimates and ground truth microphone locations. The results show that combining the estimates from the two arrays leads to an improvement in the speaker detection rates in addition to an improvement in the azimuth estimates for the located speakers.

3.3   Speech Recognition Experiments

Speech recognition experiments were conducted on the final post-filtered output of the separation system using the recogniser supplied by the SSC2 challenge organisers. The system is an HTK based HMM recogniser with accoustic models trained on the WSJCAM0 database, a dictionary derived from that used in the AMI LVCSR system [2] and the ARPA bigram and trigram language models developed for the WSJ0 corpus. Because the acoustic models are trained on data from close talking microphones, MLLR is used to adapt the models to the output of the separation system. Due to the small amounts of data available for adaptation, a leave one out cross-validation approach is used for adaptation. The supplied recognition system also includes scoring scripts which automatically assign the separated speech to the correct speaker. To do this, the scoring script compares both sentences against transcripts for each of the read utterances, and selects the one which gives the highest score. Tables 6 and 7 show the recogniser word error rates for the development and evaluation sets for a number of different conditions. Headset and lapel microphones were worn by the speakers during the database recording and results from these microphones are given, in addition

Table 6. WER on development data set of the proposed system

	Both	Best
Headset Microphone	11.7	5.2
Lapel Microphone	32.2	20.5
Single Distant Microphone	98.8	92.5
GT Beamformer	68.8	47.1
GT Postfiler	57.0	36.2
Beamformer	66.3	43.4
Post-filter	54.8	35.1

Table 7. WER on evaluation data set of the proposed system

	Both	Best
Headset Microphone	16.4	8.4
Lapel Microphone	46.3	26.5
Single Distant Microphone	103.3	91.0
Beamformer	73.9	50.7
Postfiler	58.0	38.3

to a single distant microphone condition in which a single microphone from sub-array 1 is used. 'GT-Beamformer' and 'GT-Post-filter' indicate results from using ground truth microphone and speaker locations for the beamformer both without and with the post-filter. Finally 'Beamformer' and 'Post-filter' results are for the full system without and with post-filter, but using automatically derived microphone locations. Results are presented for the combined score for both speakers (the 'Both' column) and for scoring only the best scoring speaker for each utterance. This shows that in many cases one speaker is significantly easier to recognise than the other - possibly because they are either louder, clearer speakers, or they read longer sentences.

On the development set, using automatically derived microphone and speaker locations results in slightly lower WER compared with the use of ground truth data. While this result is encouraging it should be noted that the ground truth speaker locations are only approximate (as explained in Section 3.2) and that the results would be expected to improve if the ground-truth speaker locations were known more accurately. The beamforming system gives much higher WER than close talking microphones, however it is closer in performance to the lapel microphones, with which microphone array systems have previously been competitive [14]. A similar pattern of results is seen in the evaluation set, which word error rates even higher than for the development data, indicating that this is a difficult task.

4   Conclusions and Future Works

We have presented an approach to speech separation based on microphone array beamforming techniques which avoids the usual problems of beamforming (that of requiring detailed knowledge of the recording environment) by automatically estimating microphone and speaker locations. The technique achieves closer performance to the lapel microphones and shows large improvements in recognition accuracy compared with the single distant microphone case. The data supplied with for the PASCAL SSC2, on which the algorithm was evaluated, is a 'worst case scenario' in terms of speaker overlap in that every recording contains overlapped speech. This may explain the difference between our approach and recognition results using lapel microphones with which, in other task, beamforming approaches are usually competitive.

There are several areas in which we plan to improve the system. The array calibration algorithm relies on the microphones being in a diffuse noise field. While this is a good model for many practical situations it may not always be appropriate to assume the diffuse noise case, particularly when the microphones are widely spaced. We plan to investigate methods of automatic microphone localisation in other noise fields, and to improve the clustering techniques to ensure localisation is not affected by widely spaced microphones.

Whilst all utterances have some overlap, most also contain segments in which one or other of the speakers is solely speaking. In order to improve speaker localisation accuracy, we plan to segment the utterance into regions of silence, single speaker activity and overlap, and use different localisation strategies in each of these cases. We also plan to more closely integrate the array calibration and source localisation to minimise errors in estimating positions.

Experiments will also be performed to adapt the post filter to not only cancel cross talk, but also extraneous noise from with the recording environment.

Acknowledgements

This work was partly supported by the European Union 6th FWP IST Integrated Project AMI (Augmented Multi-party Interaction, FP6-506811).

References

[1] Janin, A., et al.: The ICSI-SRI Spring 2006 Meeting Recognition System. In: Proc. of the Rich Transcription 2006 Spring Meeting Recognition Evaluation, Washington, USA (2006)

[2] Hain, T., et al.: The AMI system for the transcription of speech in meetings. In: Proc. IEEE Int. Conf. on Acoustics, Speech, and Signal Processing, vol. 4, pp. 357–360 (2007)

[3] Morgan, N., et al.: The meeting project at ICSI. In: Proc. Human Language Technology Conf. (2001)

[4] McCowan, I., Lincoln, M., Himawan, I.: Microphone array calibration in diffuse noise fields. IEEE Trans. on Acoustics, Speech, and Signal Processing (to appear, 2008)

[5] Cook, R.K., et al.: Measurement of correlation coefficients in reverberant sound fields. The Journal of the Acoustical Society of America 27, 1072–1077 (1955)

[6] Torgerson, W.: Theory and Methods of Scaling. Wiley, New York (1958)

[7] Cox, M.F., Cox, M.A.A.: Multidimensional Scaling. Chapman and Hall (2001)

[8] Di Biase, J.H., Silverman, H.F., Brandstein, M.S.: Robust localization in reverberant rooms. In: Brandstein, M.S., Ward, D.B. (eds.) Microphone Arrays, pp. 157–180. Springer, Heidelberg (2001)

[9] Bitzer, J., Simmer, K.U.: Superdirective microphone arrays. In: Brandstein, M.S., Ward, D.B. (eds.) Microphone Arrays, pp. 19–38. Springer, Heidelberg (2001)

[10] Cox, H., Zeskind, R., Owen, M.: Robust adaptive beamforming. IEEE Trans. on Acoustics, Speech, and Signal Processing 35, 1365–1376 (1987)

[11] Roweis, S.T.: Factorial models and refiltering for speech separation and denoising. In: Proc. of Eurospeech, pp. 1009–1012 (2003)

[12] Maganti, H.K., Gatica-Perez, D., McCowan, I.: Speech enhancement and recognition in meetings with an audio-visual sensor array. IEEE Trans. on Acoustics, Speech, and Signal Processing 15, 2257–2269 (2007)

[13] Lincoln, M., McCowan, I., Vepa, J., Maganti, H.K.: The multi-channel wall street journal audio visual corpus (mc-wsj-av): Specification and initial experiments. In: Proc. ASRU, pp. 357–362 (2005)

[14] Moore, D., McCowan, I.: Microphone array speech recognition: Experiments on overlapping speech in meetings. In: Proc. IEEE Int. Conf. on Acoustics, Speech, and Signal Processing, vol. 5, pp. 497–500 (2003)

Author Index

Lecture Notes in Computer Science

Sublibrary 3: Information Systems and Application, incl. Internet/Web and HCI

For information about Vols. 1– 4537
please contact your bookseller or Springer